Enzymatic Reactions
in Organic Media

Enzymatic Reactions in Organic Media

Edited by

A.M.P. KOSKINEN
Professor of Organic Chemistry
University of Oulu
Finland

and

A.M. KLIBANOV
Professor of Chemistry
Massachusetts Institute of Technology
USA

LIVERPOOL
JOHN MOORES UNIVERSITY
AVRIL ROBARTS LRC
TITHEBARN STREET
LIVERPOOL L2 2ER
TEL. 0151 231 4022

BLACKIE ACADEMIC & PROFESSIONAL
An Imprint of Chapman & Hall

London · Glasgow · Weinheim · New York · Tokyo · Melbourne · Madras

Published by
Blackie Academic & Professional, an imprint of Chapman & Hall,
Wester Cleddens Road, Bishopbriggs, Glasgow G64 2NZ

Chapman & Hall, 2–6 Boundary Row, London SE1 8HN, UK

Blackie Academic & Professional, Wester Cleddens Road, Bishopbriggs, Glasgow G64 2NZ, UK

Chapman & Hall GmbH, Pappelallee 3, 69469 Weinheim, Germany

Chapman & Hall USA, Fourth Floor, 115 Fifth Avenue, New York, NY 10003, USA

Chapman & Hall Japan, ITP-Japan, Kyowa Building, 3F, 2-2-1 Hirakawacho, Chiyoda-ku, Tokyo 102, Japan

DA Book (Aust.) Pty Ltd, 648 Whitehorse Road, Mitcham 3132, Victoria, Australia

Chapman & Hall India, R. Seshadri, 32 Second Main Road, CIT East, Madras 600 035, India

First edition 1996

© 1996 Chapman & Hall

Typeset in 10/12pt Times by EJS Chemical Composition, Bath

Printed in Great Britain by St Edmundsbury Press, Bury St Edmunds, Suffolk

ISBN 0 7514 0259 1

Apart from any fair dealing for the purposes of research or private study, or criticism or review, as permitted under the UK Copyright Designs and Patents Act, 1988, this publication may not be reproduced, stored, or transmitted, in any form or by any means, without the prior permission in writing of the publishers, or in the case of reprographic reproduction only in accordance with the terms of the licences issued by the Copyright Licensing Agency in the UK, or in accordance with the terms of licences issued by the appropriate Reproduction Rights Organization outside the UK. Enquiries concerning reproduction outside the terms stated here should be sent to the publishers at the Glasgow address printed on this page.

The publisher makes no representation, express or implied, with regard to the accuracy of the information contained in this book and cannot accept any legal responsibility or liability for any errors or omissions that may be made.

A catalogue record for this book is available from the British Library
Library of Congress Catalog Card Number: 95–81445

∞ Printed on permanent acid-free text paper, manufactured in accordance with ANSI/NISO Z39.48-1992 (Permanence of Paper).

Preface

LIVERPOOL
JOHN MOORES UNIVERSITY
AVRIL ROBARTS LRC
TITHEBARN STREET
LIVERPOOL L2 2ER
TEL. 0151 231 4022

The outlook of organic synthesis has changed many times during its tractable history. The initial focus on the synthesis of substances typical of living matter, exemplified by the first examples of organic chemistry through the synthesis of urea from inorganic substances by Liebig, was accepted as the birth of organic chemistry, and thus also of organic synthesis. Although the early developments in organic synthesis closely followed the pursuit of molecules typical in nature, towards the end of the 19th century, societal pressures placed higher demands on chemical methods appropriate for the emerging age of industrialization. This led to vast amounts of information being generated through the discovery of synthetic reactions, spectroscopic techniques and reaction mechanisms.

The basic organic functional group transformations were discovered and improved during the early part of this century. Reaction mechanisms were elucidated at a growing pace, and extremely powerful spectroscopic tools, such as infrared, nuclear magnetic resonance and mass spectrometry were introduced as everyday tools for a practising organic chemist. By the 1950s, many practitioners were ready to agree that almost every molecule could be synthesized. Some difficult stereochemical problems were exceptions; for example Woodward concluded that erythromycin was a "hopelessly complex target". This frustration led to a hectic phase of development of new and increasingly more ingenious protecting group strategies and functional group transformations, and also saw the emergence of asymmetric synthesis.

The last two decades have brought about a stunning wealth of new asymmetric reactions, initially as stoichiometrically controlled processes, where the chiral information is introduced from natural sources through chiral pool molecules (internal asymmetric induction), or through the use of chiral auxiliaries (relayed asymmetric induction). These rapid changes in synthesis technology have placed more and more emphasis on the development of economical routes of introducing the chiral information into the synthetic products. Currently, it is accepted that catalytic asymmetric reactions are the most economical to perform in terms of atom and chiral economy; the chiral information of the catalyst is transferred in a most efficient way from the catalyst to the substrate (external asymmetric induction). Synthetic chemistry is thus trying to mimic one of the pretexts of enzymatic transformations.

It is no wonder that chemists have turned to the nature-evolved enzymes from which to learn and utilize these efficient transformations. Although the subject area of enzyme reactions is still rather unfamiliar territory to the average practitioner of organic synthesis, we firmly believe that such processes and transformations do provide methods which are not easily achieved by classical organic transformations. Enzyme engineering through site-directed mutagenesis will provide additional avenues for refining the enzymes' specificities. Solvent engineering, as described in this book, will provide a very powerful tool to fine-tune enzymatic properties to suit particular needs in synthesis.

Transition metal cayalyzed processes have recently proved to be very efficient, with phenomenal catalytic turnover numbers. Similarly, catalytic antibodies are beginning to provide tailor-made catalysts capable of performing arduous catalytic reactions in an enzyme-like fashion. Combining these technologies with enzymatic transformations will provide the practitioner of synthesis with an armoury unprecedented in the history of chemistry. For example, the synthesis of complex oligosaccharides in a few synthetic transformations, without the need for protecting groups, is one of the dreams coming true at this very moment. It remains to be seen how far one can take this happy marriage between the different fields of chemistry and natural sciences.

It has been a joy to compile this volume, and although the schedule has at times been rather stringent, the contributors have done an excellent job in conveying the most important facets of each topic. It has been a pleasure working with the authors, without whose diligent and painstaking efforts this volume would not have emerged. We also would like to thank the publishers for their help in the practical matters of this task.

A.M.P. Koskinen
A.M. Klibanov
October 1995

Contributors

P. Adlercreutz Department of Biotechnology, Chemical Center, Lund University, PO Box 124 S-22100 Lund, Sweden

C.-S. Chen College of Pharmacology, University of Kentucky, Lexington KY40536, USA

I. Gill Department of Biotechnology and Enzymology, BBSRC Institute of Food Research, Earley Gate, Whiteknights Road, Reading RG6 2EF, UK

G. Girdaukas School of Pharmacy, University of Wisconsin–Madison, Madison WI 53706, USA

L.T. Kanerva Department of Chemistry, University of Turku, Vatselankatu 2, SF-20500 Turku, Finland

H. Kitaguchi Ashigara Research Laboratories, Fuji Photo Film Co. Ltd, 210 Nakanuma, Minamiashigara, Kanagawa, 250-01 Japan

A.M. Klibanov Department of Chemistry, Massachusetts Institute of Technology, Cambridge, MA 02139, USA

A.M.P. Koskinen Department of Chemistry, University of Oulu, Linnanmaa, FIN-90570, Oulu, Finland

S. Riva Istituto di Chimica degli Ormoni, Via Mario Bianco 9, 20131 Milano, Italy

A.J. Russell Center for Biotechnology and Bioengineering, University of Pittsburgh, 300 Technology Drive, Pittsburgh PA 15219, USA

D.B. Sarney Department of Biotechnology and Enzymology, BBSRC Institute of Food Research, Earley Gate, Whiteknights Road, Reading RG6 2EF, UK

R.A. Sheldon Laboratory for Organic Chemistry and Catalysis, Delft University of Technology, Julianalaan 136, 2628 BL Delft, The Netherlands

C.J. Sih   School of Pharmacy, University of Wisconsin–Madison, Madison WI 53706, USA

J.C. Sih   The Upjohn Company, Kalamazoo, MI 49001, USA

E.N. Volfson   Department of Biotechnology and Enzymology, BBSRC Institute of Food Research, Earley Gate, Whiteknights Road, Reading RG6 2EF, UK

Z. Yang   Department of Chemical Engineering, University of Pittsburgh, 300 Technology Drive, Pittsburgh PA 15219, USA

A. Zaks   Schering-Plough Research Institute, 1011 Morris Avenue, U-SA-19, Union NJ 07083, USA

Contents

4 New enzymatic properties in organic media 70

A. ZAKS

5 Enzymatic resolutions of alcohols, esters, and nitrogen-containing compounds 94

C.J. SIH, G. GIRDAUKAS, C.-S. CHEN and J.C. SIH

9 Productivity of enzymatic catalysis in non-aqueous media: New developments   244

E.N. VULFSON, I. GILL and D.B. SARNEY

10 Large-scale enzymatic conversions in non-aqueous media   266

R.A. SHELDON

1 Enzymes in organic solvents: meeting the challenges

A.M.P. KOSKINEN

Enantiopure compounds have undoubtedly gained a central role in the development of modern chemical technology. This is most evident from the changes in the drug market, where single-enantiomer drugs currently occupy a share of some 35 billion US dollars (1). Two-thirds of the world market of the top 25 selling drug wholesale final forms is covered by enantiopure compounds, and the sales are steadily growing at an average annual rate of nearly 7% (see Table 1.1) (2). At the same time, the sales of racemic drugs are diminishing by nearly 30% per year, currently covering only some 1.4 billion dollars. The quest for more efficient, more specifically targeted drugs will place a growing demand on enantiopure materials. The rapid pace of development is evident from the fact that of the 95 best selling drugs in the USA in 1994, 36 single-isomer drugs were new chemical entities approved between 1990 and 1993.

Similar trends can be seen in other specialty and consumer chemical sectors. Environmental issues and the demand for sustainable technology require new chemical materials which are environmentally benign, can be produced without extra burden on the ecosystem, and are truly biodegradable. Polymers, detergents, consumer products, and industrial chemicals all fall prey to these new requirements.

Enzymatic transformations have long been known to be highly stereoselective, but their wider utilization has been slowed down by the commonly held opinion that the substrate-specificity of enzymes cannot be altered easily. Despite some early non-productive suggestions that the course of enzymatic reactions could be reversed, no serious efforts towards this direction were made until the early 1980s, when Alexander Klibanov showed that enzymes do indeed function in

Table 1.1 Worldwide drug market for 25 top-selling prescription drugs

Drug category	Sales, $ billion				% Average annual change
	1993	1994	1995	1996	
Achiral	8.4	9.5	9.8	10.2	7.0
Racemic	3.9	3.0	2.2	1.4	−29.5
Enantiomeric	22.1	23.0	24.8	26.7	6.6

From Stinson (ref. 2).

LIVERPOOL
JOHN MOORES UNIVERSITY
AVRIL ROBARTS LRC
TEL. 0151 231 4022

organic solvents (3), and through this modification one can alter the course the enzymatic reactions take (4). The seminal observations soon led to a fast-growing interest in the possibilities of such a powerful new technology, and only a decade later one can find numerous commercial applications.

The first industrially applied microbiological fermentation, paving the way to the development of the subject area of the present treatise, was the introduction of the 11-hydroxy group into a progesterone nucleus, achieved by the Upjohn Company in 1951 (5). Thus 11α-progesterone could be transformed into cortisone in nine steps, allowing the synthesis of cortisone in only 14 steps from diosgenin (equation 1.1) (6).

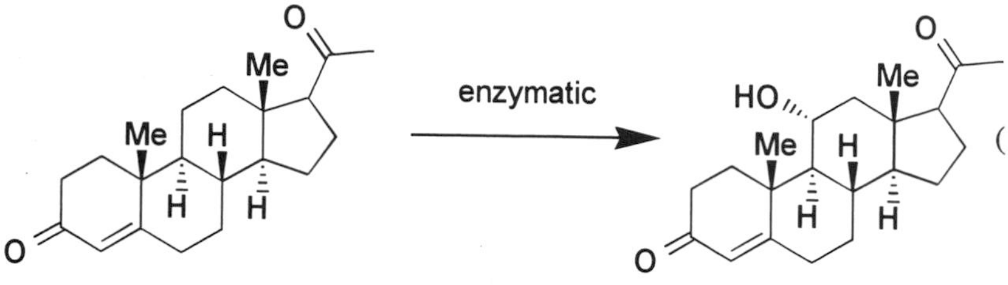

(1.1)

Recent industrial applications can be illustrated by the following two examples. The synthesis of a highly effective HIV-inhibitor, 7-butyroyl-castanospermine, has been achieved through a selective transformation of the 7-hydroxy group into its butyroyl ester by a sequence of three consecutive enzymatic steps (Scheme 1.1). Castanospermine is first transformed into the 1-butyroyl ester with transesterification with subtilisin under non-aqueous conditions. The monoester is converted into the 1,7-dibutyroyl ester with a lipase

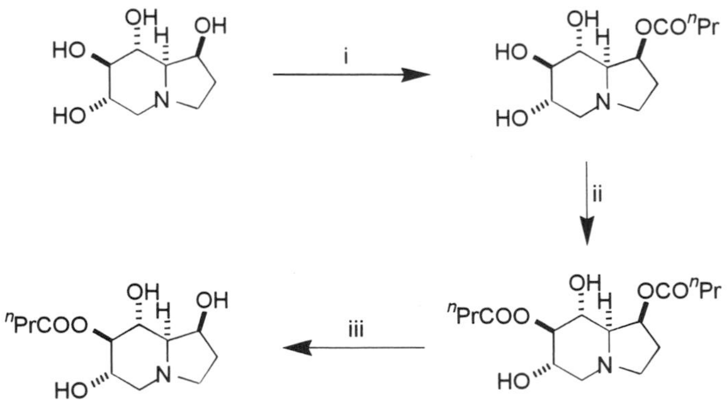

Scheme 1.1 Reagents: (i) Subtilisin, n-PrCOOCH$_2$CCl$_3$, pyridine, 92 h, 84%; (ii) Lipase CV, n-PrCOOCH$_2$CCl$_3$, THF, 72%; (iii) Subtilisin, phosphate buffer, pH 6.0, 64%.

in tetrahydrofuran (THF), and finally the 1-butyroyl ester is cleaved with subtilisin (7, 8).

The Merck synthesis of a leukotriene D_4 antagonist developed for the treatment of asthma, relies on the use of the 'meso trick' (equation 1.2). Treatment of the diester with a lipase releases only one of the acid groups from its methyl ester protecting functions, despite the fact that the chiral center is rather distant from the two enantiotropic functional groups (9).

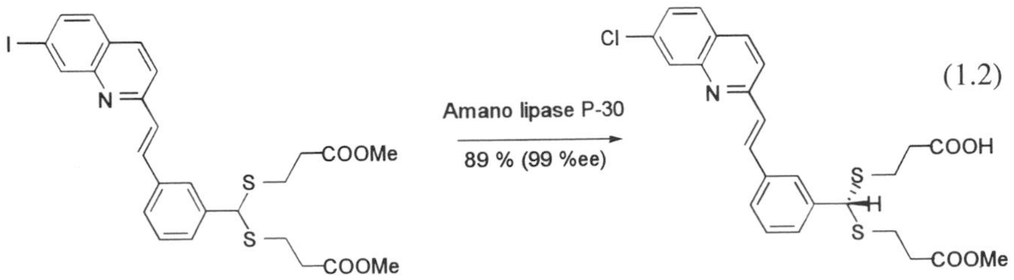

$$(1.2)$$

In the course of the total synthesis of taxol, a promising anti-cancer drug, Nicolaou et al. developed an expeditious synthesis of the functionalized C-ring precursor utilizing a boronic acid tethered Diels–Alder reaction between the hydroxyester 1 and 3-hydroxypyrone 2 (Scheme 1.2) (10, 11). The Diels–Alder product 3 was then converted to the corresponding primary alcohol 4 through a sequence of protection, reduction and deprotection steps in high overall yield. Due to the nature of the initial Diels–Alder reaction, the C-ring precursor 4 was obtained in a racemic form.

The synthesis requires a resolution step, which in the original synthetic sequence was performed at a relatively late stage in the synthesis, only after the construction of the tricyclic taxoid skeleton (compound 5). The resolution, with the associated derivatization and cleavage steps, brought about a major material loss and thus reduced the overall appeal of the synthesis.

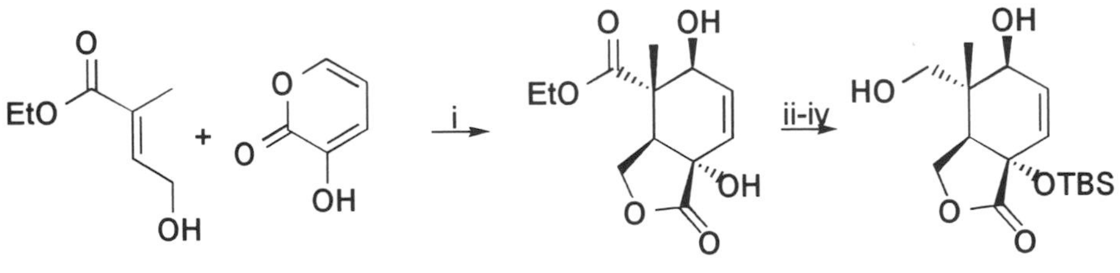

Scheme 1.2 Reagents: (i) PhB(OH)$_2$, PhH, 90°C, then 2,2-dimethyl-1,3-propanediol, 61%; (ii) TBSOTf, 2,6-lutidine, DMAP, CH$_2$Cl$_2$, 0°C, 4 h, 95%; (iii) LiAlH$_4$, Et$_2$O, 0°C, 1 h, 94%; (iv) CSA, MeOH, CH$_2$Cl$_2$, 0°C, 1 h, 90%.

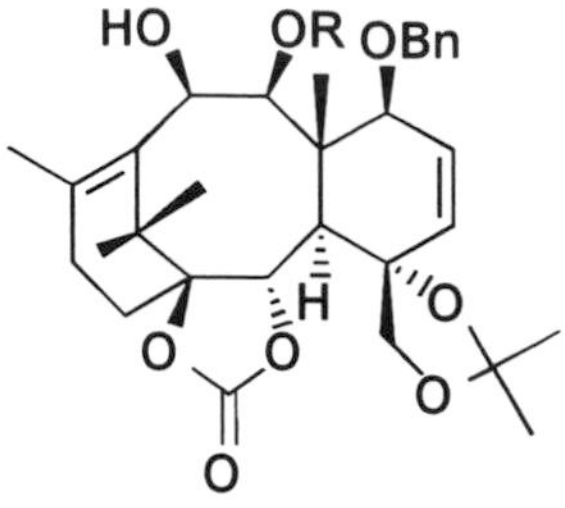

R = camphanoyl

5

Johnson's laboratory at Wayne State University has met with considerable success in the development of enzymatic transesterifications of sterically encumbered neopentyl alcohols (12), and this method was tested for the particular problem at hand (13). Thus, the racemic intermediate **4** was subjected to two consecutive enzymatic kinetic resolution steps with recombinant *Candida antarctica* lipase B (SP-435) in isopropenyl acetate–hexane (Scheme 1.3). The enzyme was produced, after transfer of the genetic coding, by *Aspergillus oryzae*, and immobilized by adsorption on an acrylic resin. Treatment of the racemic alcohol rac-**4** at 50°C in isopropenyl acetate–hexane (1:2.5) for 24 h gave the monoacetate (90%ee) in 51% yield. Separation, and treatment with Hünig's base gave the enantiomeric alcohol (+)-**4**, whose enantiopurity could be enhanced by resubjection to the same enzymatic transesterification conditions

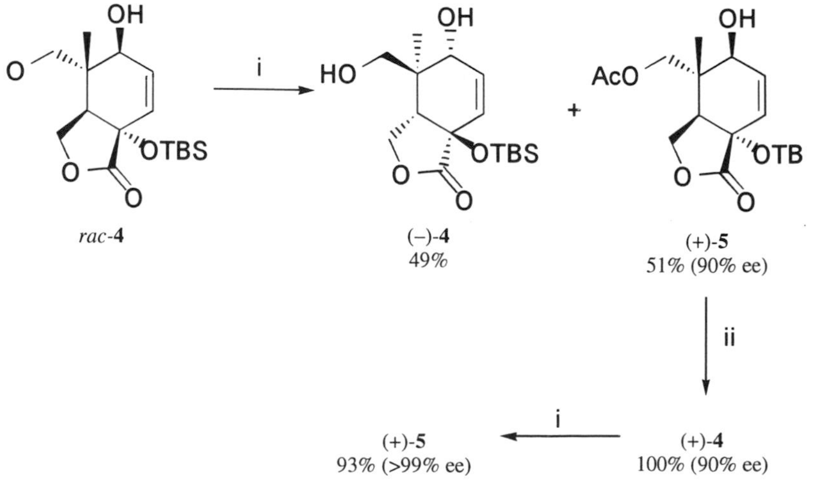

rac-**4**

(−)-**4**
49%

(+)-**5**
51% (90% ee)

ii

(+)-**5**
93% (>99% ee)

(+)-**4**
100% (90% ee)

Scheme 1.3 Reagents: (i) SP-435, isopentenyl acetate–hexane (1 : 2.5), 50°C, 24 h; (ii) i-Pr₂NEt, MeOH, 25°C.

(93% yield, >99%ee). The overall efficiency of enantiopure acetate (+)-**5** was 47% (out of a theoretical 50%).

An interesting application of a combination of microbial oxidation, enzymatic asymmetrization and more classical organic synthetic methods has been utilized in the synthesis of conduritols C (cyclohex-5-ene-1,2,3,4-tetraols), compounds of current interest because of their reported ability to inhibit glycosidases (Scheme 1.4) (14). The synthesis began by microbial oxidation of benzene to the corresponding *meso*-cyclohexa-3,5-diene-1,2-diol with a *Pseudomonas putida* mutant (15). After protection as the acetonide followed by singlet-oxygen oxidation and reduction to the corresponding diol, the stage was set for the enzymatic asymmetrization of the *meso* compound. This was effected by treatment with the crude lipase from *Pseudomonas cepacia* (Amano P-30 lipase) in isopropenyl acetate at 55°C for two days to yield the monoacetate in

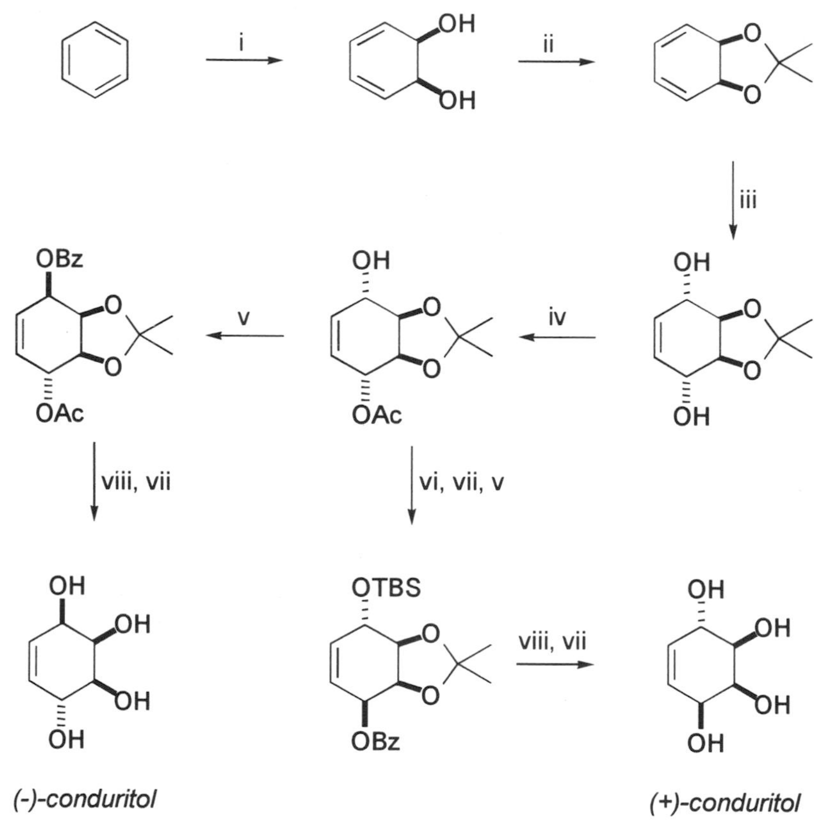

Scheme 1.4 Reagents: (i) Psdeudomonas putida; (ii) Me$_2$C(OMe)$_2$, CH$_2$Cl$_2$, *p*-TsOH, 82%; (iii) O$_2$, *meso*-tetraphenylporphin, CH$_2$Cl$_2$, MeOH, 0°C, hv, 4–6 h, then thiourea (12 h), 65%; (iv) Amano P-30 lipase, isopropenyl acetate, 55°C, 2 days, 90%; (v) PhCOOH, DEAD, Ph$_3$P, THF, 0°C, 30 min, 100%; (vi) TBSCl, imidazole, DMF; (vii) K$_2$CO$_3$, MeOH, 89%; (viii) *p*-TsOH, MeOH, 84% for (–)-conduritol, 66% for (+)-conduritol.

90% yield and >95%ee. The monoacetate could then be converted using standard chemical transformations to either (–)- or (+)-conduritol C in a straightforward manner.

A final example of enzymatically produced important advanced synthetic intermediates is 4-hydroxy-2-cyclopentenyl acetate. This intermediate has been efficiently produced by pancreatin-catalyzed acetylation of the corresponding *meso*-diol (16). The compound itself, and its derivatives shown in Figure 1.1, have gained wide applicability in the synthesis of prostaglandins (17–20), other natural cyclopentanoids (21), and azasugars (22).

The following chapters from eminent scientists in the field will provide the reader with an overview of the current status of the use of enzymatic transformations is organic solvents to attack typical synthetic organic chemistry problems. The first chapters will cover the basics of enzyme use, so as to allow the reader to comprehend what are the special features of enzymatic transformations, how enzymes behave under differing conditions, and what one can do to tame the enzymes in one's own direction. Chapter 2 by Adlercreutz highlights the role of the solvent (including small amounts of water), as well as the different forms of enzyme preparations (solids, supported enzymes and solubilized enzymes), in practical applications. Chapter 3 by Yang and Russell takes a more thermodynamic approach to these effects, discussing the effects of the water activity on enzyme activity. The effects of organic solvents on the structure and dynamics of enzymes are also included in this chapter. The discussion also introduces mathematical models for kinetics and thermodynamics. Chapter 4 by Zaks then illustrates the effects of organic solvents on enzyme specificity and thermal stability. Thermodynamic models for the dependence of substrate specificity on the solvent properties (especially log *P*) are presented.

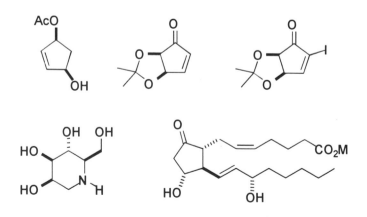

manno-1-deoxynojirimycin   **(-)-Prostaglandin E₂ methyl ester**

Figure 1.1 Structures of 4-hydroxy-2-cyclopentenyl acetate, some of its derived synthetic intermediates and synthetic targets.

After the introductory chapters, the material then covers a range of typical applications. Hydrolases and their use in synthesis are already so widespread that two chapters cover varying aspects of these transformations. Chapter 5 by Sih *et al.* contains kinetic and stereochemical models for lipase stereoselectivity, as well as numerous examples of applications. Chapter 6 by Riva provides an extensive coverage of regioselectivity of hydrolases in the acylation of poly-hydroxylated organic compounds. Chapter 7 by Kanerva contains numerous examples (including some practical experimental details) of the utility of practical kinetic resolutions using lipases and esterases. Mnemonic predictive models as well as kinetic equations and a simple approach to prediction of the extent of kinetic resolution from readily observable parameters are presented and exemplified.

Enzymatic peptide synthesis is beginning to claim victories in efficient frag-ment coupling battles of relatively large peptides. This is amply illustrated in Chapter 8 by Kitaguchi, discussing both thermodynamically and kinetically controlled peptide synthesis. Chapter 9 by Vulfson, Gill and Sarney then illustrates enzymatic solvent-free systems, transformations in heterogeneous eutectic mixtures and the design of continuous bioreactors, with an emphasis on attaining high productivities. Chapter 10 by Sheldon takes a view from the standpoint of large-scale applications. Many enzymatic transformations in organic solvents or solvent-free systems have already found their way into the oleochemical, flavor and fragrance, pharmaceutical, pesticides and polymer industries. The present and future prospects for the growth of these markets are amply illustrated in this concluding chapter.

Enzymatic synthesis definitely has an important role to play in the develop-ment of novel efficient asymmetric processes. Although the widespread use of organic solvents to affect the outcome of enzymatic transformations is barely into its teens, the successes are already numerous, in both laboratory-scale and production-scale applications. More startling future developments will undoubtedly emerge. Full use of designed engineered proteins, solvent engineering and most likely developments such as catalytic antibodies will bring further extensions to the scope of these transformations in the future.

References

1. Stinson, S.C. (1994a) Chiral drugs. *Chem. Eng. News*, Sept. 19, 38–72.
2. Stinson, S.C. (1994b) Market, environmental pressures spur changes in fine chemicals industry. *Chem. Eng. News*, May 16, 10–25.
3. Klibanov, A.M. (1986) Enzymes work in organic solvents. *Chemtech*, **16**, 354–359.
4. Klibanov, A.M. (1989) Asymmetric transformations catalyzed by enzymes in organic solvents. *Accts. Chem. Res.*, **23**, 114–120.
5. Paterson, D.H. and Murray, H.C. (1952) Microbiological oxygenation of steroids at carbon 11. *J. Amer. Chem. Soc.*, **74**, 1871–1872.
6. Mancera, O., Zaffaroni, A., Rubin, B.A., Sondheimer, F., Rosenkrantz, G. and Djerassi, C. (1952) Steroids. XXXVII. A ten step conversion of progesterone to cortisone. *J. Amer. Chem. Soc.*, **74**, 3711–3712.

7. Margolin, A.L., Delinck, D.L. and Whalon, M.R. (1990) Enzyme-catalyzed regioselective acylation of castanospermine. *J. Amer. Chem. Soc.*, **112**, 2849–2854.
8. Therisod, M. and Klibanov, A.M. (1987) Regioselective acylation of secondary hydroxyl groups in sugars catalyzed by lipases in organic solvents. *J. Amer. Chem. Soc.*, **109**, 3977–3981.
9. Hughes, D.L., Bergan, J.J., Amato, J.S., Bhupathy, M., Leazer, J.L., McNamara, J.M., Sidler, D.R., Reider, P.J. and Grabowski, E.J.J. (1990) Lipase-catalyzed asymmetric hydrolysis of esters having remote chiral/prochiral centers. *J. Org. Chem.*, **55**, 6252–6259.
10. Nicolaou, K.C., Yang, Z., Liu, J.J., Ueno, H.M., Nantermet, P.G., Guy, R.K., Claiborne, C.F., Renaud, J., Couladouros, E.A., Paulvannan, K. and Sorensen, E.J. (1994) Total synthesis of taxol. *Nature*, **367**, 630–634.
11. Nicolaou, K.C., Ueno, H., Liu, J.-J. Nantermet, P.G., Yang, Z., Renaud, J., Paulvannan, K. and Chadha, R. (1995) Total synthesis of taxol. 4. The final stages and completion of the synthesis. *J. Amer. Chem. Soc.*, **117**, 653–659, and the preceding papers.
12. Johnson, C.R. and Sakaguchi, H. (1992) Enantioselective transesterifications using immobilized, recombinant *Candida antarctica* lipase B: Resolution of 2-iodo-2-cycloalken-1-ols. *Synlett*, 813–816.
13. Johnson, C.R., Xu, Y., Nicolaou, K.C., Yang, Z., Guy, R.K., Dong, J.G. and Berova, N. (1995) Enzymatic resolution of a key stereochemical intermediate for the synthesis of (–)-taxol. *Tetrahedron Lett.*, **36**, 3291–3294.
14. Johnson, C.R., Plé, P.A. and Adams, J.P. (1991) Enantioselective syntheses of (+)- and (–)-conduritol C from benzene via microbiol oxidation and enzymatic asymmetrization. *J. Chem. Soc., Chem. Commun.*, 1006–1007.
15. Gibson, D.T., Koch, J.R. and Kallio, R.E. (1968) Oxidative degradation of aromatic hydrocarbons by microorganisms. I. Enzymic formation of catechol from benzene. *Biochemistry*, **7**, 2653–2662.
16. Theil, F., Ballschuh, S., Schick, H., Haupt, M., Häfner, B. and Schwarz, S. (1988) Synthesis of (1*S*,4*R*)-(–)-4-hydroxy-2-cyclopentyl acetate by a highly enantioselective enzyme-catalyzed transesterification in organic solvents. *Synthesis*, 540–541.
17. Takano, E., Tanigawa, K. and Ogasawara, K. (1976) Asymmetric synthesis of a prostaglandin intermediate using micro-organisms. *J. Chem. Soc., Chem. Commun.*, 189–190.
18. Nara, M., Terashima, S. and Yamada, S. (1980) Stereochemical studies—LVII. Synthesis of optically active compounds by the novel use of *meso*-compounds—1. Efficient synthesis of two structural types of optically pure prostaglandin intermediates. *Tetrahedron*, **36**, 3161–3170.
19. Noyori, R. and Suzuki, M. (1984) Prostaglandinsynthesen durch Dreikomponentenkupplung. *Angew. Chem.*, **96**, 854–882; *Angew. Chem., Intl. Ed. Engl.*, **23**, 847.
20. Johnson, C.R. and Penning, T.D. (1988) Triply convergent synthesis of (–)-prostaglandin E$_2$ methyl ester. *J. Amer. Chem. Soc.*, **110**, 4726–4735.
21. Harre, M., Raddatz, P., Walenta, R. and Winterfeldt, E. (1982) 4-Oxo-2-cyclopentenyl-acetate in Synthesebaustein. *Angew. Chem.*, **94**, 496–508; *Angew. Chem., Int. Ed. Engl.*, **21**, 480.
22. Johnson, C.R., Golebiowski, A., Schoffers, E., Sundram, H. and Braun, M.P. (1995) Chemoenzymatic synthesis of azasugars: D-*talo*- and D-*manno*-1-deoxynojirimycin. *Synlett*, 313–314.

2 Modes of using enzymes in organic media

P. ADLERCREUTZ

2.1 Introduction

The most common reason for using organic media for enzymatic reactions is that the substrate to be converted is poorly soluble in water. The addition of a moderate amount of organic solvent is a straightforward way to increase the solubility of hydrophobic substrates and thereby make the reaction feasible. Both water-miscible and water-immiscible solvents can be used. In the latter case, two-phase systems are obtained with the enzyme and other hydrophilic substances present in the aqueous phase while hydrophobic substrates and products mainly partition to the organic phase (Figure 2.1). In order for the bioconversion to occur, the substrates must be transferred to the enzyme in the aqueous phase. After the reaction, hydrophobic products are transferred back to the organic phase.

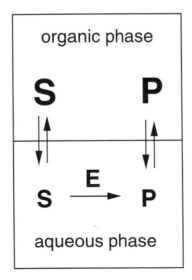

Figure 2.1 Schematic presentation of an enzymatic conversion in a two-phase system. S = substrate, P = product and E = enzyme.

The distribution of the reactants in the aqueous/organic two-phase system can be controlled by choosing a suitable solvent and to some extent by manipulations of the aqueous phase, for example by changing the pH (the pH should be suitable for the enzyme as well). The partitioning of substrates and products to the organic phase is an advantage when substrate and/or product inhibition is a problem in homogeneous systems.

For some applications, reaction media with only small proportions of organic solvents are suitable. However, in most cases it is beneficial to use a reaction medium which is predominantly organic in nature and the present chapter will focus on this type of reaction system. Often there is no macroscopic aqueous phase present; sometimes less than a monolayer of water is bound to the surface of the enzyme. Other reviews have been published on this or closely related topics (Klibanov, 1986; Khmelnitsky et al., 1988; Dordick, 1989; Gupta, 1992; Kvittingen, 1994). Special interest has been devoted to the fundamental principles governing biocatalysis in non-conventional media (Tramper et al., 1992).

Several modes of using enzymes in predominantly organic media have been developed (Figure 2.2). These can be divided into two groups. First, solid enzyme preparations can be used: lyophilized enzyme powders or enzymes on supports. Secondly, the enzyme can be solubilized in organic media. This can be achieved by covalent modification of the enzyme, by the formation of non-covalent enzyme–polymer complexes or by using surfactants either to solubilize the enzyme directly or to create micro-emulsions. The borderline between solid and solubilized enzyme preparations is not strict. In some cases very small particles are formed, and in these systems the enzyme has sometimes been classified as dissolved and sometimes as suspended. In this chapter different examples of the use of solid and dissolved enzyme preparations in organic media will be reviewed, after a short general description of the influence of the solvent and the water content on enzymes operating in predominantly organic media.

2.2 Choice of solvent

In all the solvent-containing biocatalytic systems, the nature of the solvent influences the activity and stability of the enzyme to a large extent. The polarity of the solvent plays a key role. A lowering of the dielectric constant of the medium leads to increased electrostatic interactions between charged residues in the enzyme. This can lead to decreased internal flexibility of the enzyme. Since some flexibility is needed for the catalytic events, a reduced flexibility is normally accompanied by a reduced catalytic activity of the enzyme. Accordingly, high polarity of the active site of an enzyme has been suggested as a factor increasing the catalytic activity (Affleck et al., 1992). The change in dielectric constant also alters the pK values for all titrable residues on the surface of the protein. Such changes in or near the active site can change the binding

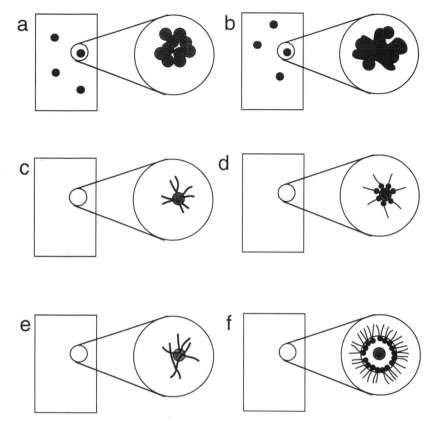

Figure 2.2 A schematic presentation of different modes of using enzymes in organic solvents: (a) enzyme powder suspended in solvent; the small circles in the enlarged figure represent enzyme molecules; (b) enzyme on a support suspended in solvent; (c) covalently modified enzyme dissolved in solvent; (d) enzyme solubilized in solvent with a surfactant; (e) enzyme solubilized in solvent with a polymer; (f) enzyme solubilized in reversed micelles (micro-emulsion).

and/or conversion of substrates, and if the change in dielectric constant is drastic the three-dimensional structure of the protein may be affected.

The presence of an organic solvent always constitutes a risk of enzyme inactivation. Much effort has been spent on trying to generalize how organic substances interact with enzyme molecules and how they influence enzyme stability. When water-miscible solvents are used, moderate concentrations can often be added without negative effects on the enzyme. However, addition of larger amounts, which are sometimes needed to dissolve the substrates, often causes enzyme inactivation. The degree of inactivation depends on which solvent is used, and for each solvent the inactivation occurs rather abruptly when the water concentration decreases below a certain value (Khmelnitsky *et al.*, 1988). Critical water concentrations, at which the enzyme expresses half of its activity in water, were measured for different solvents. The critical water concentration for a number of alcohols, diols and triols increased with increasing log P value of the solvent. Glycerol, the most hydrophilic solvent (lowest log P value) in the group, thus caused least inactivation. Log P is defined as the logarithm of the partition coefficient of a substance in the standard 1-octanol–water two-phase system. Log P values can be determined experimentally by measuring the partitioning of the solvent between octanol and water. Alternatively, log P values can be calculated on the basis of the molecular structure of the solvent by using the hydrophobic fragmental constants of Rekker (1977).

The tendency of solvents to inactivate enzymes does not depend only on their hydrophobicity. Other physicochemical characteristics of solvents, such as the solvating ability and molecular geometry, are of importance as well (Khmelnitsky *et al.*, 1991). Taking into consideration several factors, 'denaturation capacities' of solvents were calculated. For most water-miscible solvents these calculated values correlated well with experimental data of enzyme inactivation. The calculations showed that water-immiscible solvents are not soluble enough in water to reach the inactivation threshold concentration. However, this group of solvents can influence enzymes not only due to the dissolved solvent molecules (molecular toxicity); the presence of a separate organic phase constitutes another risk for enzyme inactivation (phase toxicity) (Bar, 1986). For some combinations of enzymes and solvents the molecular toxicity dominates, and for other cases the phase toxicity is a more serious problem (Ghatorae *et al.*, 1994). The effect of phase toxicity depends on both the nature of the water–organic interface and the interfacial area. A large interfacial area causes increased enzyme inactivation but on the other hand it promotes the desired mass transfer of substrates and products between the phases.

Several studies have been carried out with the aim of rationalizing the effects of water-immiscible solvents on biocatalysts. Log P has often been used as parameter for characterizing the solvents (Laane *et al.*, 1987). Normally solvents in this group with high log P values (> 4) (hydrophobic solvents) cause less inactivation of biocatalysts than more hydrophilic solvents. This trend is thus the reverse of that found for alcohols described above. Accordingly, considering the whole range of solvents, there seems to be a group with intermediate log P values (between 0 and 2) which inactivates biocatalysts more than solvents with both lower and higher log P values (Laane and Tramper, 1990). For some applications, solvents from this group are needed to dissolve the substrates, and in these cases a compromise must often be made, taking into consideration both enzyme stability and substrate solubility.

It should be borne in mind that correlations between simple parameters such as log P or even more complicated ones such as denaturation capacity can never exactly predict the effects of solvents on enzymes in general. As clearly shown in a recent study, there are large individual variations among the enzymes (Ghatorae *et al.*, 1994). For example, a lipase was more stable in hydrophobic solvents while chymotrypsin was more stable in less hydrophobic solvents.

Solid enzyme preparations can normally be used in a wide range of solvents, which gives many possibilities to optimize the choice of solvent with respect to enzyme activity, stability and other important factors. However, in the systems using solubilized enzymes in organic media, the choice of solvent is often more restricted. Enzymes covalently modified with polyethylene glycol show good solubility mainly in aromatic hydrocarbons and in chlorinated hydrocarbons. Aromatic hydrocarbons are useful solvents for enzyme–polymer complexes, as well. In other solvents the complexes form suspensions of fine particles instead of solutions. Micro-emulsions have been described with several different solvents. However, for the majority of the studies of enzymes in micro-emulsions, hydrocarbons have been used as organic solvents.

2.3 Effects of water

The amount of water present in the reaction mixture influences biocatalysis in several ways. Even after lyophilization or other drying procedures, the enzyme

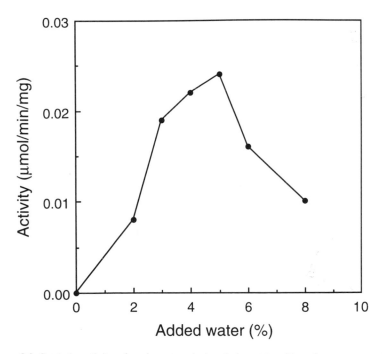

Figure 2.3 Catalytic activity of α-chymotrypsin in ethyl acetate with various amounts of water added. The enzyme was deposited on Chromosorb W/AW and catalyzed the esterification of N-acetyl-L-phenylalanine with ethanol. The solubility of water in the reaction medium was about 4.7%. Data from Wehtje et al. (1993b).

preparation contains some water, which is strongly bound. Although this comparatively dry preparation can be used in organic media, the reaction rate can often be increased considerably by the addition of more water (Figure 2.3). Often an optimal water content can be found; further water addition will cause decreased enzyme activity. Water probably activates the enzyme by increasing the internal flexibility of the enzyme molecule. However, water can also act as a substrate in the enzymatic reaction, especially in the reactions of hydrolytic enzymes. This results in side-reactions and lower product yields.

2.3.1 Quantification of water

When enzymatic reactions are carried out in organic media, water is distributed between the different phases present. Some water is bound to the enzyme and thereby has a large influence on the catalytic activity. Some water is dissolved in the solvent, and if supports, polymers or other substances are present these will bind water as well.

The amount of water in the reaction mixture can be measured in several ways. The most common way is to measure the water concentration (in % by vol. or mol/l). However, due to the distribution of water, the total water concentration does not say very much about the hydration of the enzyme. A better way to characterize the degree of hydration of a biocatalytic system in organic media is to use the thermodynamic water activity as the parameter (Halling, 1984). The water activity determines how much water is bound to the enzyme (Halling, 1990) and this in turn determines the catalytic activity to a large extent (Zaks and Klibanov, 1988a; Adlercreutz, 1991).

It is thus beneficial to work at fixed water activity, for example in studies of the influence of the solvent on enzymatic catalysis. Otherwise the effects due to differences in enzyme hydration will strongly influence the results and mask the effects sought. A typical example of this was seen when reaction rates were compared for the same reaction carried out in different solvents at varying water concentrations. In the different solvents, maximal reaction rate was observed at widely different water concentrations. However, when water was quantified in terms of water activity the optimum was observed at about the same water activity (Valivety et al., 1992).

2.3.2 Reactions at controlled water activity

Both for fundamental studies and for preparative conversions it is of interest to carry out the enzymatic reaction under controlled water activity. A simple method to achieve this is to pre-equilibrate both the enzyme preparation and the substrate solution in atmospheres of controlled water activity before starting the reaction (Goderis et al., 1987; Adlercreutz, 1991). Normally, saturated salt solutions are used to obtain atmospheres with controlled water activity. The method has been used in several kinetic studies. By keeping the reactor in the controlled atmosphere during the reaction, equilibration can take place continuously. However, for preparative conversions equilibration via the gas phase is sometimes too slow, especially if water is formed in the reaction (e.g. in esterification reactions). By pumping the saturated salt solution through a silicone tube which passes through the reactor, much more effective control of the water activity in the reactor can be achieved (Wehtje *et al.*, 1993c). Another method to control the water activity during enzymatic reactions in organic media is to add salt hydrates directly to the reactor (Kuhl and Halling, 1991). Pairs of salt hydrates with different degrees of hydration are in equilibrium at a fixed water activity. As long as both forms of the salt hydrate are present the water activity will remain the same.

2.3.3 Quantification of water in micro-emulsions

In micro-emulsion systems containing reversed micelles, the molar ratio between water and surfactant is often used as a measure to quantify water. The reason is that this ratio determines the size of the micelles and this in turn governs the catalytic activity of the enzyme. Usually maximal activity is obtained when the size of the micelles is equal to the size of the enzyme molecules (Martinek *et al.*, 1989).

2.4 Solid enzyme preparations

2.4.1 Lyophilized enzyme powders and enzyme crystals

Proteins can be dissolved in a few polar organic solvents (Chin *et al.*, 1994). However, in most solvents, enzymes and other proteins are insoluble and therefore heterogeneous catalytic systems are commonly used for enzyme catalysis in organic media. The enzyme forms a solid phase in the bulk organic phase. The most straightforward way of using solid enzymes in organic media is to suspend the solid enzyme directly in the solvent. Enzyme powders obtained by lyophilization have been used in a large number of applications. Typical examples are the resolutions of racemic mixtures using hydrolytic enzymes, for example lipases. Numerous examples have been published in which such an enzymatic step was combined with organic chemical steps to prepare a chiral substances in optically active form (Haraldsson, 1992).

2.4.1.1 'pH control' in organic media.

It is always important that the ionization state of the enzyme is suitable for catalysis. Since protonation and deprotonation of the enzyme seldom occurs to any appreciable extent in organic media, the ionization state of the enzyme must be suitable already in the lyophilized preparation. This is done by adjusting the pH value of the enzyme solution prior to lyophilization. The enzyme keeps its ionization state from the aqueous solution. This has been called the 'pH memory' of enzymes in organic media (Zaks and Klibanov, 1985). To increase the buffering capacity of the system, buffers are sometimes present in the aqueous enzyme solutions that are lyophilized. Even better buffering capacity can be obtained by dissolving buffering substances in the organic phase. Examples of useful substances are trisoctylamine and triphenyl acetic acid (Blackwood *et al.*, 1994). Even though the buffering substances were primarily present in the organic phase, they were able to control the pH in the aqueous phase.

2.4.1.2 Inactivation during lyophilization.   The lyophilization procedure might inactivate the enzyme. To avoid this and thereby increase the activity of lyophilized enzymes in dry organic solvents, the lyophilization can be carried out in the presence of lyoprotectants such as sorbitol (Dabulis and Klibanov, 1993). The inactivation is believed to be caused at least partly by a reversible conformational change in the enzyme. This process can be reversed and the enzyme reactivated by the addition of small amounts of water (Dabulis and Klibanov, 1993).

2.4.1.3 Hydration of enzyme powders.   The morphology of lyophilized enzyme powders under different degree of hydration has been observed directly using environmental scanning electron microscopy (Roziewski and Russell, 1992). Considerable swelling of the enzyme powder was seen at high water activity. The water uptakes by different globular proteins are fairly similar. At a water activity of 0.9 the uptake was around 0.3 g per protein (Bull, 1944). The water adsorption isotherms of proteins in organic media are quite similar to those in air at low water activity, but at high water activity the sorption of water is suppressed by the organic solvents (McMinn *et al.*, 1993). The largest suppression was observed in *n*-propanol while di-*n*-butyl ether and toluene had smaller effects.

2.4.1.4 Active site quantification.   An important question is how many of the active sites of the enzyme are available and active in the solid enzyme preparation when it is suspended in the organic medium. Theoretical considerations make it probable that a large number of the active sites should be accessible in protein crystals and lyophilized amorphous preparations (Faber, 1991). Considering the enzyme molecules as spheres with a monolayer of water around them, there is room for solvent channels within the enzyme preparation so that the substrate molecules can reach the active sites almost directly from the organic phase.

It has been proposed that the enzyme molecules on the surface of the particles are denatured because of the contact with the organic medium. This layer of denatured protein should then protect the rest of the enzyme from inactivation (Khmelnitsky *et al.*, 1988). The ideal way to determine the number of accessible active sites is by titration. This is a standard method for checking preparations of proteases in aqueous media, but it has been necessary to modify the procedure for determinations on enzymes in organic media (Zaks and Klibanov, 1988b; Paulaitis *et al.*, 1992). Active-site titration of lyophilized chymotrypsin and subtilisin showed that about 65% of the active sites were accessible (Zaks and Klibanov, 1988b). The rest of the active sites were probably blocked by protein–protein contacts. They were made accessible by redissolving the enzyme in water (Zaks and Klibanov, 1988b).

2.4.1.5 Enzyme crystals.   Instead of a lyophilized enzyme powder, enzyme crystals can be used as catalysts in organic media. In order to increase the stability, crosslinking with glutaraldehyde has been carried out (Lee *et al.*, 1986). Crosslinking increased the stability of crystals towards dimethoxyethane. Thermolysin was treated in a similar way which made this enzyme preparation more thermostable and stable towards exogeneous proteolysis (St Clair and Naiva, 1992).

2.4.2 Enzymes immobilized on supports

Lyophilized enzyme powders are suitable catalysts for many applications. However, the enzyme particles sometimes tend to aggregate and attach to the walls of the reactor, especially when the enzyme is hydrated to obtain catalytic activity. These problems can be reduced by immobilizing the enzyme on a solid support. Furthermore, enzyme preparations on supports often show considerably higher catalytic activity than enzyme powders. This is at least partially due to spreading of the enzyme on a large area, which makes a larger proportion of the active sites available for catalytic function and facilitates mass transfer of substrates and products. Another possible reason is that the support might protect the enzyme from reversible inactivation during drying or lyophilization and thereby keep the number of active enzyme molecules high. Still another possible explanation is that the support provides a favorable micro-environment for the enzyme during catalysis and therefore the catalytic activity increases.

Especially for large-scale applications, the immobilization of enzymes on supports has been found beneficial and is used to a large extent (Macrae, 1985). For both small- and large-scale applications, it is of importance to choose a suitable immobilization method and a suitable support. Porous supports are widely used. Both their morphological and their chemical characteristics are of importance (Table 2.1). The differences between observed catalytic activities on different supports are sometimes due to mass-transfer limitations or partition effects for water or substrates and products. For practical applications, mechanical properties of the support are of great importance. Fragile particles should not be used in stirred reactors, and in packed-bed reactors it is important that the pressure drop is not too high. Thus the compressibility of the support is of importance. Finally, the price of the support is of course an important factor, especially for large-scale applications.

LIVERPOOL
JOHN MOORES UNIVERSITY
AVRIL ROBARTS LRC
TEL. 0151 231 4022

Table 2.1 Important characteristics of a support
to be used for enzymes in organic media

Morphological characteristics:
    particle size
    pore size
    specific surface areaa

Chemical nature of the support surface:
    water partition characteristics
    substrate/product partition characteristics
    direct effects of the support on the enzyme

2.4.2.1 Immobilization methods.   Since enzymes are insoluble in most of the organic media used, there is usually no need for covalent linkages between the enzyme and the support. One common way to prepare an immobilized enzyme for use in organic media is to mix an aqueous solution of the enzyme with the support and remove the water at reduced pressure. In this way everything in the aqueous solution will be deposited on the support. Quite high enzyme loadings can be obtained with this method and it is always known how much enzyme is present in the immobilized preparation. It should be borne in mind that hydrophilic contaminants in crude enzyme preparations will remain close to the enzyme when it is used in water-immiscible solvents. In aqueous media these contaminants will dissolve and may not influence the enzyme. However, in organic media they will affect the micro-environment of the enzyme and can have a large influence on its catalytic activity. Often contaminants like proteins and carbohydrates can form a suitable 'support material' for the enzyme and thereby have a positive influence. In order to improve the control of the reaction conditions, it is better to use a pure enzyme, which can be immobilized on a suitable support with or without extra additives.

In an alternative immobilization method, the enzyme is precipitated from an aqueous solution in the presence of the support by addition of a cold, water-miscible solvent, such as acetone. Components of the aqueous solution which are soluble in the solvent will not be included in the enzyme preparation. Still another method is to let the enzyme adsorb spontaneously on the support from an aqueous solution. Only substances which interact strongly enough with the support will be included in the immobilized enzyme preparation. This method has been used successfully for making lipase preparations (Ruckenstein and Wang, 1993).

Some mixtures of water and water-miscible solvents can dissolve enzymes. In order to prevent leakage of enzyme from the support, covalent coupling can be used. This can be achieved by using one of the numerous methods developed for enzyme use in aqueous media (Rosevear *et al.*, 1987). Crosslinking of an enzyme adsorbed or deposited on a support with glutaraldehyde is another useful method (Wehtje, 1992).

2.4.2.2 Mass-transfer limitations.   When a solid enzyme preparation is used as catalyst, mass-transfer limitations may influence the overall reaction rate. Concerning the substrate in the reaction, mass transfer both from the bulk liquid phase to the particle surface (external mass transfer) and from the particle surface to the enzyme molecule (internal mass transfer) should be considered. Mass transfer of the reaction product in the opposite direction can be of importance as well. The external mass transfer can be facilitated by increasing the stirring speed in a stirred-tank reactor and by increasing the flow rate in a packed-bed reactor. A couple of studies have shown that under normal conditions for enzymatic catalysis in organic media, external mass transfer does not limit the overall reaction rate (Luck *et al.*, 1988; Ison *et al.*, 1994).

When high catalytic activity is present in a porous particle, diffusion of the substrate through the pores to the enzyme may become rate limiting. To study this effect, a lipase was immobilized on controlled-pore glass supports with the same pore size but different particle size. It was clearly shown that the reaction rate decreased with increasing particle size (Bosley and Clayton, 1994) (Figure

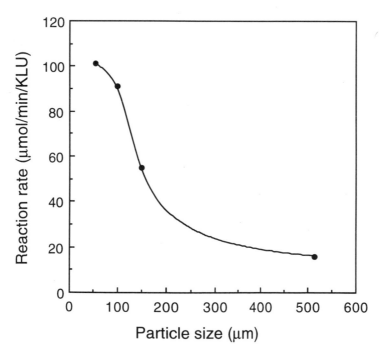

Figure 2.4 Effect of particle size on the catalytic activity of *Rhizomucor miehei* lipase adsorbed on controlled-pore glass with a mean pore diameter of 35 nm. The enzyme loading was 2200 lipase units (LU) per m^2 surface area of the support. The reaction studied was the esterification of oleic acid and 1-octanol without extra solvent. Data from Bosley and Clayton (1994).

2.4). Although few other studies on this topic have been made in organic media, it is likely that internal mass-transfer limitations will occur in other cases when high enzyme loadings and large particles are used. These phenomena have been described in detail for other cases of heterogeneous catalysis (Satterfield, 1970).

2.4.2.3 Influence of pore size.   Few materials have pores of uniform size. In most supports used for enzyme immobilization the pore diameter varies in a wide range (Ison *et al.*, 1994). This makes it difficult to study the effects of pore size on enzyme activity. However, controlled-pore glasses are ideally suited for this kind of study. When a lipase was immobilized by adsorption on controlled-pore glass, it was found that the pore diameter should be at least 35 nm for the lipase to penetrate the pores and cover the available surface area (Bosley and Clayton, 1994). The catalytic efficiency increased with increasing pore size up to about 100 nm. These requirements for large pores may seem surprising considering that the diameter of the lipase molecule is around 5 nm.

2.4.2.4 Influence of surface area.   The surface area of a support is of vital importance for its performance. A large surface area has the positive effect of spreading out the enzyme and thereby making the active sites accessible. However, if a small amount of enzyme is adsorbed on a large surface area, a

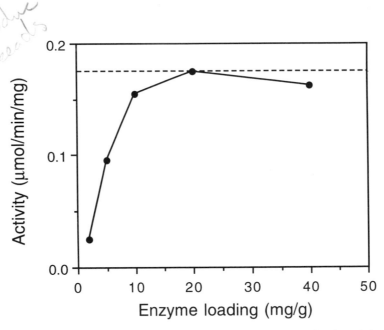

Figure 2.5 Specific activity of a *Pseudomonas* lipase (lipase P) as a function of the enzyme loading on Celite. The reaction studied was the esterification of oleic acid and ethanol in hexane (Wehtje, E., unpublished).

considerable proportion of the enzyme can be inactivated. This was revealed in a study of the influence of the enzyme loading on the specific activity of enzymes deposited on porous supports (Wehtje *et al.*, 1993a). At high enzyme loadings, a reduction in specific activity was observed because of mass-transfer limitations (Figure 2.5). However, a larger decrease in specific activity was found at low enzyme loadings. In order to obtain high specific activity at low enzyme loading, it was necessary to add another protein, such as albumin, or a polymer, such as polyethylene glycol, to the preparation. The additive should be deposited on the support prior to or at the same time as the enzyme, in order to protect it during the drying procedure. The larger the surface area of the support, the larger the amount of additive that was needed (Figure 2.6). In order to obtain full enzymatic activity, the enzyme and the other protein should form a monolayer on the accessible surface area (Wehtje *et al.*, 1993a).

2.4.2.5 Water partition effects. In the reaction mixture, water partitions between the enzyme, the support and the reaction medium. The amount of water present on the enzyme plays an important role in determining its catalytic activity. With a fixed amount of water present in the system, the solubility of water in the reaction medium is of great importance. Similarly, the water-attracting capacity of the support influences the amount of water on the enzyme

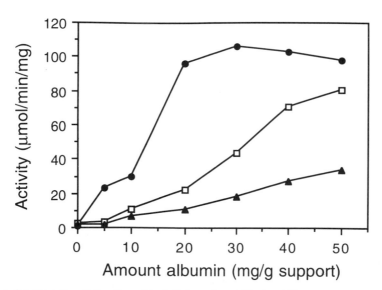

Figure 2.6 Catalytic activity of mandelonitrile lyase immobilized on different controlled-pore glass supports, as a function of the amount of albumin. The enzyme (0.25 mg/g support) was immobilized after the immobilization of albumin. The supports had different specific surface areas: 9.5 m^2/g (●), 25 m^2/g (□) and 83 m^2/g (▲). Data from Wehtje *et al.* (1993a).

and thereby its catalytic activity. A simple test of the water-absorbing capacity of supports has been developed (Reslow *et al.*, 1988). Dry support material was mixed with water-saturated di-isopropyl ether, and after equilibration the water content remaining in the solvent was measured. The amount of water taken up by the support can then be calculated. The 'aquaphilicity' was defined as the ratio of the amount of water on the support to the amount of water in the solvent in the standard system. When enzymes were immobilized on different supports and the catalytic activity was measured at a fixed water concentration, the activity decreased with increasing aquaphilicity of the support (Figure 2.7).

It was observed that the enantioselectivity of subtilisin increased with increasing aquaphilicity of the support when it was used in anhydrous aceto-nitrile (Orsat *et al.*, 1994). This was interpreted as being due to water adsorption on the supports with high aquaphilicity, thereby decreasing the hydration of the enzyme and making it less flexible and therefore more enantioselective.

2.4.2.6 Substrate partition effects. The chemical nature of the support can influence the partition of substrates and products in the system and this can sometimes be of great importance for the reaction rate. This effect has been

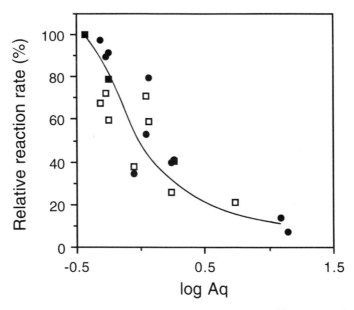

Figure 2.7 The relative reaction rate obtained with α-chymotrypsin (□) and horse liver alcohol dehydrogenase (●) when deposited on support materials with different aquaphilicity (Aq). The reaction catalyzed by α-chymotrpsin was the esterification of *N*-acetyl-L-phenylalanine with ethanol. Alcohol dehydrogenase catalyzed the reduction of cyclohexanone with the concomitant oxidation of ethanol. Di-isopropyl ether was used as reaction medium. The highest reaction rates obtained in each reaction were set at 100%. Data from Reslow *et al.* (1988).

observed in a steroid conversion catalyzed by whole cells immobilized in polyurethanes with varying hydrophobicity (Omata et al., 1979). The reaction rate increased with increasing hydrophobicity of the polyurethane. The increase in reaction rate correlated well with the increase in partition coefficient for the substrate between the polyurethane and the reaction medium. Thus, the high reaction rates were interpreted as being due to an enrichment of the hydrophobic substrate in the hydrophobic gel thereby increasing the local substrate concentration (Omata et al., 1979). It is quite possible that the most important aspect of the substrate enrichment was an increase in mass transfer of the substrate to the cells.

In other cases an enrichment of the substrate in the support is not favorable. The epoxidation of alkenes by *Nocardia corallina* cells is subject to substrate inhibition. The cells were immobilized in mixtures of alginate and silicone with hexadecane as the reaction medium (Kawakami et al., 1992). In the epoxidation of 1-tetradecene, no substrate inhibition occurred and the optimal matrix contained 80–90% silicone. In this matrix the substrate was enriched to a high concentration in the gel, giving a high reaction rate. With 1-octene as substrate, moderate substrate inhibition occurred and it was therefore beneficial to use more alginate in the matrix, so that the local substrate concentration around the cells was somewhat lower. Styrene caused severe substrate inhibition, and as a result pure alginate was the optimal matrix giving a low local substrate concentration (Kawakami et al., 1992).

2.4.2.7 Direct effects of the support on the enzyme. Most of the effects caused by the support on enzymatic activity in organic media can be explained either by differences in support morphology or by effects of partition of water or substrates as described above. However, it is possible that the chemical nature of the support can influence the activity of the enzyme directly. Such effects could be due to the presence of different enzyme conformations on different supports. Surprising effects were observed when an alcoholysis reaction was carried out by chymotrypsin on different supports (Adlercreutz, 1991). The reactions were carried out at fixed water activities to avoid effects of water partition. Large differences between the supports were observed in the relative rates of alcoholysis and the competing hydrolysis reaction. By using a polyamide support and a low water activity (0.33), it was possible to suppress hydrolysis effectively while still maintaining a high alcoholysis activity. At this low water activity, very low catalytic activity was detected on the other supports (Adlercreutz, 1991).

2.4.2.8 Effects of additives. It is obvious that the support is a major factor determining the micro-environment of the enzyme. However, it is possible to modulate the micro-environment by adding substances which partition to the support. These additives may change the catalytic properties of the enzyme. It was possible to achieve high alcoholysis activity for chymotrypsin on Celite at low water activity by using sorbitol as an additive (Adlercreutz, 1993). Sorbitol was co-immobilized with the enzyme on the support. Similar catalytic function of chymotrypsin at low water activity was thus achieved by either immobilization on polyamide (Adlercreutz, 1991) or immobilization on Celite in the presence of sorbitol (Adlercreutz, 1993). The effect of sorbitol can be due either to a direct influence of sorbitol on the enzyme or to an indirect effect due to extra water present in the vicinity of the enzyme due to sorbitol.

In another study of chymotrypsin on Celite, it was found beneficial to use ethylcellulose as an additive (Otamiri et al., 1991). The main effect of the ethylcellulose was probably to act as a buffer of the water activity. This was of prime importance since the reaction studied was an esterification reaction producing water.

In some cases it has been shown useful to add polar solvents to enzymes operating in non-polar solvents (Zaks and Klibanov, 1988a; Reslow et al., 1992). Although the polar solvent often partly dissolves in the non-polar one, at least part of it partitions to the enzyme. The polar solvent thus modifies the microenvironment of the enzyme. It can partly replace water in the activation of the enzyme, but does not act as a substrate. Thus, synthetic reactions are favored and hydrolysis is suppressed.

2.4.2.9 Supports used for enzymes in organic media. Several materials of quite different types and with different properties have been successfully used for different applications of enzymes in organic media (Malcata et al., 1990).

2.4.2.9.1 Inorganic supports Controlled-pore glass is a very useful support for fundamental studies of biocatalysis in organic media (Bosley and Clayton, 1994), but it is too expensive for most practical applications. Another type of inorganic support that has been widely used as enzyme support is diatomaceous earth (kieselguhr). Different types of diatomaceous earth are available at reasonable prices under a variety of names, e.g. Celite. Natural diatomaceous earth contains both large pores (1–20 μm) and much smaller ones. Some types of Celite have been subjected to calcination, a process which reduces the number of small pores. Celite has been used for both large- (Macrae, 1985) and small-scale reactions (Adlercreutz, 1991). In a comparison of several supports, Celite provided the highest reaction rates for chymotrypsin and horse liver alcohol dehydrogenase (Reslow et al., 1988). Other studies have shown that it is suitable for lipases as well (Wisdom et al., 1984). Chromosorb is a calcinated variety of diatomaceous earth that has been used both in systems with quite low water content (Wehtje et al., 1993b) and in applications where a macroscopic aqueous phase was absorbed in its pores (Cambou and Klibanov, 1984). Other frequently used inorganic supports for enzymes in organic media are silica gel and alumina (Brady et al., 1988).

2.4.2.9.2 Synthetic polymers. Several kinds of synthetic polymeric materials have been shown to be suitable supports for enzymes. In a comparison between several organic and inorganic supports, hydrophobic polymers like polyethene and polypropene provided the highest catalytic activities of a *Candida* lipase (Brady *et al.*, 1988). Another good support for lipases is crosslinked polystyrene (Ruckenstein and Wang, 1993). On all these hydrophobic supports, lipases can be effectively adsorbed. The adsorption of other enzymes has been improved by covalent modification of the enzyme with a hydrophobic imidoester before adsorption (Ampon and Means, 1988). With this modification, trypsin and yeast alcohol dehydrogenase were immobilized on an acrylic polyester.

The hydrolytic activity of *Candida* lipase decreased after modification with hydrophobic imidoesters, but the activity in an esterification reaction increased considerably (Basri *et al.*, 1992). The esterification activity increased with increasing degree of modification of the lipase and also with increasing hydrophobicity of the modifying reagent. It was not evaluated whether the increase in activity was due to K_m or V_{max} effects, but the latter alternative seems more probable. In a comparison between different covalent modification methods it was found that modification with monomethoxypolyethylene glycol was the most effective in providing a lipase with high esterification activity and this enzyme was easily adsorbed on polymer beads (Basri *et al.*, 1994). The highest activity of the immobilized lipase was observed on the acrylic polyester XAD-7.

Another way to immobilize enzymes in acrylic polymers is to link the enzyme to acrylic monomers which are then polymerized (Fulcrand *et al.*, 1991). In this particular case, polyethylene glycol was included in the preparation as well, to provide a suitable micro-environment for the enzyme. Several peptide synthesis reactions catalyzed by chymotrypsin were carried out in *t*-amyl alcohol in good yields.

Polyurethane was found to be a suitable support for thermolysin in the synthesis of Aspartame in ethyl acetate (Yang and Su, 1988) and for whole-cell catalyzed steroid conversions (Omata *et al.*, 1979). The hydrophobicity of polyurethanes can be varied by the use of different proportions of the prepolymers (Omata *et al.*, 1979). Polyvinyl alcohol has been used as support for chymotrypsin catalyzing peptide synthesis reactions in acetonitrile (Noritomi *et al.*, 1989).

Ion-exchange resins have been used as supports for lipases to be used in organic media. On suitable materials of this type, the enzyme can easily be immobilized by adsorption, and with supports having pores of proper sizes, high enzyme loadings can be applied (Ison *et al.*, 1994).

2.4.2.9.3 Polysaccharide supports. Agarose gels and other polysaccharide gels are frequently used for covalent immobilization of enzymes for use in aqueous media and to a minor extent they are used in organic media as well. In organic media it was found to be favorable to use multipoint attachment of the enzyme to the support to increase the stability of the enzyme (Blanco *et al.*, 1989, 1992). The agarose beads contained large amounts of water and often a macroscopic aqueous phase was present outside the beads as well.

Alginate is a polysaccharide containing carboxyl groups. Gels of alginate crosslinked with calcium ions have been widely used for the immobilization of whole cells. Recently, calcium alginate gels have been used also for immobilizing enzymes to be used in organic media (Hertzberg *et al.*, 1992). As with the agarose gels, two-phase systems were formed with the aqueous phase inside the beads. The polysaccharide gels are quite hydrophilic. For some applications it is advantageous with a more hydrophobic support. For this reason immobilization of biocatalysts in mixtures of calcium alginate and silicone have been used (Kawakami *et al.*, 1992).

In order to prevent hydrophilic support particles from aggregating in a hydrophobic reaction medium, they can be entrapped in a hydrophobic polymer. In this way, horse liver alcohol dehydrogenase in an aqueous solution was first absorbed in Sephadex particles which were entrapped in silicone before use in hexane (Kawakami *et al.*, 1992).

Chitin, which is built up of 2-amino-2-deoxy-D-glucose residues with about 80% of the amino groups acetylated, was found useful as a support for chymotrypsin catalysing peptide synthesis reaction in acetonitrile (Kise *et al.*, 1987).

2.4.2.9.4 Enzyme–lipid aggregates.   Lipase–lipid aggregates have been formed by using a large number of phospholipid analogues of the dialkyl ether type (Akita *et al.*, 1992). One reason for using lipids containing ether bonds instead of ester bonds was that lipases can hydrolyse ester bonds in phospholipids as well as in the desired substrate molecules. The lipase–lipid aggregates were not soluble in the solvents but showed good catalytic activity compared to other types of enzyme preparations (Akita *et al.*, 1992). The phospholipid can be regarded as a kind of support material for the enzyme. Di-isopropyl ether was the best solvent.

2.5 Solubilized enzyme preparations

2.5.1 Covalently modified enzymes soluble in organic media

2.5.1.1 Modification with polyethylene glycol.   Enzymes can be covalently modified to make them soluble in organic solvents; polyethylene glycol (PEG) has been used extensively for this purpose (Inada *et al.*, 1986). Covalent modification of enzymes with PEG was originally carried out to make the enzymes non-immunogenic (Matsushima *et al.*, 1980), but later it was found that the same

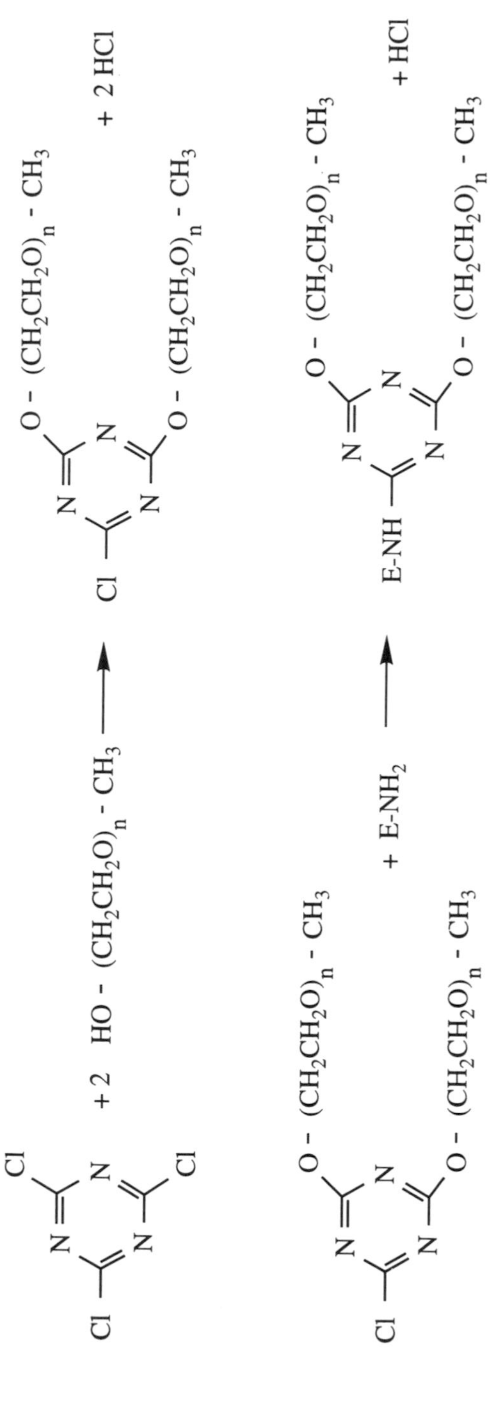

Figure 2.8 Formulae for the activation of monomethoxy poly(ethylene glycol) with cyanuric chloride and the subsequent coupling of the activated polymer with the amino groups of an enzyme.

technique was suitable for making the enzymes soluble in organic solvents (Takahashi *et al.*, 1984b). Normally, the polyethylene glycol is first activated and then the activated derivative is allowed to react with the enzyme. The original activation method employed cyanuric chloride as reagent (Figure 2.8). A monomethoxy-PEG was allowed to react with cyanuric chloride so that two PEG molecules were bound to each cyanuric chloride residue. Amino groups on the enzyme made a nucleophilic attack in the third activated position of cyanuric chloride. In this way two PEG chains were linked per amino group modified. Although several procedures have been described for the coupling of PEG to enzymes, only one other method has seen widespread use to prepare organo-soluble enzymes, and that uses chloroformate activation (Veronese *et al.*, 1985). Still another method involves the oxidation of monomethoxy-PEG to the aldehyde derivative, with subsequent Schiff's base formation with the amino groups of the enzyme and reduction with borohydride (Wirth *et al.*, 1991).

In general, the solubility of the modified enzyme in organic solvent increases with increasing degree of modification. On the other hand, inactivation of the enzyme occurs during the derivatization, and this effect increases with increasing degree of modification (Figure 2.9). Often a compromise with about half of the amino groups modified provides useful preparations.

The question of the size of the enzyme-containing entities has been addressed by Halling and coworkers (Khan *et al.*, 1992). It was shown that in organic solvents, PEG-modified subtilisin was present as a suspension of microparticles. The size of the particles was estimated by using dynamic light scattering to be at least 300 nm. The presence of particles in the preparations means that mass-transfer limitations may limit the overall reaction rate. However, in a comparison of the catalytic activities of different kinds of preparations of a *Pseudomonas* lipase, the highest V_{max} values were obtained with the PEG-modified preparation (at high water activity) (Bovara *et al.*, 1993). K_m and V_{max} values for a large number of substrates of a *Pseudomonas* lipase have been determined (Nishio *et al.*, 1988).

PEG-modified enzymes in organic media bind water (Takahashi *et al.*, 1984a). The water concentration or the water activity in the reaction medium greatly influences the catalytic activity of the enzyme, as in other types of enzyme preparations in organic media. In protease-catalysed peptide synthesis reactions, the V_{max} values increased water concentration (Gaertner and Puigserver, 1989; Ljunger *et al.*, 1993).

2.5.1.1.1 Synthetic applications.   PEG-modified enzymes have been used to a limited extent for synthetic applications. Compared to solid enzyme preparations, PEG-modified enzymes are slightly more difficult to recover and reuse. However, there are at least two rather convenient ways of doing this. Either the modified enzymes are precipitated with a non-polar solvent and redissolved in a new batch of substrate solution or they can be retained in the reactor with a suitable membrane.

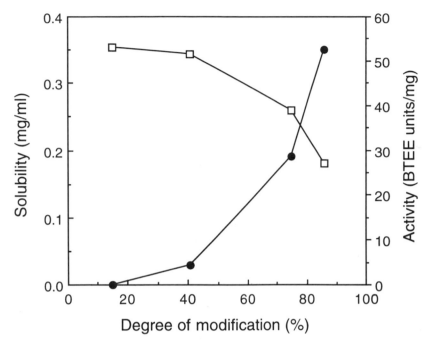

Figure 2.9 Solubility in benzene (●) and catalytic activity (□) of α-chymotrypsin after coupling with the polyethylene glycol 5000. The degree of modification is defined as the relative amount of amino groups that have been modified. The reaction studied was the hydrolysis of benzoyltyrosine ethyl ester (BTEE). Data from Ljunger *et al.* (1993).

Solubilized enzymes have advantages for the conversion of solid substrates. This effect has been utilized in peptide synthesis. Using PEG-modified enzymes, an enzymatic solid-phase synthesis method was developed in which the peptide chain is allowed to grow on a solid resin (Sakurai *et al.*, 1990), in analogy with the widely used methods for chemical solid-phase peptide synthesis.

2.5.1.2 Modification with polystyrene. Polystyrene-linked trypsin has been prepared in a two-step process (Ito *et al.*, 1993). Azobis(4-cyanovaleric acid) was coupled with the amino groups of trypsin (about 50% of the amino groups were modified). Polymerization of styrene was carried out in the presence of the modified enzyme, thus yielding polystyrene-linked trypsin. This preparation expressed catalytic activity in chloroform while unmodified trypsin was inactive. Using similar procedures, a *Pseudomonas* lipase was covalently linked to poly(*N*-vinylpyrrolidone), polystyrene and poly(methyl methacrylate) with polymer chains of molecular weights between 27 000 and 160 000 (Ito *et al.*, 1992). Transparent solutions were formed with all the polymers and the solubility increased with increasing molecular weight of the polymers. The three polymers provided modified enzymes with approximately the same specific activity.

2.5.1.3 Modification with polyacrylates.   Enzymes have been linked to microgels or latex particles. The particles can be dissolved in water or in organic solvents, depending on the properties of the polymers. In one example, a polymer was formed using acrylic acid, methyl methacrylate and 2-ethoxyethyl methacrylate as monomers (Davey *et al.*, 1989). The particles were practically monodisperse with a diameter of 44.1 nm in the dry state and swelled both in aqueous buffer and in organic solvents. Chymotrypsin and a couple of other enzymes were linked to the carboxylic acid groups of the polymer using carbodi-imide coupling, so that about 29% of the wet microgel surface was covered by enzyme. The kinetic constants for hydrolysis reactions were determined in methanol and tetrahydrofuran with low water content.

2.5.1.4 Other covalent modification methods.   Covalent coupling of a variety of hydrophobic groups has been used for making enzymes more suitable for catalysis in organic media. Increased solubility or dispersibility was achieved with reagents like dinitrofluorobenzene (Scott *et al.*, 1993), palmitoyl chloride (Dordick *et al.*, 1986) and *p*-dimethyl sulfonio phenyl ester (Murakami *et al.*, 1993). The latter two reagents introduce acyl groups, mainly on amino groups of the lysine side chains. When trypsin and chymotrypsin were modified by using this method, the optimal hydrolytic activity in mixtures of water and dimethylformamide was observed after modification with decanoyl groups (Kawasaki *et al.*, 1994). The modification of the enzymes caused a decrease in K_m but had no significant effect on k_{cat}.

Reductive alkylation with aldehydes such as acetaldehyde or octaldehyde increased the activity of trypsin (Ampon *et al.*, 1991). The reaction rate in anhydrous dimethylformamide was 5–6 times that of unmodified enzyme in the esterification of glucose with oleic acid. The optimal degree of modification was about 15% for the octyltrypsin and 50% for the ethyltrypsin (Ampon *et al.*, 1991).

Covalent modification of enzymes in order to improve adsorption on supports was discussed in the section dealing with enzymes immobilized on supports.

2.5.2 Non-covalent complexes soluble in organic media

2.5.2.1 Enzyme–surfactant complexes.   Surface-active lipids have been used to solubilize lipases in organic solvents. Non-ionic or cationic surfactants were able to form complexes with the lipase, but anionic and zwitterionic surfactants were ineffective (Okahata and Ijiro, 1988). Especially good results were obtained with didodecyl glucosyl glutamate (Figure 2.10). Upon mixing of aqueous solutions of the enzyme and the surfactant, the complex precipitated (Okahata and Ijiro,

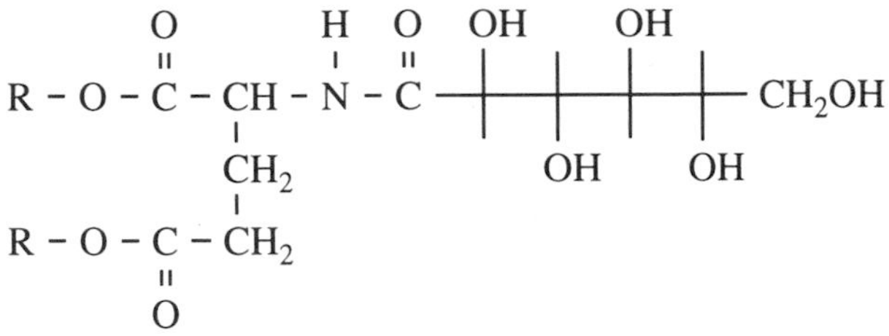

Figure 2.10 Dialkyl glucosyl glutamate. This group of surfactants is used to solubilize enzymes in organic media. Typically R = dodecyl.

1992). Good solubility of the complex and enzymatic activity was observed in di-isopropyl ether, benzene, toluene and hexane. In chlorinated hydrocarbons, no activity was detected although the complexes were soluble. In this procedure about 150 surfactant molecules were attached to each lipase molecule, which resulted in complexes with a molecular weight of about 130 kDa (Tsuzuki *et al.*, 1991a). The catalytic activity was found to increase with increasing alkyl chain length (between C8 and C16) in dialkyl glucosyl glutamates, but it was not investigated whether this was due to differences in K_m or V_{max} (Tsuzuki *et al.*, 1991b). The slightly different surfactant dioleyl ribityl glutamate has been used to solubilize lipases in organic media as well (Goto *et al.*, 1993). The preparation showed catalytic activity when dissolved in hydrocarbons (Goto *et al.*, 1994).

It has been shown that subtilisin Carlsberg can be solubilized in octanol by *n*-octyl α-D-glucoside (Blinkovsky *et al.*, 1994). The solubilized enzyme was four times as active as the suspended enzyme powder. Even somewhat higher activity was obtained after solubilization with palmitoylraffinose.

The surfactant Aerosol OT (dioctyl sulfosuccinate, sodium salt), which is widely used in the preparation of micro-emulsions (see below) has been used to solubilize chymotrypsin in iso-octane (Paradkar and Dordick, 1994). The number of surfactant molecules participating in the solubilization was only about 30 per enzyme molecule, which is much less than in the case of reversed micelles.

2.5.2.2 Enzyme–polymer complexes. Enzymes can be dissolved in organic media using polymers as solubilizing agents. Complexes of ethylcellulose and chymotrypsin were soluble and catalytically active in toluene (Otamiri *et al.*, 1992a). The most efficient way to prepare the complexes was to lyophilize the enzyme and the polymer from an aqueous solution and then dissolve the preparation in the solvent. Salts in the aqueous solution were of importance for the complex formation. Other polymers, such as poly(vinyl butyral), poly(methyl methacrylate), poly(vinyl methyl ketone) and poly(ethylene glycol) were able to form soluble complexes with enzymes as well (Otamiri *et al.*, 1992b). The ethylcellulose complexes of chymotrypsin have been further characterized by light-scattering measurements. The apparent molecular weights were 65–149 kDa, indicating that the complexes were small, containing just a few enzyme and polymer molecules. An increase in the total concentration of ethyl-cellulose–chymotrypsin preparation resulted in an increase in apparent molecular weight of the complexes (Otamiri et al., 1992a). The enzyme–polymer complexes showed higher denaturation temperatures in toluene in differential scanning calorimetric studies than polymer-free preparations (Otamiri et al., 1994).

From a practical point of view it is often important that the enzyme and the product can be separated, so that a pure product can be isolated and the enzyme can be reused. When one is using the enzyme–polymer complexes, this can easily be achieved by just adding water to the system (Otamiri et al., 1992b). Addition of water causes phase separation, and the enzyme can be recovered from the aqueous phase while hydrophobic products partition to the organic phase.

Enzyme–polymer complexes have also been formed using alkylated poly-(ethyleneimine) as solubilizing agent (Vakurov et al., 1994). In the dry state the enzyme–polymer complexes had good storage stability.

With palmitoyl poly(sucrose acrylate) as solubilizing agent the activity of subtilisin Carlsberg was increased 75 times compared to the suspended enzyme (Blinkovsky et al., 1994). Light-scattering measurements on the enzyme–polymer complexes indicated hydrodynamic radii of 40–50 nm, small enough to include single enzyme molecules forming complexes with multiple polymer molecules.

2.5.2.3 Surfactant-coated nanogranules. Micro-emulsions (see below) have been used to prepare surface-modified nanogranules containing entrapped enzymes (Khmelnitsky et al., 1989). A micro-emulsion was formed containing the enzyme and a monomer in the aqueous microdroplets. In addition to the normal surfactant Aerosol OT, a polymeric surfactant was used. The monomers were polymerized forming small polymer spheres (nanogranules) containing entrapped enzyme. Aerosol OT was removed by washing and the nanogranules surrounded by the polymeric surfactant (e.g. Pluronic F-108) were obtained. The diameter of the polyacrylamide core and the overall (hydrodynamic) diameter of the nanogranules were about 8 and 58 nm, respectively. The nanogranules were soluble in a range of organic solvents including toluene, carbon tetrachloride, acetone and ethanol (Khmelnitsky et al., 1989).

2.5.3 Enzymes in micro-emulsions

One way to dissolve enzymes in low water systems is to solubilize them in micro-emulsions. Micro-emulsions are macroscopically homogeneous and isotropic mixtures of water, organic solvent and surfactant. Since micro-emulsions are thermodynamically stable, they can be formed by just mixing the components. In a typical example an aqueous solution of the enzyme to be solubilized is mixed with a solution of a surfactant in the organic solvent. After mixing a clear solution is formed.

On a microscopic scale, micro-emulsions are structured into aqueous and oil microdomains separated by a surfactant-rich film. Under some conditions reversed micelles (Figure 2.2) are formed and with other proportions of the components both the aqueous and the organic phase are continuous. Enzymes can be solubilized in the aqueous microdomains with high retention of catalytic activity. Numerous studies of enzymatic catalysis have been carried out in micro-emulsions, predominantly in systems containing reversed micelles. Here, only a minor part of this work can be presented. More thorough reviews have been made (Luisi and Laane, 1986; Martinek et al., 1989; Sánches-Ferrer and García-Carmona, 1994).

2.5.3.1 Surfactants and solvents used in micro-emulsions. The most commonly used surfactant in studies of enzymes in micro-emulsions is Aerosol OT (dioctyl sulfosuccinate, sodium salt). This anionic surfactant promotes the formation of reversed micelles because it has two hydrophobic chains, which furthermore are branched. The hydrophobic part is thus considerably bulkier than the hydrophilic head group and this favors the formation of reversed micelles. Other commonly used surfactants are cetyl trimethyl ammonium bromide (CTAB) and various non-ionic surfactants like those in the Brij, Triton and Tween series. The organic solvent in the micro-emulsion is often a hydrocarbon or a chlorinated hydrocarbon e.g. chloroform.

2.5.3.2 Spectroscopic studies. Since micro-emulsions are optically transparent they are well suited to spectroscopic investigations. Detailed studies of fluorescence and circular dichroism (CD) spectra of lysozyme in various reversed micellar systems have been carried out (Steinmann et al., 1986). In most systems the enzyme had a CD spectrum quite similar to that in water, showing that the solubilization of the enzyme in the micro-emulsions did not cause any drastic structural changes (Steinmann et al., 1986).

2.5.3.3 Preparative conversions. Mass-transfer limitations do not normally occur in micro-emulsions due to the large interfacial area. Under some conditions, large amounts of water can be included in the micro-emulsions, which is an advantage in reactions involving both hydrophilic and hydrophobic reactants. In preparative conversions, the presence of the surfactant can create problems during the isolation of the product. However, a few approaches towards solving this problem have been presented. One approach is to use membrane reactors. The membrane should be permeable to the reaction product but retain the enzyme and the surfactant within the reactor (Luisi and Laane, 1986). However, it is difficult to avoid severe losses of surfactant through the membrane. One way to solve this problem is to use a polymeric surfactant. Poly(ethyleneimine) modified with cetyl bromide and ethyl bromide has been used for this purpose (Khmelnitsky et al., 1992). With this surfactant, enzymes were solubilized in reversed micelles having sizes in the range 22–51 nm, using benzene as organic phase and n-butanol as cosurfactant. In addition to being suitable for membrane reactors, the polymeric surfactant also functioned as a pH buffer in the system (Khmelnitsky et al., 1992).

Another way to isolate the reaction product and reuse the enzyme has been presented for reactions in bicontinuous micro-emulsions (Larsson et al., 1990a). The reaction was carried out in the micro-emulsion. After the desired reaction time, the temperature was changed by a few degrees, which caused the conversion of the micro-emulsion into a two-phase system. One phase consisted of the organic solvent in which hydrophobic compounds were dissolved and the other phase contained all the enzyme and almost all the surfactant. By replacing the organic phase with a new substrate solution it was possible to repeat the conversion process without adding new enzyme or surfactant.

In case the substrate is hydrophobic and the product hydrophilic, it can be advantageous to carry out the reaction in a Winsor II system which consists of a water-in-oil micro-emulsion coexisting with an aqueous phase. Under proper conditions the substrate and the enzyme partition mainly to the micro-emulsion phase while the product partitions mainly to the aqueous phase. This behavior has been demonstrated in the chymotrypsin-catalyzed resolution of DL-phenyl-alanine methyl ester (Towey et al., 1994). It was possible to speed up the reaction by vigorous stirring of the reaction mixture, thus creating an emulsion. When stirring was interrupted, phase separation occurred. A drawback was that even under optimal conditions a considerable fraction of the enzyme (20%) partitioned to the aqueous phase. In order to solve this problem, immobilized enzyme in a 'tea-bag' was used for a similar chymotrypsin-catalyzed resolution of amino acids (Bhalerao et al., 1994).

2.5.3.4 Detergentless micro-emulsions. Since the surfactant causes problems in product isolation and sometimes causes enzyme inactivation, detergentless micro-emulsions have been tried as reaction media for enzymes (Khmelnitsky et al., 1987). When water, hexane and isopropanol were mixed in proper proportions for micro-emulsion formation, the catalytic activity of the enzyme (trypsin) was much higher than in other mixtures of the same components. The isopropanol probably functioned as a surfactant, stabilizing microdroplets of water.

2.5.4 Other surfactant-containing systems

Mixtures of water, organic solvent and surfactant can form several types of phases. Besides the reversed micellar phases, different liquid crystalline phases with lamellar, hexagonal and cubic structures can be formed. The last two groups can have either 'normal' or 'reversed' molecular orientation. Enzymes can express catalytic activity when solubilized in all these phases (Klyachko et al., 1986). The enzymatic activity varied with the water content in all cases and when a change in water content caused a transition from one type of structure to another, an abrupt change in activity was sometimes observed. Because of their high viscosity, the pure liquid crystalline systems are of little practical interest for biocatalysis. However, under some conditions a liquid crystalline phase can co-exist with an organic solvent phase, and this has been shown to be a useful system (Miethe et al., 1989). Mass-transfer limitations were not encountered to a large extent. The enzyme (yeast alcohol dehydrogenase) showed catalytic activity similar to that observed in water, and after the reaction the liquid crystalline phase was removed by simple filtration.

2.6 Other non-conventional reaction media

Although not covered in detail in this review, a couple of other non-conventional media for enzymatic reactions are mentioned because they have much in common with organic media. Supercritical and near-supercritical fluids have been used for enzymatic reactions in a large number of studies. For a review see Nakamura (1990). The most widely used medium in this group is supercritical carbon dioxide. A main advantage is that the supercritical fluid can be removed easily after the reaction by just decreasing the pressure. Furthermore, supercritical carbon dioxide has low toxicity compared to organic solvents. For this reason it has already replaced organic solvents in a few large-scale extraction processes. As solvent, supercritical carbon dioxide resembles hexane. Consequently, hydrophobic compounds can be dissolved, but the solubility of more polar compounds is low. To improve the solubility of moderately hydrophobic compounds, organic cosolvents can be added. Enzymes are normally not soluble in the supercritical fluids, and therefore heterogeneous systems with solid enzyme preparations are normally used.

A potential advantage with supercritical media is the high diffusivity and low surface tension which leads to low mass-transfer resistance. However, most kinetic comparisons of enzymatic reactions have shown quite similar reaction rates in organic solvents and supercritical fluids (Dumont et al., 1992). A drawback with supercritical fluids is the high pressures needed, making special equipment necessary.

Another interesting approach is to carry out enzymatic reactions with gaseous substrates. This topic has recently been reviewed (Lamare and Legoy, 1993). The enzyme preparation is present in the form of a solid, which can be packed in a column through which the gaseous substrate is passed (Lamare and Legoy, 1995). Enzymes used in this way are hydrogenase, alcohol oxidase and lipases. It is often advantageous to use a thermostable enzyme, since rather high reaction temperatures are used to keep the substrate in the gas phase (Robert et al., 1992).

As with organic reaction media, the water activity in the reactor plays a major role in determining the catalytic activity of the enzyme (Barzana *et al.*, 1989; Lamare and Legoy, 1995).

2.7 Comparisons between different modes of using enzymes in organic media

When it has been recognized that it is preferable to carry out a certain reaction in an organic medium, the question arises as to which type of reaction system to choose. The following points should be considered:

Enzyme preparation procedure
Product isolation and enzyme reuse
Enzymatic activity
Enzymatic stability.

The most easily prepared enzyme catalyst is the enzyme powder, but most of the other preparations mentioned here do not involve complicated procedures either. The exception is covalent modification which often involves one derivatization step and one purification step before the modified enzyme can be used.

Product isolation and enzyme reuse are very easy when solid enzyme preparations are used. When solubilized enzymes are applied, enzyme recovery is considerably more complicated. It often involves precipitation of the enzyme or a change of conditions so that a two-phase system is formed. Preferably, the enzyme and other compounds (surfactants, etc.) which should be reused should partition to one phase and a substantial portion of the product to the other phase. An alternative to these methods is to use membranes to retain the enzyme in the reactor, but this method has not been widely used for organic reaction media.

During the preparation of the enzyme catalyst, inactivation might occur. Examples of procedures that can cause inactivation are covalent modification, lyophilization and immobilization on supports with large surface areas. The observed rate of the enzymatic reactions depends both on the catalytic activity of the enzyme and on mass-transfer limitations.

An important point in the comparison of different enzyme preparations is that the influence of the enzyme hydration should be considered. Preferably, the comparison should be made at fixed water activity or in a range of water activities/water concentrations. A study carried out in this way indicated that support-immobilized enzyme is preferable at low water activity and PEG-modified enzyme at high water activity (Bovara *et al.*, 1993). The directly added lipase powder expressed low activity in the entire range of water activities. Another study showed that both chymotrypsin and horse liver alcohol dehydrogenase expressed the highest catalytic activity when solubilized in a microemulsion (Larsson *et al.*, 1990b). The enzymes expressed intermediate activity when deposited on Celite and considerably lower activity when the enzymes were used directly in the solvent without support. This indicates that the solubilization of enzymes is beneficial when a maximal reaction rate is the main goal. Furthermore, the use of a proper support material can increase the activity of enzymes in solid preparations.

Comparisons of the stability of enzymes in different forms in organic media are not straightforward to carry out. Changes in mass-transfer conditions or enzyme hydration can easily dominate over the real changes in enzymatic activity during operation. The operational stability of horse liver alcohol dehydrogenase was better when solubilized in a micro-emulsion or used directly in the solvent, compared to when it was deposited on Celite (Larsson *et al.*, 1990b). On the other hand, chymotrypsin was more stable when it was deposited on Celite or used directly in the solvent, than in a micro-emulsion (Larsson *et al.*, 1990b). More comparative data are needed before any generalizations concerning the influence of type of enzyme preparation on the stability in organic media can be made.

Considering all factors, it is clear that there is no single best mode of using enzymes in organic media. The use of enzymes on supports has important advantages in large-scale synthetic processes and is therefore used to a large extent in the present industrial applications. The solid enzyme preparations can be recommended as a first choice for synthetic applications on a small scale as well. The solubilized preparations are advantageous for spectroscopic studies and have a potential for providing the highest reaction rates.

References

Adlercreutz, P. (1991) On the importance of the support material for enzymatic synthesis in organic media. Support effects at controlled water activity. *Eur. J. Biochem.*, **199**, 609–614. —

Adlercreutz, P. (1993) Activation of enzymes in organic media at low water activity by polyols and saccharides. *Biochim. Biophys. Acta*, **1163**, 144–148.

Affleck, R., Xu, Z.-F., Suzawa, V., Focht, K., Clark, D.S. and Dordick, J.S. (1992) Enzymatic catalysis and dynamics in low-water environments. *Proc. Natl. Acad. Sci. USA*, **89**, 1100–1104.

Akita, H., Umezawa, I., Matsukura, H. and Oishi, T. (1992) A lipid–lipase aggregate with ether linkage as a new type of immobilized enzyme for enantioselective hydrolysis in organic solvents. *Chem. Pharm. Bull.*, **40**, 318–324.

Ampon, K. and Means, G.E. (1988) Immobilization of proteins on organic polymer beads. *Biotechnol. Bioeng.*, **32**, 689–697.

Ampon, K., Salleh, A.B., Teoh, A., Yunus, W.M.Z.W., Razak, C.N.A. and Basri, M. (1991) Sugar esterification catalyzed by alkylated trypsin in dimethylformamide. *Biotechnol. Lett.*, **13**, 25–30.

Bar, R. (1986) Phase toxicity in multiphase biocatalysis. *Trends Biotechnol.*, **4**, 167.

Barzana, E., Karel, M. and Klibanov, A.M. (1989) Enzymatic oxidation of ethanol in the gaseous phase. *Biotechnol. Bioeng.*, **34**, 1178–1185.

Basri, M., Ampon, K., Yunus, W.M.Z.W., Razak, C.N.A. and Salleh, A.B. (1992) Amidination of lipase with hydrophobic imidoesters. *J. Amer. Oil Chem. Soc.*, **69**, 579–583.

Basri, M., Ampon, K., Yunus, W.M.Z.W., Razak, C.N.A. and Salleh, A.B. (1994) Immobilization of hydrophobic lipase derivatives on to organic polymer beads. *J. Chem. Tech. Biotechnol.*, **59**, 37–44.

Bhalerao, U.T., Sreenivas Rao, A. and Fadnavis, N.W. (1994) A novel tea-bag methodology for enzymatic resolutions of α-amino acid derivatives in reverse micellar media. *Synth. Commun.*, **24**, 2109–2118.

Blackwood, A.D., Curran, L.J., Moore, B.D. and Halling, P.J. (1994) 'Organic phase buffers' control biocatalyst activity independent of initial aqueous pH. *Biochim. Biophys. Acta*, **1206**, 161–165.

Blanco, R.M., Guisán, J.M. and Halling, P.J. (1989) Agarose–chymotrypsin as a catalyst for peptide and amino acid ester synthesis in organic media. *Biotechnol. Lett.*, **11**, 811–816.

Blanco, R.M., Halling, P.J., Bastida, A., Cuesta, C. and Guisán, J.M. (1992) Effect of immiscible organic solvents on activity/stability of native chymotrypsin and immobilized-stabilized derivatives. *Biotechnol. Bioeng.*, **39**, 75–84.

Blinkovsky, A.M., Khmelnitsky, Y.L. and Dordick, J.S. (1994) Organosoluble enzyme–polymer complexes: a novel type of biocatalyst for nonaqueous media. *Biotechnol. Techniques*, **8**, 33–38.

Bosley, J.A. and Clayton, J.C. (1994) Blueprint for a lipase support: Use of hydrophobic controlled-pore glasses as model systems. *Biotechnol. Bioeng.*, **43**, 934–938.

Bovara, R., Carrea, G., Ottolina, G. and Riva, S. (1993) Effects of water activity on V_{max} and K_m of lipase catalyzed transesterification in organic media. *Biotechnol. Lett.*, **15**, 937.

Brady, C., Metcalfe, L., Slaboszewski, D. and Frank, D. (1988) Lipase immobilized on a hydrophobic, microporous support for the hydrolysis of fats. *J. Amer. Oil Chem. Soc.*, **65**, 917–921.

Bull, H.B. (1944) Adsorption of water by proteins. *J. Amer. Chem. Soc.*, **66**, 1499–1507.

Cambou, B. and Klibanov, A.M. (1984) Preparative production of optically active esters and alcohols using esterase-catalyzed stereospecific transesterification in organic media. *J. Amer. Chem. Soc.*, **106**, 2687–2692.

Chin, J.T., Wheeler, S.L. and Klibanov, A.M. (1994) On protein solubility in organic solvents. *Biotechnol. Bioeng.*, **44**, 140–145.

Dabulis, K. and Klibanov, A.M. (1993) Dramatic enhancement of enzymatic activity in organic solvents by lyoprotectants. *Biotechnol. Bioeng.*, **41**, 566–571.

Davey, J.P., Pryce, R.J. and Williams, A. (1989) Microgels as soluble enzyme supports in organic media. *Enz. Microb. Technol.*, **11**, 657–661.

Dordick, J. (1989) Enzymatic catalysis in monophasic organic solvents. *Enz. Microb. Technol.*, **11**, 194–211.

Dordick, J.S., Marletta, M.A. and Klibanov, A.M. (1986) Peroxidases depolymerize lignin in organic media but not in water. *Proc. Natl. Acad. Sci. USA*, **83**, 6255–6257.

Dumont, T., Barth, D., Corbier, C., Branlant, G. and Perrut, M. (1992) Enzymatic reaction kinetics; comparison in an organic solvent and in supercritical carbon dioxide. *Biotechnol. Bioeng.*, **39**, 329–333.

Faber, K. (1991) A rationale to explain the catalytic activity of solid enzymes in organic solvents. *J. Mol. Catalysis*, **65**, L49–L51.

Fulcrand, V., Jacquier, R., Lazaro, R. and Viallefont, P. (1991) Enzymatic peptide synthesis in organic solvent mediated by gels of copolymerized acrylic derivatives of α-chymotrypsin and polyoxyethylene. *Int. J. Peptide Protein Res.*, **38**, 273–277.

Gaertner, H. and Puigserver, A. (1989) Kinetics and specificity of serine proteases in peptide synthesis catalyzed in organic solvents. *Eur. J. Biochem.*, **181**, 207–213.

Ghatorae, A.S., Guerra, M.J., Bell, G. and Halling, P.J. (1994) Immiscible organic solvent inactivation of urease, chymotrypsin, lipase, and ribonuclease: separation of dissolved solvent and interfacial effects. *Biotechnol. Bioeng.*, **44**, 1355–1361.

Goderis, H., Ampe, G., Feyton, M., Fouwe, B., Guffens, W., Van Cauwenbergh, S. and Tobback, P. (1987) Lipase-catalyzed ester exchange reactions in organic media with controlled humidity. *Biotechnol. Bioeng.*, **30**, 258–266.

Goto, M., Kameyama, H., Goto, M., Miyata, M. and Nakashio, F. (1993) Design of surfactants suitable for surfactant-coated enzymes as catalysts in organic media. *J. Chem. Eng. Jpn*, **26**, 109–111.

Goto, M., Kamiya, N., Miyata, M. and Nakashio, F. (1994) Enzymatic esterification by surfactant-coated lipase in organic media. *Biotechnol. Prog.*, **10**, 263–268.

Gupta, M.N. (1992) Enzyme function in organic solvents. *Eur. J. Biochem.*, **203**, 25–32.

Halling, P.J. (1984) Effects of water on equilibria catalysed by hydrolytic enzymes in biphasic reaction systems. *Enz. Microb. Technol.*, **6**, 513–516.

Halling, P.J. (1990) High-affinity binding of water by proteins is similar in air and in organic solvents. *Biochim. Biophys. Acta*, **1040**, 225–228.

Haraldsson, G.G. (1992) The application of lipases in organic synthesis. In *The Chemistry of Acid Derivatives* (Patai, S., ed.). John Wiley & Sons Ltd, Chichester, pp. 1395–1473.

Hertzberg, S., Kvittingen, L., Anthonsen, T. and Skjåk-Braek, G. (1992) Alginate as immobilization matrix and stabilizing agent in two-phase liquid systems: application in lipase-catalysed reactions. *Enz. Microb. Technol.*, **14**, 42–47.

Inada, Y., Takahashi, K., Yoshimoto, T., Ajima, A., Matsushima, A. and Saito, Y. (1986) Application of polyethylene glycol-modified enzymes in biotechnological processes: organic solvent-soluble enzymes. *Trends Biotechnol.*, **4**, 190–194.

Ison, A.P., Macrae, A.R., Smith, C.G. and Bosley, J. (1994) Mass transfer effects in solvent-free fat interesterification reactions: influences on catalyst design. *Biotechnol. Bioeng.*, **43**, 122–130.

Ito, Y., Fujii, H. and Imanishi, Y. (1992) Lipase modification by various synthetic polymers for use in chloroform. *Biotechnol. Lett.*, **14**, 1149–1152.

Ito, Y., Fujii, H. and Imanishi, Y. (1993) Catalytic peptide synthesis by trypsin modified with polystyrene in chloroform. *Biotechnol. Prog.*, **9**, 128–130.

Kawakami, K., Abe, T. and Yoshida, T. (1992) Silicone-immobilized biocatalysts effective for bioconversions in nonaqueous media. *Enz. Microb. Technol.*, **14**, 371–375.

Kawasaki, Y., Takahashi, Y., Nakatani, M., Murakami, M., Dosako, S. and Okai, H. (1994) Enzymatic functions of fatty acid-modified chymotrypsin and trypsin in aqueous–organic media. *Biosci. Biotech. Biochem.*, **58**, 512–516.

Khan, S.A., Halling, P.J., Bosley, J.A., Clark, A.H., Peilow, A.D., Pelan, E.G. and Rowlands, D.W. (1992) Polyethylene glycol-modified subtilisin forms microparticulate suspensions in organic solvents. *Enz. Microb. Technol.*, **14**, 96–100.

Khmelnitsky, Y.L., Zharinova, I.N., Berezin, I.V., Levashov, A.V. and Martinek, K. (1987) Detergentless microemulsions. A new microheterogeneous medium for enzymatic reaction. In *Enzyme Engineering 8* (Laskin, A.I., Mosbach, K., Thomas, D. and Wingard, L.B.J., eds) New York Academy of Sciences, New York, pp. 161–164.

Khmelnitsky, Y., Levashov, A., Klyachko, N. and Martinek, K. (1988) Engineering biocatalytic systems in organic media with low water content. *Enz. Microb. Technol.*, **10**, 710–724.

Khmelnitsky, Y.L., Neverova, I.N., Momtcheva, R., Yaropolov, A.I., Belova, A.B., Levashov, A.V. and Martinek, K. (1989) Surface-modified polymeric nanogranules containing entrapped enzymes: a novel biocatalyst for use in organic media. *Biotechnol. Lett.*, **3**, 275–280.

Khmelnitsky, Y., Mozhaev, V., Belova, A., Sergeeva, M. and Martinek, K. (1991) Denaturation capacity: a new qualitative criterion for selection of organic solvents as reaction media in biocatalysis. *Eur. J. Biochem.*, **198**, 31–41.

Khmelnitsky, Y.L., Gladilin, A.K., Roubailo, V.L., Martinek, K. and Levashov, A.V. (1992) Reversed micelles of polymeric surfactants in nonpolar organic solvents. A new microheterogeneous medium for enzymatic reactions. *Eur. J. Biochem.*, **206**, 737–745.

Kise, H., Hayakawa, A. and Noritomi, H. (1987) Enzymatic reactions in aqueous-organic media. IV. Chitin-α-chymotrypsin complex as a catalyst for amino acid esterification and peptide synthesis in organic solvents. *Biotechnol. Lett.*, **9**, 543–548.

Klibanov, A.M. (1986) Enzymes that work in organic solvents. *Chemtech*, **16**, 354–359.

Klyachko, N.L., Levashov, A.V., Pshezhetsky, A.V., Bogdanova, N.G., Berezin, I.V. and Martinek, K. (1986) Catalysis by enzymes entrapped into hydrated surfactant aggregates having lamellar or cylindrical (hexagonal) or ball-shaped (cubic) structure in organic solvents. *Eur. J. Biochem.*, **161**, 149–154.

Kuhl, P. and Halling, P. (1991) Salt hydrates buffer water activity during chymotrypsin-catalysed peptide synthesis. *Biochim. Biophys. Acta*, **1078**, 326–328.

Kvittingen, L. (1994) Some aspects of biocatalysis in organic solvents. *Tetrahedron*, **50**, 8253–8274.

Laane, C., Boeren, S., Vos, K. and Veeger, C. (1987) Rules for optimization of biocatalysis in organic solvents. *Biotechnol. Bioeng.*, **30**, 81–87.

Laane, C. and Tramper, J. (1990) Tailoring the medium and reactor for biocatalysis. *Chemtech*, **20**, 502–506.

Lamare, S. and Legoy, M.-D. (1993) Biocatalysis in the gas phase. *Trends Biotechnol.*, **11**, 413–418.

Lamare, S. and Legoy, M.-D. (1995) Working at controlled water activity in a continuous process: the gas/solid system as a solution. *Biotechnol. Bioeng.*, **45**, 387–397.

Larsson, K.M., Adlercreutz, P. and Mattiasson, B. (1990a) Enzymatic catalysis in microemulsions. Enzyme reuse and product recovery. *Biotechnol. Bioeng.*, **36**, 135–141.

Larsson, K.M., Janssen, A., Adlercreutz, P. and Mattiasson, B. (1990b) Three systems used for biocatalysis in organic solvents — a comparative study. *Biocatalysis*, **4**, 163–175.

Lee, K.M., Blaghen, M., Samama, J.-P. and Biellmann, J.-F. (1986) Crosslinked crystalline horse liver alcohol dehydrogenase as a redox catalyst: activity and stability toward organic solvent. *Bioorg. Chem.*, **14**, 202–210.

Ljunger, G., Adlercreutz, P. and Mattiasson, B. (1993) Reactions catalyzed by PEG-modified α-chymotrypsin in organic solvents. Influence of water content and degree of modification. *Biocatalysis*, **7**, 279–288.

Luck, T., Kiesser, T. and Bauer, W. (1988) Engineering parameters for the application of immobilized lipases in a solvent-free system. In *Biotechnology for Fats and Oils Industry* (Applewhite, T.H., ed.). Amer. Oil Chem. Soc., Champaign, Illinois, USA, pp. 343–345.

Luisi, P.L. and Laane, C. (1986) Solubilization of enzymes in apolar solvents via reverse micelles. *Trends Biotechnol.*, **4**, 153–161.

Macrae, A.R. (1985) Interesterification of fats and oils. In *Biocatalysts in Organic Synthesis* (Tramper, J., van der Plas, H.C. and Linko, P., eds). Elsevier, Amsterdam, pp. 195–208.

Malcata, F.X., Reyes, H.R., Garcia, H.S., Hill, C.G.J. and Amundson, C.H. (1990) Immobilized lipase reactors for modification of fats and oils — a review. *J. Amer. Oil Chem. Soc.*, **67**, 890–910.

Martinek, K., Klyachko, N.L., Kabanov, A.V., Khmelnitsky, Y.L. and Levashov, A.V. (1989) Micellar enzymology: its relation to membranology. *Biochim. Biophys. Acta*, **981**, 161–172.

Matsushima, A., Nishimura, H., Ashihara, Y., Yokota, Y. and Inada, Y. (1980) Modification of *E. coli* aspariginase with 2,4-bis(*O*-methoxypolyethylene glycol)-6-chloro-*S*-triazine (activated PEG2); disappearance of binding ability towards anti-serum and retention of enzymic activity. *Chem. Lett.*, **1980**, 773–776.

McMinn, J.H., Sowa, M.J., Charnick, S.B. and Paulaitis, M.E. (1993) The hydration of proteins in nearly anhydrous organic solvent suspensions. *Biopolymers*, **33**, 1213–1224.

Miethe, P., Gruber, R. and Voss, H. (1989) Enzymes in lyotropic liquid crystals — a new method of bioconversion in non-aqueous media. *Biotechnol. Lett.*, **11**, 449–454.

Murakami, M., Kawasaki, Y., Kawanari, M. and Okai, H. (1993) Transesterification of oil by fatty acid-modified lipase. *J. Amer. Oil Chem. Soc.*, **70**, 571–574.

Nakamura, K. (1990) Biochemical reactions in supercritical fluids. *Trends Biotechnol.*, **8**, 288–292.

Nishio, T., Takahashi, K., Tsuzuki, T., Yoshimoto, T., Kodera, Y., Matsushima, A., Saito, Y. and Inada, Y. (1988) Ester synthesis in benzene by polyethylene glycol-modified lipase from *Pseudomonas fragi* 22.39B. *J. Biotechnol.*, **8**, 39–44.

Noritomi, H., Watanabe, A. and Kise, H. (1989) Enzymatic reactions in aqueous-organic media VII. Peptide and ester synthesis in organic solvents by α-chymotrypsin immobilized through non-covalent binding to poly(vinyl alcohol). *Polymer J.*, **21**, 147–153.

Okahata, Y. and Ijiro, K. (1988) A lipid-coated lipase as a new catalyst for triglyceride synthesis in organic solvents. *J. Chem. Soc., Chem. Commun.*, 1392–1394.

Okahata, Y. and Ijiro, K. (1992) Preparation of a lipid-coated lipase and catalysis of glyceride ester in homogeneous organic solvents. *Bull. Chem. Soc. Jpn*, **65**, 2411–2420.

Omata, T., Iida, T., Tanaka, A. and Fukui, S. (1979) Transformation of steroids by gel-entrapped *Nocardia rhodocrous* cells in organic solvent. *Eur. J. Appl. Microbiol. Biotechnol.*, **8**, 143–155.

Orsat, B., Drtina, G.J., Williams, M.G. and Klibanov, A.M. (1994) Effect of support material and enzyme pretreatment on enantioselectivity of immobilized subtilisin in organic solvents. *Biotechnol. Bioeng.*, **44**, 1265–1269.

Otamiri, M., Adlercreutz, P. and Mattiasson, B. (1991) Effects on ester synthesis in toluene by immobilized chymotrypsin by addition of polymers to reaction medium. *Biotechnol. Appl. Biochem.*, **13**, 54–64.

Otamiri, M., Adlercreutz, P. and Mattiasson, B. (1992a) Complex formation between chymotrypsin and ethyl cellulose as a means to solubilize the enzyme in active form in toluene. *Biocatalysis*, **6**, 291–305.

Otamiri, M., Adlercreutz, P. and Mattiasson, B. (1992b) Complex formation between chymotrypsin and polymers as a means to improve exposure of the enzyme to organic solvents. In *Biocatalysis in Non-Conventional Media* (Tramper, J., Vermüe, M., Beeftink, H.H. and von Stockar, U., eds). Elsevier, Amsterdam, pp. 363–369.

Otamiri, M., Adlercreutz, P. and Mattiasson, B. (1994) A differential scanning calorimetric study of chymotrypsin in the presence of added polymers. *Biotechnol. Bioeng.*, **43**, 73–78.

Paradkar, V.M. and Dordick, J.S. (1994) Mechanism of extraction of chymotrypsin into isooctane at very low concentrations of Aerosol OT in the absence of reversed micelles. *Biotechnol. Bioeng.*, **43**, 529–540.

Paulaitis, M.E., Sowa, M.J. and McMinn, J.H. (1992) Effect of enzyme hydration on the catalytic activity of chymotrypsin in nearly anhydrous organic suspensions. *Ann. N.Y. Acad. Sci.*, **672**, 278–282.

Rekker, R.F. (1977) *The Hydrophobic Fragmental Constant*. Elsevier, Amsterdam.

Reslow, M., Adlercreutz, P. and Mattiasson, B. (1988) On the importance of the support material for bioorganic synthesis. Influence of water partition between solvent, enzyme and solid support in water-poor reaction media. *Eur. J. Biochem.*, **172**, 573–578.

Reslow, M., Adlercreutz, P. and Mattiasson, B. (1992) Modification of the microenvironment of enzymes in organic solvents. Substitution of water by polar solvents. *Biocatalysis*, **6**, 307–318.

Robert, H., Lamare, S., Parvaresh, F. and Legoy, M.-D. (1992) The role of water in gaseous biocatalysis. In *Biocatalysis in Nonconventional Media* (Tramper, J., Vermüe, M., Beeftink, H.H. and von Stockar, U., eds). Elsevier, Amsterdam, pp. 85–91.

Rosevear, A., Kennedy, J.F. and Cabral, J.M.S. (1987) *Immobilized Enzymes and Cells*. Adam Hilger, Bristol.

Roziewski, K. and Russell, A.J. (1992) Effect of hydration on the morphology of enzyme powder. *Biotechnol. Bioeng.*, **39**, 1171–1175.

Ruckenstein, E. and Wang, X. (1993) Lipase immobilized on hydrophobic porous polymer supports prepared by concentrated emulsion polymerization and their activity in the hydrolysis of triacylglycerides. *Biotechnol. Bioeng.*, **42**, 821–828.

Sakurai, K., Kashimoto, K., Kodera, Y. and Inada, Y. (1990) Solid phase synthesis of peptides with polyethylene glycol-modified protease in organic solvents. *Biotechnol. Lett.*, **12**, 685–688.

Sánches-Ferrer, A. and García-Carmona, F. (1994) Biocatalysis in reverse self-assembling structures: reverse micelles and reverse vesicles. *Enz. Microb. Technol.*, **16**, 409–415.

Satterfield, C.N. (1970) *Mass Transfer in Heterogeneous Catalysis*. MIT Press, Cambridge, Massachusetts.

Scott, C.D., Scott, T.C. and Woodward, C.A. (1993) Use of dinitrofluorobenzene to modify enzymes for enhanced solubilization and activity in organic solvents. *Appl. Biochem. Biotechnol.*, **39/40**, 279–287.

St Clair, N.L. and Naiva, M.A. (1992) Cross-linked enzyme crystals as robust biocatalysts. *J. Amer. Chem. Soc.*, **114**, 7314–7316.

Steinmann, B., Jäckle, H. and Luisi, P.L. (1986) A comparative study of lysozyme conformation in various reverse micellar systems. *Biopolymers*, **25**, 1133–1156.

Takahashi, K., Nishimura, H., Yoshimoto, T., Okada, M., Ajima, A., Matsushima, A., Tamaura, Y., Saito, Y. and Inada, Y. (1984a) Polyethylene glycol-modified enzymes trap water on their surface and exert enzymatic activity in organic solvents. *Biotechnol. Lett.*, **6**, 765–770.

Takahashi, K., Nishimura, H., Yoshimoto, T., Saito, T. and Inada, Y. (1984b) A chemical modification to make horseradish peroxidase soluble and active in benzene. *Biochem. Biophys. Res. Commun.*, **121**, 261–265.

Towey, T.F., Rees, G.D., Steytler, D.C., Price, A.L. and Robinson, B.H. (1994) Winsor-II microemulsion systems in bioseparations — a model reaction for extractive bioconversions. *Bioseparation*, **4**, 139–147.

Tramper, J., Vermüe, M., Beeftink, H.H. and von Stockar, U. (eds) (1992) *Biocatalysis in Nonconventional Media*. Elsevier, Amsterdam.

Tsuzuki, W., Okahata, Y., Katayama, O. and Suzuki, T. (1991a) Preparation of organic-soluble enzyme (lipase B) and characterization by gel permeation chromatography. *J. Chem. Soc., Perkin Trans. 1*, 1245–1247.

Tsuzuki, W., Sasaki, T. and Suzuki, T. (1991b) Effect of detergent attached to enzyme molecules on the activity of organic-solvent-soluble lipases. *J. Chem. Soc., Perkin Trans. 12*, 1851–1854.

Vakurov, A.V., Gladilin, A.K., Levashov, A.V. and Khmelnitsky, Y.L. (1994) Dry enzyme-polymer complexes: stable organosoluble biocatalysts for nonaqueous enzymology. *Biotechnol. Lett.*, **16**, 175–178.

Valivety, R.H., Halling, P.J. and Macrae, A.R. (1992) Reaction rate with suspended lipase catalyst shows similar dependence on water activity in different organic solvents. *Biochim. Biophys. Acta*, **1118**, 218–222.

Veronese, F.M., Largajolli, R., Boccu, E., Benassi, C.A. and Schiavon, O. (1985) Surface modification of proteins. Activation of monomethoxypolyethylene glycols by phenylchloroformates and modification of ribonuclease and superoxide dismutase. *Appl. Biochem. Biotechnol.*, **11**, 141–152.

Wehtje, E. (1992) *Parameters influencing enzyme activity in organic media*. PhD thesis, Lund University, Lund, Sweden.

Wehtje, E., Adlercreutz, P. and Mattiasson, B. (1993a) Improved activity retention of enzymes deposited on solid supports. *Biotechnol. Bioeng.*, **141**, 171–178.

Wehtje, E., Adlercreutz, P. and Mattiasson, B. (1993b) Reaction kinetics of immobilized α-chymotrypsin in organic media. 1. Influence of solvent polarity. *Biocatalysis*, **7**, 149–161.

Wehtje, E., Svensson, I., Adlercreutz, P. and Mattiasson, B. (1993c) Continuous control of water activity during biocatalysis in organic media. *Biotechnol. Lett.*, **7**, 873–878.

Wirth, P., Souppe, J., Tritsch, D. and Biellmann, J.-F. (1991) Chemical modification of horseradish peroxidase with ethanal–methoxypolyethylene glycol: solubility in organic solvents, activity, and properties. *Bioorg. Chem.*, **19**, 133–142.

Wisdom, R.A., Dunnill, P. and Lilly, M.D. (1984) Enzymatic interesterification of fats: factors influencing the choice of support for immobilized lipase. *Enz. Microb. Technol.*, **6**, 443–446.

Yang, C.-P. and Su, C.-S. (1988) Synthesis of aspartame precursor: α-L-aspartyl-L-phenylalanine methyl ester in ethyl acetate using thermolysin entrapped in polyurethane. *Biotechnol. Bioeng.*, **32**, 595–603.

Zaks, A. and Klibanov, A.M. (1985) Enzyme-catalyzed processes in organic solvents. *Proc. Natl. Acad. Sci. USA*, **82**, 3192–3196.

Zaks, A. and Klibanov, A.M. (1988a) The effect of water on enzyme action in organic media. *J. Biol. Chem.*, **263**, 8017–8021.

Zaks, A. and Klibanov, A.M. (1988b) Enzymatic catalysis in nonaqueous solvents. *J. Biol. Chem.*, **263**, 3194–3201.

3 Fundamentals of non-aqueous enzymology

Z. YANG and A.J. RUSSELL

3.1 Introduction

Non-aqueous enzymology is such a rapidly developing research area that it has attracted interest from chemists, biochemists and chemical engineers. This has been reflected by an explosive growth of the literature (for reviews see Dordick, 1989; Klibanov, 1989; Blinkovsky et al., 1992; Russell et al., 1992; Halling, 1994). It is generally accepted that when enzymes are placed in organic media, they exhibit altered properties such as enhanced thermostability (Ayala et al., 1986; Wheeler and Croteau, 1986), altered specificity (see review by Wescott and Klibanov, 1994), molecular memory (Ståhl et al., 1991; Dabulis and Klibanov, 1993), and the ability to catalyze reactions that are kinetically or thermodynamically impossible in aqueous solution (Kuhl et al., 1990; West et al., 1990). In addition, the industrial utility of biocatalysts is enhanced because of the increased solubility of hydrophobic substrates, ease of product and enzyme recovery, and reduced risk of microbial contamination of reactors. In order to take full advantage of these perceived benefits of non-aqueous enzymology, we must first understand the fundamental interaction between a solvent and an enzyme. The structural and mechanistic integrity of the protein, the role of water and solvents on activity, and the specific solvent effects on the kinetics and thermodynamics of enzyme-catalyzed processes, are all central to the development of our understanding of non-aqueous enzymology and its applications.

3.2 Structural integrity

Perhaps the most obvious question regarding enzymes in organic solvents is whether a protein maintains its native conformation when transferred from the tranquility of an aqueous buffer to the harsh realities of a non-aqueous process environment. A naive expectation would be to predict that the protein would 'turn inside out' and thus lose its activity. This preconception actually inhibited the growth of non-aqueous enzymology for many years, and even today, in the face of industrial processes catalyzed by enzymes in organic solvents, scientists do not uniformly agree that enzymes are active in solvents.

The current hypothesis is that when an appropriately prepared enzyme is placed in an anhydrous organic solvent, it is kinetically (but not thermodynamically) 'trapped' in its native-like conformation (Zaks and Klibanov,

1988a). The result of this hypothesis (a native structure) has been supported directly by NMR, EPR and X-ray crystallography. Earlier, Guinn *et al.* (1991a) studied the mobility of the active site of horse liver alcohol dehydrogenase, using EPR spectroscopy. The data indicate that the enzyme is conformationally restricted in the anhydrous environment, but not unfolded. In order to elminate ambiguity from exogenous probes necessary for EPR studies, Burke *et al.* (1993) performed a solid-state ^{2}H-NMR investigation of protein dynamics using unmodified α-lytic protease. The flipping rate of Tyr 123 (at the active center) was extremely slow ($\leq 10^3$ s^{-1}) in the enzyme powder, whether dry or suspended in *tert*-butyl methyl ether, as compared to the flipping rate observed in aqueous crystals ($\geq 10^7$ s^{-1}). This experiment confirmed that proteins exhibit reduced mobility in organic solvents. In a study of the active site integrity of α-chymotrypsin in organic solvents using ^{13}C-NMR, it was clear that the active site was similar in both lyophilized powder and organic solvents (Burke *et al.*, 1992). Using ^{15}N-NMR to probe the hydrogen-bonded network at the active site of α-lytic protease, Klibanov and coworkers (Burke *et al.*, 1989) have also shown that the tautomeric structure and hydrogen bonding catalytic triad in the active center are identical in acetone, octane, and water. Since the catalytic triad involves residues remote from each other in the polypeptide chain, its intactness suggests that there is no significant distortion of the protein's native, folded structure when the enzyme is placed in the dry solvents. In addition, the three-dimensional structure of subtilisin Carlsberg in acetonitrile is identical to that in water (at 2.3 Å resolution), as shown in non-aqueous X-ray crystallography experiments (Fitzpatrick *et al.*, 1993). Recently, Yennawar *et al.* (1994, 1995) have undertaken a structural study of crystalline γ-chymotrypsin soaked in aqueous and non-aqueous solutions by X-ray crystallographic methods at a resolution of 2.2 Å, and their results have also revealed that the overall conformation of chymotrypsin is not changed by replacing water with hexane (see Figure 3.1), although changes in the side chain positions do occur, which may be responsible for the altered substrate specificity when the enzyme is placed in organic media. Finally, protein structure and dynamics in organic solvents have also been predicted by molecular dynamics simulations of bovine pancreatic trypsin inhibitor (Hartsough and Merz, 1992). The results show that the amino acid side chains of the protein are more rigid in chloroform than in water, but the backbone flexibility of the protein in each medium is similar.

The structural rigidity of enzymes in organic solvents has been indirectly substantiated by enzyme activity assays. For instance, the activities of lipase in tributyrin/heptanol (Zaks and Klibanov, 1985), and α-chymotrypsin in octane (Zaks and Klibanov, 1988a) are dependent upon the pH of the aqueous solution from which the enzymes have been lyophilized, with the 'pH dependence' resembling that in water. The explanation is that the ionogenic groups of the enzyme acquire the ionization state corresponding to the pH of the aqueous solution, and this state is then retained during lyophilization and dispersal in the organic solvent due to the conformational rigidity of the enzyme.

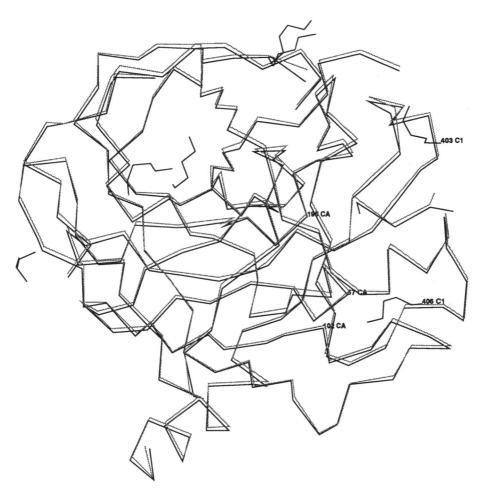

Figure 3.1 Superposition of the α-carbon trace of τ-chymotrypsin in the hexane (dark) and native (light) structures (kindly provided by Prof. G.K. Farber from Penn State University).

The rigidity of enzymes in organic solvents also results in altered substrate specificity. In spite of the wide substrate specificity a lipase exhibits in aqueous solution, dry lipase cannot accept bulky tertiary alcohols as substrates in trans-esterification reactions in organic solvents. This has been explained by the fact that lipase lacks the conformational mobility needed to accommodate large substrate molecules in its active site when in the dry state (Zaks and Klibanov, 1984). It has also been suggested that many enzymes, such as terpene cyclase (Wheeler and Croteau, 1986), cytochrome oxidase and ATPase (Ayala *et al.*,

1986), and lipase (Zaks and Klibanov, 1984) are more thermostable in organic solvents than in water. This enhanced thermostability could also be related to structural rigidity of enzymes in the absence of a macro-aqueous phase (Volkin *et al.*, 1991).

In conclusion, we reiterate that there is ample evidence that proteins suspended in organic solvents do not unfold. Rather, the protein retains a native-like conformation with altered flexibility. The subtle effects of solvent on protein structure and mobility will be the basis of improving our understanding of how to alter enzyme function by 'solvent engineering'.

3.3 Mechanistic integrity

Even though enzymes are structurally intact in organic media, an important consideration is whether the solvent will cause an alteration in the mechanism for a given biocatalytic reaction. Dastoli and Price (1967) demonstrated classical Michaelis–Menten kinetics for xanthine oxidase-catalyzed oxidation of croton-aldehyde in 14 anhydrous organic solvents. Decades later the impact of non-aqueous enzyme kinetics became more apparent when it was discovered that lipases and serine proteases could catalyze transesterification reactions in the absence of bulk water. In aqueous solution these enzymes catalyze ester hydrolyses via the acyl-enzyme mechanism (Jencks, 1987). A typical mechanism for ester hydrolysis involves the formation of a non-covalent enzyme–substrate complex, which is then transformed to a tetrahedral intermediate by the attack of the hydroxyl group of an active-site serine residue upon the carbonyl of the substrate. This intermediate collapses, leading to the release of the alcohol product and the formation of a stable covalent acyl-enzyme intermediate (acylation). The nucleophilic attack of water on the acyl-enzyme then results in the formation of a tetrahedral enzyme–product complex which collapses to release the free enzyme and the product (deacylation).

Engineering of enzyme activity in organic solvents is dependent on our under-standing of the catalytic process in totality. It is clearly important to determine whether the acyl-enzyme mechanism is also followed when transesterification between an ester alcohol is catalyzed by lipases and serine proteases in organic solvents. This issue has been addressed in detail for esterases in several laboratories.

Klibanov and coworkers have analyzed the mechanism of transesterifications catalyzed by three lipases (Zaks and Klibanov, 1985), subtilisin and α-chymo-trypsin (Zaks and Klibanov, 1988a) in various organic solvents via the double-reciprocal parallel-lines method. The observed kinetic behavior indicates that all the tested enzymes follow classical Michaelis–Menten kinetics. Although it is always an experimental challenge to obtain k_{cat} and K_m values separately for an enzyme in organic media, because the K_m is generally so high that substrate

solubility becomes limiting, k_{cat}/K_m is usually straightforward to determine from initial rate determinations at relatively low substrate concentrations.

The mechanistic similarity of serine proteases in aqueous and non-aqueous media has also been substantiated by conducting Hammett analyses of the cleavage of para-substituted phenyl acetates catalyzed by subtilisin Carlsberg (Kanerva and Klibanov, 1989). When the specificity constant (k_{cat}/K_m) was plotted against the substituent constant (σ^{-1}) of the substrate, linear correlations were obtained in six solvents (including water) differing widely in hydrophobicity and polarity. The reaction constants (ρ) determined from the above linear plots are essentially independent of the nature of the solvent. The ρ value, which reflects the charge distribution in the transition state, is very sensitive to enzyme mechanism, and therefore the data suggest that the structure of the transition state formed upon acylation of subtilisin with phenyl acetates is the same regardless of the solvent.

The transition state for the deacylation step has also been compared in different solvents by investigating kinetic isotope effects on a model transesterification reaction between vinyl butyrate and 1-butanol (BuOH or BuOD) catalyzed by subtilisin or lipase (Adams $et\ al.$, 1990). A similar deuterium kinetic isotope effect was observed for both enzymes, both $^{H}V_{max}/^{D}V_{max}$ and $^{H}K_m^{alcohol}/^{D}K_m^{alcohol}$ are similar in various solvents and equal $^{H}k_3/^{D}k_3$ (k_3 is the deacylation rate constant), indicating that the transition state structure for deacylation is also independant of the nature of the reaction medium.

Chatterjee and Russell (1993) have successfully demonstrated the presence of the stable acyl-enzyme intermediate by conducting a steady-state kinetic analysis of subtilisin-catalyzed transesterifications in a variety of organic solvents. In support of the ping-pong acyl-enzyme mechanism (Figure 3.2), a partial reaction between enzyme and p-nitrophenol butyrate in the absence of an alcohol has been observed in hexane, acetonitrile, and dioxane. The predicted linear dependence of both $1/k_{cat}$ and $1/K_m$ on $1/[S_2]$, and the predicted independence of k_{cat}/K_m on $[S_2]$ have all been experimentally confirmed, as expected, for the acyl-enzyme mechanism. Further, at 43°C when deacylation is rate-controlling, the transesterification reaction between Ac-Phe-OMe or Ac-Phe-OEt and a propanol/butanol mixture results in the same ratio of propyl to butyl ester product, and transesterification of the methyl and ethyl esters of phenylalanine by 30 mM propanol gives equivalent rates of formation of the propyl ester, indicating that deacylation is taking place from a common acyl-enzyme intermediate. All these experimental data confirm the maintenance of the acyl-enzyme mechanism in organic environments.

In addition to the kinetic studies listed above, Chatterjee and Russell (1992a) have also investigated the mechanism of subtilisin by comparing the activities of native subtilisin and a protein engineered variant, thiol-subtilisin, in which the active site serine residue is replaced by a thiol-containing cysteine. The data indicate that changing the solvent does not significantly affect the difference between the stability of the transition states for the two

LIVERPOOL
JOHN MOORES UNIVERSITY
AVRIL ROBARTS LRC
TEL. 0151 231 4022

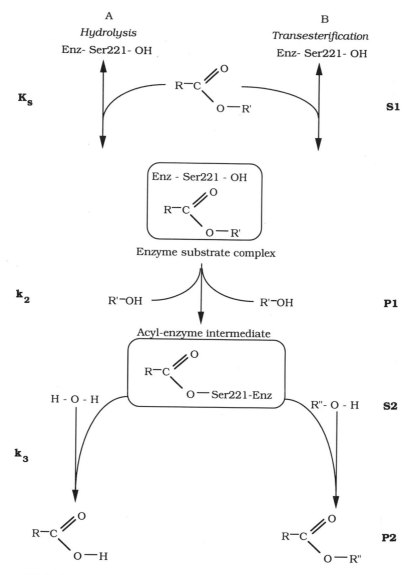

Figure 3.2 Acyl-enzyme mechanism for hydrolysis (A) and transesterification (B) catalyzed by subtilisin. K_s is the dissociation constant for the binding of the first substrate (S1) to the enzyme, and k_2 and k_3 are the rates of acylation and deacylation, respectively.

enzymes. Thus the active site serine is as catalytically important in water as in organic solvents.

In conclusion, for those enzymes studied, the catalytic mechanism is not affected during transition from water to solvents.

3.4 Water

A common thread in all studies of enzymes in organic media is that the amount of water associated with the enzyme is a key determinant of the properties (e.g. activity, stability, and specificity) that the enzyme exhibits. Controlling the amount and location of water is critical to the successful application of non-aqueous biocatalytic systems. Water is necessary for the acquisition and maintenance of the enzymes' catalytically active conformation in essentially anhydrous organic solvents, but water is also involved in many enzyme inactivation processes (Volkin *et al.*, 1991).

3.4.1 Effect of water on enzyme activity

Although the water content in a typical non-aqueous enzymatic system is usually as low as 0.01%, small variations in water content result in large changes in enzyme catalytic activity. While an enzyme is essentially inactive in completely 'dry' systems, its activity accelerates as the degree of enzyme hydration increases. Enzyme activity has been observed in truly dehydrated enzyme systems, but it is likely that this results from unfolded protein acting as a stabilizer for a small amount of native enzyme. Too much water will, however, facilitate enzyme aggregation, which can lead to a decrease in enzyme activity. The mechanism of such water-induced aggregation in organic media has not been studied in detail, although it has been suggested that intermolecular disulphide bond formation could be important (Liu *et al.*, 1991).

When water is added to a solvent–enzyme system, it partitions between the solvent and the enzyme. The activity of the enzyme is dependent on the amount of water associated with the enzyme, and to a lesser degree on the water content in the whole system (Zaks and Klibanov, 1988b). As long as a minimal amount of water is associated with the enzyme, its activity in organic media is retained. Naturally, the water content–activity dependence for different enzymes will be subtly distinct, and this rule may not always apply.

The amount of water required for catalysis is dependent on the enzyme. Lipases are highly active when only a few molecules of water are associated with the protein molecule (Valivety *et al.*, 1992a, b, c). Subtilisin and chymotrypsin require less than 50 molecules of water per enzyme molecule to exhibit their minimal activities (Zaks and Klibanov, 1988a). Alcohol oxidase, alcohol dehydrogenase and polyphenol oxidase, however, are only active when hundreds of water molecules are bound per enzyme molecule (enough for a monolayer to form) (Zaks and Klibanov, 1988b). It has been suggested that making an enzyme more hydrophobic by chemical modification can reduce the water requirement for the enzyme. This suggestion is substantiated by the increase in the activity of horseradish peroxidase after deglycosylation, or benzyl- and poly(ethylene glycol)-modification (Vazquez-Duhalt *et al.*, 1992). Once again, however, the rule is not general since there is no good relationship between enzyme surface charge and minimum water content. Maximal activity in nonaqueous media undoubtedly requires not only an appropriate concentration of water, but also its correct location. The pretreatment of subtilisin by ultrasound irradiation results in a substantial increase of enzyme activity, possibly due to the re-distribution of water molecules in subtilisin (Vulfson *et al.*, 1991).

3.4.2 *Effect of water on protein mobility*

The EPR spectra of spin labels attached to the active sites of alcohol dehydro-genase (Guinn *et al.*, 1991a) and subtilisin (Affleck *et al.*, 1992a) have shown that active site mobility in organic solvents increases significantly upon the addition of water. A study of the structure and dynamic motions of cytochrome c in tetrahydrofuran by two-dimensional NH-exchange NMR spectroscopy has revealed that addition of water to the solvent can enhance the flexibility of all regions of the native protein structure (Wu and Gorenstein, 1993). For subtilisin, the increase in active site flexibility is accompanied by an increase in active site polarity (Affleck *et al.*, 1992b). This is supported by dielectric studies on collagen, cytochrome c, elastin and lysozyme powders (Bone and Pethig, 1985). The most likely explanation of increased mobility is that water acts as a plasticizer to increase the flexibility, and hence the polarizability, of the protein structure (Bone and Pethig, 1985).

3.4.3 *Relationship between enzyme activity and protein mobility*

The addition of water to solid enzyme preparations in organic solvents may increase activity via enhancement of the polarity and flexibility of an enzyme's active site. This effect has been quantified for subtilisin in tetrahydrofuran (Affleck *et al.*, 1992a; Xu *et al.*, 1994), although the addition of too much water led to the formation and growth of a water cluster within the active site, thereby changing the enzyme structure and reducing enzyme activity (Affleck *et al.*, 1992a).

Hydration of the charged and polar groups of protein molecules appears to be a prerequisite for enzyme catalysis (Rupley *et al.*, 1983). It is possible that in the dry state, the charged and polar groups of the enzyme molecule interact with each other and produce an inactive locked conformation (Zaks and Klibanov, 1988b). The role of water may be to form hydrogen bonds with these functional groups (Zaks and Klibanov, 1988b), thereby dielectrically screening the electro-static interactions between ionized groups, and neutralizing the dipole–dipole interactions between peptide units and polar side-groups in the folded poly-peptide chain (Bone and Pethig, 1985). Electrostatic interactions may be dominant in controlling the enzyme's catalytic behavior in an organic medium. This hypothesis is supported by introducing various 'hydrogen-bond formers', such as glycerol, ethylene glycol and formamide, to the solvent (Zaks and

Klibanov, 1988b). For example, addition of 3% formamide results in a 35-fold increase in the activity of polyphenol oxidase in octanol (with 1% water).

Some of the novel properties exhibited by enzymes in organic media may be due to their high conformational rigidity in the absence of water, and these effects will be altered by water-enhanced flexibility increases. A recent study (Yang *et al.*, 1993) has shown that the apparent pH dependence of subtilisin in organic media is dependent upon the solvent used and the amount of water added, and that the apparent pK_a values of the enzyme can be reduced by about 1 pH unit simply by enzyme hydration. This effect appears to be related to the increase in the polarity of the enzyme's active site. Russell and Klibanov (1988) have reported that a 100-fold enhancement of subtilisin activity in anhydrous organic solvents was obtained when the enzyme was lyophilized from aqueous solution containing competitive inhibitors, compared to the enzyme lyophilized in the absence of such ligands. Addition of water, however, removed this 'molecular memory' by increasing protein flexibility. Although dry lipase is extremely thermostable at 100°C (half-life 12 h) in heptanol/tributyrin, its activity vanishes when only 0.8% water is added (Zaks and Klibanov, 1984).

3.4.4 Water activity

When water is added to a non-aqueous enzymatic system, it is distributed between enzyme, solvent, substrate and support (if present). As discussed above, the amount of water bound to enzyme molecules is central in fixing the catalytic activity of the enzyme. Halling (1989) has proposed that thermodynamic water activity (a_w) is a parameter which can be used to quantify the water level associated with the enzyme. Indeed, for lysozyme and chymotrypsin in suspensions of four organic solvents, McMinn *et al.* (1993) have shown that at low a_w levels, the water adsorption isotherms (water adsorbed by the protein versus water activity) of the proteins are similar in the solvent and gas phase, suggesting that organic solvents do not directly affect the tight binding of water by enzymes. At higher a_w levels, water adsorption by the proteins in the organic media is suppressed relative to the sorption of water vapor, which might be the evidence for direct interactions between water and solvent at the protein–solvent interface. These results were confirmed by Halling (1990a), who obtained the water adsorption isotherms of six proteins suspended in a variety of organic solvents using the published literature data. Therefore, it has been suggested that enzyme kinetics in different solvents are best studied at a constant water activity, so as to ensure similar enzyme hydration levels and in turn to avoid the influence of water. This supposition may, however, be enzyme-dependent. For example in Figure 3.3, the activity of a lipase was measured in 23 organic solvents when the solvents were dry, water-saturated, and when the water activity was fixed at 0.59 by addition of a salt hydrate, $Na_4P_2O_7 \cdot 10H_2O$. It is striking that for this enzyme the trend in the data was not dependent on water activity (S. Kamat and A.J. Russell, unpublished results).

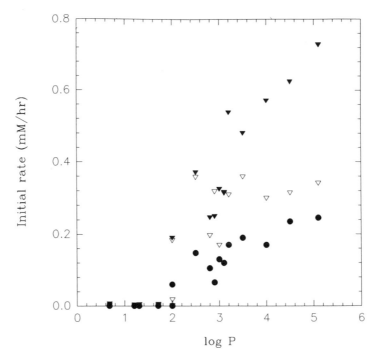

Figure 3.3 Initial rates of transesterification between methyl methacrylate and 2-ethyl hexanol catalyzed by *Candida cylindracea* lipase suspended in organic solvents dried with molecular sieves (●); in solvents saturated with water (▽); and in solvents in the presence of $Na_4P_2O_7 \cdot 10H_2O$ ($a_w = 0.59$) (▼) (S. Kamat and A.J. Russell, unpublished results).

Water activity can be measured from the ratio of water vapor pressure over a reaction system divided by the vapor pressure of pure water under the same conditions. A constant water activity in different organic solvents can be achieved in at least three ways:

(1) Separate pre-equilibration of substrate solution and enzyme preparation with a saturated aqueous solution of a salt (Valivety *et al.*, 1992b)
(2) Direct addition of a salt hydrate to the reaction system
(3) Addition of fixed, but differing, amounts of water to each solvent.

The second method was introduced by Halling (1992), who noted that the equilibrium of a salt hydrate pair can provide a constant vapor pressure at a fixed temperature. This method is effective with α-chymotrypsin (Kuhl and Halling, 1991), lipase (Kvittingen *et al.*, 1992), polyphenol oxidase (Yang *et al.*, 1992) and subtilisin (Yang *et al.*, 1993).

3.4.5 Effect of water activity on enzyme activity

A consideration of water activity is beneficial in predicting and investigating both water availability and catalytic activity for enzymes in organic solvents (Halling, 1994). This has been demonstrated by investigating the relationship between catalytic activity and a_w while changing solvents (Valivety et al., 1992c) and substrates (Kvittingen et al., 1992). When water activity is not controlled, there is usually an increase in the optimal water content (to enable a maximal catalytic activity) as solvent polarity increases. However, in solvents varying in polarity from hexane to pentanone, a lipase shows a similar optimal water activity at approximately 0.55 (Valivety et al., 1992c). The same phenomenon was observed by Martins et al. (1994). Substrates also have a strong affinity to water, but a constant optimal a_w of 0.78 was obtained for the esterification of butanol and butanoic acid in hexane catalyzed by a lipase, as the reactant concentrations were increased up to 0.37 M (Kvittingen et al., 1992). These results clearly show that there is a relationship between enzyme activity and water activity, and in some circumstances this dependence is hardly affected by the other components involved in the reaction systems. In many systems, however, other solvent and substrate properties also have a direct and/or indirect influence on enzyme activity. This can be reflected in a change in enzyme-catalyzed reaction rates even when water activity is kept constant (Halling, 1994).

The dependence of enzyme activity on water activity varies from enzyme to enzyme. For example, polyphenol oxidase (Yang and Robb, 1991) and α-chymotrypsin (Blanco et al., 1992) promote their highest rate enhancements at a 'complete' hydration level ($a_w = 1$, at which point a monolayer of water around the enzyme molecules is ensured), while Rhizomucor miehei lipase remains highly active at a_w below 0.0001 (Valivety et al., 1992a). Optimal a_w values for other lipases vary widely from 0.12 to 1.0 (Valivety et al., 1992b). α-Chymotrypsin, suspended in organic solvents, can catalyze peptide synthesis when water activity is controlled by a salt hydrate ($Na_2CO_3 \cdot 10H_2O$), although all the components involved (enzyme, substrates, products, and salt hydrate) are undissolved in the bulk solvent (Kuhl et al., 1990). The most plausible explanation is that the reaction occurs in the water layer surrounding the enzyme molecules.

It is well known that a low water concentration (low a_w) shifts the thermodynamic equilibria of hydrolysis reactions toward synthesis (for example in esterifications and peptide syntheses). For instance, the equilibrium constant for lipase-catalyzed esterification increased with a decrease in a_w (Svensson et al., 1994). Therefore, water activity can be used to control the equilibrium position. In order to compromise for the low reaction rates observed at low water activity levels, however, it is beneficial to initiate the reaction at high a_w (to acquire the optimal reaction rate) and then shift to a low a_w toward the end of the reaction to obtain a high yield (Cassell and Halling, 1988; Svensson et al., 1994).

3.5 Solvent

The nature of the organic solvent is another important factor in non-aqueous enzymology. Solvent not only directly or indirectly affects the enzyme activity and stability, but also changes the enzyme specificity (including substrate specificity, enantioselectivity, prochiral selectivity, regioselectivity and chemo-selectivity). Generally, a solvent affects enzyme activity via interaction with water, enzyme, substrates, and products.

3.5.1 Effect of solvent on the water associated with the enzyme

Although some solvents seem to have little effect on water molecules which are tightly bound to the enzyme (as discussed in section 3.4.1), relatively hydrophilic solvents are capable of stripping off even the essential water from the enzyme surface, leading to an insufficiently hydrated enzyme molecule and in turn to a decrease in enzyme activity. The water-stripping effect has been studied by Gorman and Dordick (1992), who measured the release of T_2O from three enzymes (chymotrypsin, subtilisin, and peroxidase) dispersed in various solvents. They found that T_2O desorption occurred in all enzymes and solvents tested, and was completed within minutes. The stripping of water from an enzyme by a solvent has been correlated with solvent polarity ($1/\varepsilon$) and to a lesser extent with solvent hydrophobicity (log P). For instance, methanol desorbed about 60% of the water bound to the enzymes, while hexane only desorbed 0.5%.

Taking this into consideration, it is not surprising that a good linear correlation between optimal water content (to achieve the optimal activity) and log P of the solvent was observed for a papain-catalyzed esterification (Stevenson and Storer, 1991). The optimal water content varied from 4% in acetonitrile to 0.05% in tetrachloromethane. Since enzyme and solvent compete for water, the optimal water content is also related to the amount of enzyme and the concentration of substrate (Hirata et al., 1990).

A recent study on the effect of pressure on enzyme function in organic media has revealed that, due to an increase in the amount of water being removed from the enzyme, enzyme activity is depressed by increasing the pressure (Kim and Dordick, 1993). This provides further evidence that enzyme activity in organic media is dominated by the interaction of enzyme-bound water with the solvent.

Removal of water from an enzyme by organic solvents can be minimized by hydrophilization of the enzyme's surface. In this way, the hydration shell of the enzyme can be retained. For example, covalent modification of α-chymotrypsin with pyromellitic dianhydride results in a strong stabilization of the enzyme against denaturation by organic solvents (Khmelnitsky et al., 1991a).

3.5.2 Effect of solvent on the enzyme

3.5.2.1 Influence of solvent on protein dynamics. Solvent can also affect enzyme activity by causing changes in the dynamic motion and conformation of the protein. EPR spectroscopy and molecular dynamics simulations of α-chymotrypsin in a variety of organic solvents have shown that the motion in the active-site region decreased with decreasing solvent dielectric constant (Figure 3.4) (Affleck *et al.*, 1992b). This is in accord with changes in the electrostatic force between charged residues on the protein, but contrasts with the increase in alcohol dehydrogenase activity upon a decrease in solvent dielectric constant (Guinn *et al.*, 1991a). Using solid-state NMR to investigate the solvent dependence of tyrosyl ring motion in α-lytic protease, Burke *et al.* (1993) have shown that there are no clear correlations between protein mobility and either enzyme activity or solvent dielectric constant. For example, the specific activities of α-lytic protease were almost the same in *tert*-butyl methyl ether and dioxane, where the enzyme showed quite different dynamic levels; while in dioxane and acetonitrile, where the enzyme showed similar tyrosine ring flipping rates, the enzyme activities were entirely different.

3.5.2.2 Influence of solvent on protein conformation. There is some evidence that the native conformation of an enzyme may be partially altered when suspended in organic solvents. The effect of solvent on peroxidase structure has been studied by Ryu and Dordick (1992a). A decrease in the difference between the fluorescence of peroxidase's buried tryptophan residue and free L-tryptophan in dioxan indicates that peroxidase denatured in the solvent. Further, an increase in the absorbance of the Soret band (403 nm) in a water/dioxan mixture suggests the solvent-induced exposure of the prosthetic heme to the reaction mixture. However, peroxidase appears to maintain its active-site structural integrity in most of the other organic solvents tested. α-Chymotrypsin is also denatured in water–2,3-butanediol, as indicated by both fluorescence intensity changes and shifts in the wavelength of maximal emission (Mozhaev *et al.*, 1989).

3.5.2.3 Influence of solvent on enzyme active center. The solvent can also affect enzyme activity by disrupting some proportion of the total number of enzyme active sites. The active site concentration of α-chymotrypsin in water is not affected by addition of three dipolar aprotic solvents up to 32% dioxan, 14% acetone and 13% acetonitrile (Clement and Bender, 1963), but only two-thirds of the total chymotrypsin molecules are catalytically active in dry octane, as determined by one approach to non-aqueous active site titration (Zaks and Klibanov, 1988a). Later experiments have shown that this loss of active sites is actually not caused by octane but by lyophilization during enzyme preparation. When solid-state NMR was used to assess the active site structure of α-chymotrypsin in organic media, it was found that lyophilization disrupted around 42%

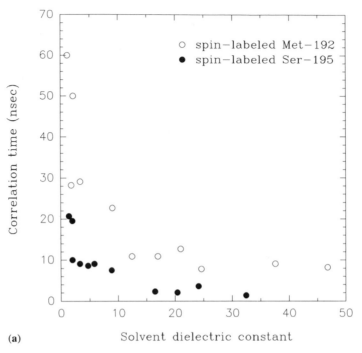

(a)

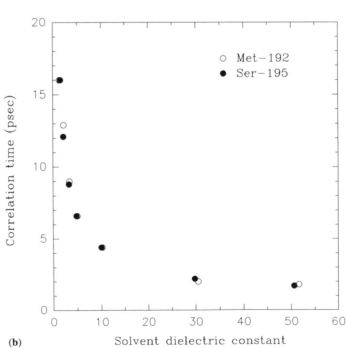

(b)

of the active centers, but addition of anhydrous solvents caused 0–50% additional disruption, depending on solvent hydrophobicity (Burke *et al.*, 1992). For example, the additional disruptions caused by octane and dioxane were 0 and 29%, respectively. The solvent-induced active center disruption could be the result of a combination of dehydration and/or unfolding of the enzyme. Although the low enzyme activity in organic media compared to that in water can be explained in part by active center disruption, it is still not clear whether the reason for the loss of activity is due to denaturation or a reduction in molecular dynamics (Burke *et al.*, 1993).

An alternative mode of interaction between solvent and enzyme is via competition with the substrate for binding at the active site. This effect may be more significant when less polar solvents are used (Kasche *et al.*, 1991). Additionally, solvent molecules can penetrate into the enzyme's active site and in turn decrease the local polarity of the active site, thereby increasing the electrostatic repulsion between the enzyme and the substrate, hence decreasing the binding. This competitive inhibition can account for the profound increase in the K_m values for substrates with enzymes in organic solvents such as dioxane and acetonitrile (Clement and Bender, 1963).

3.5.3 Effect of solvent on substrates and products

Solvents can interact directly or indirectly with substrates and products, thus affecting enzyme activity. Solvents can alter the concentrations of substrates or products in the aqueous layer around the enzyme, and since substrates must penetrate into this water layer for the reaction to occur, and products must partition out of the layer to drive the reaction forward, the solvent affects activity. This effect has been demonstrated by Yang and Robb (1994) and Kawakami and Nakahara (1994), and will be discussed in section 3.6.4.

3.6 Kinetics and thermodynamics of non-aqueous enzyme-catalyzed processes

3.6.1 Models

Since enzyme activity in different organic solvents varies significantly (for example, the rate of transesterification catalyzed by chymotrypsin increased more than 43 000 times when the solvent used was changed from dimethyl

---

Figure 3.4 Flexibility of α-chymotrypsin active site in different organic solvents: (a) Rotational correlation times of enzyme-bound spin labels determined from EPR spectra; (b) correlation times of α-carbon atoms determined from molecular dynamics simulations (replotted from data in Affleck *et al.* (1992b)).

sulfoxide to hexadecane; Zaks and Klibanov, 1988a), there are likely to be quantitative correlations between solvent properties and enzyme activity. The search for such correlations has been actively debated in the literature.

The most popular solvent descriptor which has been considered as a key determinant of enzyme activity is log P (Laane *et al.*, 1987), the logarithm of the partition coefficient of a solvent in an octanol/water mixture. Log P reflects the hydrophobicity of the solvent, and many enzymes have been found to be more active in solvents with a higher log P (Laane *et al.*, 1987). The dielectric constant of water may not be sufficient to screen the electrostatic interactions between the protein and a very polar solvent, allowing the solvent to penetrate the essential water layer and interact with the enzyme's active site. Also, as discussed before, hydrophilic solvents have a strong tendency actually to remove the essential water from the vicinity of the enzyme. Both processes lead to a loss in enzyme activity. A hydrophobic solvent (with a higher log P) will, however, enable the enzyme to retain its native conformation. Many exceptions to the 'log P rule', however, have also been reported (Cantarella *et al.*, 1991; Bonneau *et al.*, 1993). Indeed, when water activity was controlled in the reaction system, no relationship was observed between the activity of polyphenol oxidase and log P of the solvent (Yang and Robb, 1994).

Other solvent properties have also been related to enzyme activity. For example, the activity of alcohol dehydrogenase increased with a decrease in the solvent dielectric (Guinn *et al.*, 1991a), and a linear correlation was observed between lipase activity and log $S_{w/o}$ (molar solubility of water in a solvent) (Martins *et al.*, 1994). However, since any single-parameter approach is limited in its ability to reflect the possible enzyme–solvent interactions, a three-dimensional solubility parameter approach, which combines solvent hydrophobicity, Hildebrand solubility parameter and a hydrogen-bonding parameter, is useful in correlating and predicting enzyme activity in nonaqueous media (Schneider, 1991).

In spite of numerous studies, it is still unclear what characteristics of the organic solvent are the key determining factors in terms of enzyme activity. It is worth mentioning that water-immiscibility and apolarity of the solvent are not necessarily useful in predicting enzyme efficiency (Narayan and Klibanov, 1993). In addition to the physicochemical properties of the solvent, the solvent geometry can also influence enzyme activity in organic media. Ottolina *et al.* (1994) have shown that in a chiral solvent, carvone, the maximal rate of lipase-catalyzed transesterification in the (S)-solvent was twice as high as in the (R)-solvent, while the K_m remained the same. Polyphenol oxidase was also more active in (R)-solvent than in (S)-solvent, but the initial transesterification rate for subtilisin in (S)-carvone was lower than in (R)-carvone.

Generation of quantitative models is required for the fast growing field of non-aqueous enzymology to develop further. Based on the generally accepted viewpoint that disruption of the hydration shell surrounding a protein is mainly responsible for the denaturation of the protein by organic solvents, van Erp *et al.*

(1991) have developed a theoretical model to describe quantitatively the effect of water content and solvent nature on enzyme function in organic media. This model has been tested by studying the catalytic behavior of laccase suspended in water/organic mixtures.

To explain the reversible denaturation of several proteins via solvent-induced water stripping, Khmelnitsky et al. (1991b) have developed a thermodynamic model which relates the threshold concentration of the organic solvent (the concentration at which an enzyme in water begins to inactivate) with its hydrophobicity, solvating ability and molecular geometry. Based on this model, a quantitative parameter, 'denaturation capacity', was proposed as a measure of the denaturing strength of the solvent. This model can successfully predict the denaturation thresholds of many enzymes.

3.6.2 Apparent kinetic constants versus individual rate constants

The effects of water and solvent on enzyme activity are most often described by changes in K_m and V_{max} for an enzyme–substrate pair. When laccase was suspended in nine different organic solvents, an increase in water content always resulted in a sharp enhancement of both K_m and V_{max} (van Erp et al., 1991). However, a decrease in K_m was observed for subtilisin BPN' in tetrahydrofuran when 5 µl/ml of water was added (Xu et al., 1994). When α-chymotrypsin was used to catalyze an esterification in a variety of organic solvents, V_{max} increased markedly, but K_m decreased slightly, upon an increase in the log P of the solvent (Reslow et al., 1987). An opposite effect was observed when porcine pancreatic lipase was used to catalyze the transesterification between trifluoroethyl butanoate and (±)-sulcatol (Secundo et al., 1992). For both enantiomeric substrates, V_{max} values were slightly affected by the solvent, whereas K_m values dramatically increased with increasing solvent polarity. Clearly, there is no general relationship between K_m or V_{max} (k_{cat}) and the water content and nature of the solvent.

Interestingly, an investigation of trypsin kinetics shows that enhancing the content of a water-miscible solvent (9:1 mix of dimethyl formamide and dimethyl sulfoxide) reduced the k_{cat} for the enzyme's amidase activity relative to that for its esterase activity, while K_m increased sharply for both reactions (Guinn et al., 1991b). A depression of amidase activity, while maintaining a high esterase activity, in the presence of certain concentrations of water-miscible organic solvents was also observed with chymotrypsin, papain and subtilisin (Barbas et al., 1988). These enzymes can therefore be used as peptide ligases in peptide synthesis, and the peptide products will be free from secondary hydrolysis.

The K_m and k_{cat} (or V_{max}) values reported above are actually all apparent kinetic constants. Undoubtedly, measuring these apparent kinetic constants and correlating them with the nature of reaction medium can yield important information about enzyme structure and function. However, apparent kinetic constants can be a function not only of the individual rate constants for a particular mechanism, but also of any other factors affecting the enzyme activity (such as rate-determining step changes, mass-transfer limitations, and substrate concentrations). It is therefore necessary to measure individual rate constants, which will help us understand in detail how a solvent interacts with an enzyme, and thus how to use 'solvent engineering' to control enzyme activity.

Chatterjee and Russell (1992b) have determined individual rate constants for enzyme reactions in nearly anhydrous environments. Using the added nucleophile method, they have measured the effect of changing solvent on the binding and catalytic steps for subtilisin-catalyzed transesterifications (Figure 3.5). Their results indicate that the decreased overall activity of subtilisin in organic solvents is mainly due to the high value of the enzyme-substrate binding constant (K_s),

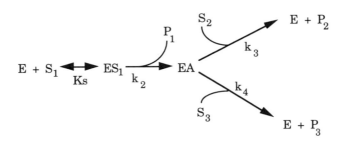

$$V_{P1} = \frac{\dfrac{k_2\,(k_3[S_2] + k_4[S_3])}{k_2 + k_3[S_2] + k_4[S_3]}\,[E_0]\,[S_1]_0}{[S_1]_0 + K_s\dfrac{k_3\,[S_2] + k_4\,[S_3]}{k_2 + k_3\,[S_2] + k_4\,[S_3]}} \tag{1}$$

$$V_{P2} = \frac{\dfrac{k_2\,k_3\,[S_2]}{k_2 + k_3\,[S_2] + k_4\,[S_3]}\,[E_0]\,[S_1]_0}{[S_1]_0 + K_s\dfrac{k_3\,[S_2] + k_4\,[S_3]}{k_2 + k_3\,[S_2] + k_4\,[S_3]}} \tag{2}$$

$$V_{P3} = \frac{\dfrac{k_2\,k_4\,[S_3]}{k_2 + k_3[S_2] + k_4[S_3]}\,[E_0]\,[S_1]_0}{[S_1]_0 + K_s\dfrac{k_3[S_2] + k_4[S_3]}{k_2 + k_3[S_2] + k_4[S_3]}} \tag{3}$$

Figure 3.5 Kinetic scheme and equations for determinations of individual rate constants for subtilisin-catalyzed transesterification in organic solvents using the added nucleophile method (Chatterjee and Russell, 1992b).

which decreases upon an increase in log P of the solvent. The rate constants for acylation (k_2) and deacylation (k_3) increased with an increase in log P and dielectric constant of the solvent, respectively. The variation of k_3 upon solvent dielectric substantiates that the tetrahedral intermediate involved in the deacylation step in organic solvents is charged.

Chatterjee and Russell (1992b) also showed that solvent changes could result in changes in the rate-determining step of subtilisin Carlsberg. In a recent kinetic study using subtilisin BPN′ to catalyze a transesterification in tetrahydrofuran, Wangikar *et al.* (1993) have also found that the rate-determining step in the dry solvent is acylation, but upon addition of 2% water, subtilisin reverts to a deacylation rate-limiting reaction, primarily due to an increase in acylation rate. The rate-determining step for hydrolysis of the same ester in aqueous solution is deacylation. Therefore, the rate-determining step can be switched by both solvent and the addition of water to a given solvent.

3.6.3 Transition state stabilization

Wangikar *et al.* (1993) have also demonstrated that the reactivity of an enzyme is very much dependent on the polarity of solvent, substrate, and enzyme active site. Since subtilisin-catalyzed transesterifications in organic media follow the acyl-enzyme mechanism involving a charged and polar tetrahedral transition state, it is not surprising that the transition state is stabilized by a polar environment. Indeed, the transition state of subtilisin BPN′ is intrinsically more stable in acetone (polar) than in hexane (non-polar) (the difference in intrinsic free energies of transition-state stabilization between hexane and acetone is 1.75 kcal/mol). A polar active-site mutation (Gly166 → Asn) can compensate for the transition-state destabilization upon going from acetone to hexane, and thus stabilize the intrinsic transition state of the enzyme in hexane by 1.22 kcal/mol as compared to acetone. The polar mutant also stabilizes the transition state of a polar substrate (*N*-acetyl-L-serine ethyl ester) as compared to that of a nonpolar substrate (*N*-acetyl-L-phenylalanine ethyl ester) in a range of different organic solvents. The fact that the transition-state stabilization correlates well with solvent polarity rather than solvent hydrophobicity suggests that the solvent effect is primarily electrostatic, as expected for a charged transition state. In this way, subtilisin BPN′ is significantly activated (up to 178-fold) by a polar active-site mutant in going from acetone to hexane. Addition of water can also improve the degree of enzyme catalysis by increasing the acylation rate and enzyme-substrate binding (i.e. decreasing the intrinsic K_m value). Xu *et al.* (1994) continued their investigations on transition state stabilization of subtilisins in organic media, and again confirmed that polarity is an important factor in stabilizing the charged transition state, and that the rate of an enzymatic reaction proceeding through a charged transition state can be enhanced by increasing the active-site polarity (for example, by the addition of water).

3.6.4 Substrate solvation

3.6.4.1 Effect of substrate solvation on kinetics. Dordick and coworkers (1993) have suggested that it is important to distinguish the effects of solvent on enzyme function in both kinetic and thermodynamic terms. A solvent can affect enzyme activity and specificity by altering the partitioning of substrates and products between the bulk organic phase and the active-site micro-aqueous phase (Halling, 1994). This partitioning effect can be used to explain the general increase in apparent K_m values when enzymes are placed in an organic solvent rather than in aqueous solution (Secundo *et al.*, 1992; Ulbrich-Hofmann and Selisko, 1993; Yang and Robb, 1993). For example, enhancements in K_m values for lipase in five organic solvents were accompanied by an increase in the substrate partition coefficient (solvent versus water) (Secundo *et al.*, 1992).

Ryu and Dordick (1989, 1992a,b) have investigated linear free-energy relationships between horseradish peroxidase activity (catalyzing the oxidation of phenols) and the properties of substrates and solvents. The following equation successfully modelled the experimental data:

$$\log (V_{max}/K_m) \propto \delta\pi + \rho\sigma \qquad (3.1)$$

where π and σ represent the hydrophobicity and electronic properties of the substrate, and δ and ρ are the solvent-dependent sensitivity to substrate hydrophobicity and electronic properties, respectively:

$$\delta = -0.21 (\log P) - 0.86 \qquad (3.2)$$

$$\rho = -1.75 - 0.012 \, \varepsilon \qquad (3.3)$$

Dordick's work shows that catalytic efficiency depends on the physicochemical characteristics of both substrate (hydrophobicity and electronic properties) and solvent (hydrophobicity and polarity). These results are related to the partitioning behavior of phenols between the bulk reaction medium and the peroxidase active site. This partitioning is reduced by increasing substrate and solvent hydrophobicities, thereby requiring a higher concentration of phenols to saturate the enzyme. This leads to an increase in apparent K_m in organic versus aqueous media.

The effect of substrate partitioning on enzyme activity is further demonstrated by the dependence of enzyme activity on substrate concentration in the aqueous phase. When polyphenol oxidase (Yang and Robb, 1994) and whole cells of *Pichia pastoris* (Kawakami and Nakahara, 1994) were used to catalyze the oxidation of catechols and benzyl alcohol, respectively, reaction rates increased with substrate concentration to a maximum, but then gradually decreased due to substrate inhibition. The optimal substrate concentration (in the aqueous phase), at which the reaction rate in organic solvents reached a maximum, was identical to that observed in water. The concentration of the substrate partitioned into the aqueous phase associated with the catalyst is obviously solvent-dependent, and these data further substantiate the need for a careful consideration of the meaning of any kinetic data for enzymes in organic solvents.

Indeed, a bell-shaped relationship was observed between the activity of polyphenol oxidase and substrate partition coefficients ($P_s = [S]_{solvent}/[S]_{H_2O}$) with two catechol substrates (Yang et al., 1992). Actually, it is reasonable to expect that there is an optimum P_s for each substrate, because at low P_s values so much substrate may be present in the enzyme hydration layer that substrate inhibition may occur, and at high P_s values so little may enter that V_{max} is not retained. Thermodynamically, the reduction in enzyme activity upon an increase in P_s can be understood by a combination of the following steps:

(1) The substrate dissolved in the bulk organic phase must partition into the active-site micro-aqueous phase;
(2) the substrate located in the active site can then react with the enzyme to form the enzyme–substrate complex;
(3) decomposition of the enzyme–substrate complex will lead to the formation of the product in aqueous phase around the enzyme;
(4) the product must partition out to the bulk reaction medium to drive the reaction.

The reaction rate is determined by the sum of the free energy changes in the first two steps. The ΔG^0 for step (1) is equal to RT (ln P_s), and it is reasonable to assume that the change in free energy during step (2) is not greatly affected by changing the solvent. Therefore, enzyme activity in this case is mainly controlled by the partitioning of the substrate, determining the substrate ground state level. Indeed, by using a transfer free energy method, Ryu and Dordick (1992a) have demonstrated that with horseradish peroxidase the substantial increase in the K_m as both substrate and solvent hydrophobicities increased is a result of the reduced binding efficiency of phenolic substrates due to the enhanced stabilization of substrates in the ground state, rather than the destabilization of the enzyme–substrate complex. Practically, it is essential to select an organic solvent which provides the optimal partitioning of the substrate into the aqueous phase as well as the preferential extraction of the product into the organic phase.

Because of the effect of substrate solvation, Halling (1994) suggests that it is necessary to distinguish two types of Michaelis constants: the true K_m (K_{ma}, based on concentrations measured in an aqueous phase) and the apparent K_m (K_{mo}, based on concentrations in the bulk organic phase). Only after discounting the solvation differences (by dividing the K_{mo} by the substrate partition coefficient P_s or the substrate solubility in the solvent S_s), can one assess the real K_m which reflects the intrinsic effect of solvent on the enzyme catalysis (Halling, 1994; Reimann et al., 1994). For the subtilisin-catalyzed transesterification between N-CBZ-L-alanine (or leucine) p-nitrophenyl ester and 1-butanol, the apparent K_m values for the ester increased with an increase in substrate solubility in the solvent, but the corrected K_m (K_m/S_s) gave the entirely opposite trend (Reimann et al., 1994).

3.6.4.2 Effect of substrate solvation on equilibrium position. It is well known that a low-water organic medium is favorable in shifting the hydrolytic equilibria to the synthetic direction. Reduced water concentration and partitioning of substrates and products account for the degree to which a shift is possible (Halling, 1994).

Compared to the numerous studies on the effect of solvent on enzyme activity, the effect of solvent on equilibrium has received much less attention. Using group contribution correlations, Halling (1990b) has made predictions on the effect of solvent on the equilibrium position of enzymatic reactions, using data for liquid–liquid distributions of reactants. Esterification reactions are $> 10^4$-fold more favored in non-polar solvents (such as hexane) over polar solvents (such as ethyl ether). The equilibrium constant correlates well with the solubility of water in the solvent. This prediction appears to apply for other equilibrium reactions, such as alcohol oxidation and chloride hydrolysis.

The effectiveness of Halling's prediction has been verified by Valivety *et al.* (1991), who studied the effect of 20 organic solvents on the equilibrium positions of the esterification between dodecanol and decanoic acid catalyzed by a lipase. The equilibrium constant is correlated to the solubility of water in the solvent, which in turn correlates well with the sum of Guttman's donor number (DN) and the electron pair acceptance index number (E_T) of the solvent. DN and E_T can be taken as measures of hydrogen-bond donating and accepting capacities of a given solvent (Marcus, 1985). Therefore, the data suggest that solubility of water in the solvent is an excellent predictor of the equilibrium position. The relationship between water solubility in a solvent and the sum of DN and E_T can be understood when one considers that the solvation of water in an organic solvent requires both donation and acceptance of hydrogen bonds or other dipole–dipole interactions. Hence, it is likely that DN and E_T for a solvent can be used to determine both the solubility of water and the equilibrium position of esterification in that solvent.

3.7 Concluding remarks

Above we have described what is currently known about the fundamentals of non-aqueous enzymology. To enable this new and fast-developing field to mature, and also to facilitate the development of its applications, a number of fundamental issues must be addressed. First, a detailed understanding of the relationship between the structure of a biocatalyst (dynamics, conformation, and active site structure) and its catalytic function (activity, stability, and specificity) in organic media must be developed. Second, the role of water and solvent on enzyme action in organic media must be investigated further. For example, why do different enzymes have distinct water-content dependencies? As far as solvent is concerned, the two major effects which a solvent has are to alter the distribution of the essential water on the enzyme, and to alter the substrate concentration in the vicinity of the enzyme by substrate solvation. Both effects are clearly important in defining the eventual activity of an enzyme in a given solvent. However, even when these key effects have been discounted (for example either by controlling water activity or by considering substrate partitioning), one can still observe the individual effects of particular organic solvents. This clearly indicates that solvents also have intrinsic effects on enzyme function in non-aqueous environments. Quantitative mathematical modelling related to the kinetics and thermodynamics of non-aqueous enzyme-catalyzed reactions should also continue to be generated. It would be naive to think that the catalytic efficiency of an enzyme in organic solvents is controlled by only one mechanism. A combination of the amount and location of water, the nature and configuration of the solvent, the characteristics of the enzyme, and the mutual interactions between water, solvent and enzyme are central to producing a given enzyme activity, specificity and stability.

Undoubtedly, answers to these fundamental questions will help us understand why enzymes are generally less active in organic solvents than in aqueous solution, and based on this understanding, we can improve our ability to tailor rationally enzyme activity by 'solvent engineering'.

References

Adams, K.A.H., Chung, S.-H. and Klibanov, A.M. (1990) Kinetic isotope effect investigation of enzyme mechanism in organic solvents. *J. Amer. Chem. Soc.*, **112**, 9418–9419.

Affleck, R., Xu, Z.-F., Suzawa, V., Focht, K., Clark, D.S. and Dordick, J.S. (1992a) Enzymatic catalysis and dynamics in low-water environments. *Proc. Natl. Acad. Sci. USA*, **89**, 1100–1104.

Affleck, R., Haynes, C.A. and Clark, D.S. (1992b) Solvent dielectric effects on protein dynamics. *Proc. Natl. Acad. Sci. USA*, **89**, 5167–5170.

Ayala, G., Gómez-Puyou, T., Gómez-Puyou, A. and Darszon, A. (1986) Thermostability of membrane enzymes in organic solvents. *FEBS Lett.*, **203**, 41–43.

Barbas, C.F., Matos, I.J.R., West, J.B. and Wong, C.-H. (1988) A search for peptide ligase: cosolvent-mediated conversion of proteases to esterases for irreversible synthesis of peptides. *J. Amer. Chem. Soc.*, **110**, 5162–5166.

Blanco, R.M., Rakels, J.L.L., Guisán, J.M. and Halling, P.J. (1992) Effect of thermodynamic water activity on amino-acid ester synthesis catalyzed by agarose–chymotrypsin in 3-pentanone. *Biochim. Biophys. Acta*, **1156**, 67–70.

Blinkovsky, A.M., Martin, B.D. and Dordick, J.S. (1992) Enzymology in monophasic organic media. *Current Opinion in Biotechnology*, **3**, 124–129.

Bone, S. and Pethig, R. (1985) Dielectric studies of protein hydration and hydration-induced flexibility. *J. Mol. Biol.*, **181**, 323–326.

Bonneau, P.R., Eyer, M., Graycar, T.P., Estell, D.A. and Jones, J.B. (1993) The effects of organic solvents on wild-type and mutant subtilisin-catalyzed hydrolyses. *Bioorg. Chem.*, **21**, 431–438.

Burke, P.A., Smith, S.O., Bachovchin, W.W. and Klibanov, A.M. (1989) Demonstration of structural integrity of an enzyme in organic solvents by solid-state NMR. *J. Amer. Chem. Soc.*, **111**, 8290–8291.

Burke, P.A., Griffin, R.G. and Klibanov, A.M. (1992) Solid-state NMR assessment of enzyme active center structure under nonaqueous conditions. *J. Biol. Chem.*, **267**, 20057–20064.

Burke, P.A., Griffin, R.G. and Klibanov, A.M. (1993) Solid-state nuclear magnetic resonance investigation of solvent dependence of tyrosyl ring motion in an enzyme. *Biotechnol. Bioeng.*, **42**, 87–94.

Cantarella, M., Cantarella, L. and Alfani, F. (1991) Hydrolytic reactions in two-phase systems. Effect of water-immiscible organic solvents on stability and activity of acid phosphatase, β-glucosidase, and β-fructofuranosidase. *Enzyme Microb. Technol.*, **13**, 547–553.

Cassells, J.M. and Halling, P.J. (1988) Effect of thermodynamic water activity on thermolysin-catalysed peptide synthesis in organic two-phase systems. *Enzyme Microb. Technol.*, **10**, 486–491.

Chatterjee, S. and Russell, A.J. (1992a) Activity of thiolsubtilisin in organic solvents. *Biotechnol. Prog.*, **8**, 256–258.

Chatterjee, S. and Russell, A.J. (1992b) Determination of equilibrium and individual rate constants for subtilisin-catalyzed transesterification in anhydrous environments. *Biotechnol. Bioeng.*, **40**, 1069–1077.

Chatterjee, S. and Russell, A.J. (1993) Kinetic analysis of the mechanism for subtilisin in essentially anhydrous organic solvents. *Enzyme Microb. Technol.*, **15**, 1022–1029.

Clement, G.E. and Bender, M.L. (1963) The effect of aprotic dipolar organic solvents on the kinetics of α-chymotrypsin-catalyzed hydrolyses. *Biochemistry*, **2**, 836–843.

Dabulis, K. and Klibanov, A.M. (1993) Dramatic enhancement of enzymatic activity in organic solvents by lyophotectants. *Biotechnol. Bioeng.*, **41**, 566–571.

Dastoli, F.R. and Price, S. (1967) Catalysis by xanthine oxidase suspended in organic media. *Arch. Biochem. Biophys.*, **118**, 163–165.

Dordick, J.S. (1989) Enzymatic catalysis in monophasic organic solvents. *Enzyme Microb. Technol.*, **11**, 194–211.

Fitzpatrick, P.A., Steinmetz, A.C.U., Ringe, D. and Klibanov, A.M. (1993) Enzyme crystal structure in a neat organic solvent. *Proc. Natl. Acad. Sci. USA*, **90**, 8653–8657.

Gorman, L.A.S. and Dordick, J.S. (1992) Organic solvents strip water off enzymes. *Biotechnol. Bioeng.*, **39**, 392–397.

Guinn, R.M., Skerker, P.S., Kavanaugh, P. and Clark, D.S. (1991a) Activity and flexibility of alcohol dehydrogenase in organic solvents. *Biotechnol. Bioeng.*, **37**, 303–308.

Guinn, R.M., Blanch, H.W. and Clark, D.S. (1991b) Effect of a water-miscible organic solvent on the kinetic and structural properties of trypsin. *Enzyme Microb. Technol.*, **13**, 320–326.

Halling, P.J. (1989) Organic liquids and biocatalysts: theory and practice. *Trends Biotechnol.*, **7**, 50–51.

Halling, P.J. (1990a) High-affinity binding of water by proteins is similar in air and in organic solvents. *Biochim. Biophys. Acta*, **1040**, 225–228.

Halling, P.J. (1990b) Solvent selection for biocatalysis in mainly organic systems: predictions of effects on equilibrium position. *Biotechnol. Bioeng.*, **35**, 691–701.

Halling, P.J. (1992) Salt hydrates for water activity control with biocatalysts in organic media. *Biotechnol. Techniques*, **6**, 271–276.

Halling, P.J. (1994) Thermodynamic predictions for biocatalysis in nonconventional media: theory, tests, and recommendations for experimental design and analysis. *Enzyme Microb. Technol.*, **16**, 178–206.

Hartsough, D.S. and Merz, K.M. (1992) Protein flexibility in aqueous and nonaqueous solutions. *J. Amer. Chem. Soc.*, **114**, 10113–10116.

Hirata, H., Higuchi, K. and Yamashina, T. (1990) Lipase-catalyzed transesterification in organic solvent: effects of water and solvent, thermal stability and some applications. *J. Biotechnol.*, **14**, 157–167.

Jencks, W.P. (1987) *Catalysis in chemistry and enzymology.* Dover, New York.

Kanerva, L.T. and Klibanov, A.M. (1989) Hammett analysis of enzyme action in organic solvents. *J. Amer. Chem. Soc.*, **111**, 6864–6865.

Kasche, V., Michaelis, G. and Galunsky, B. (1991) Binding of organic solvent molecules influences the P_1'-P_2' stereo- and sequence-specificity of α-chymotrypsin in kinetically controlled peptide synthesis. *Biotechnol. Lett.*, **13**, 75–80.

Kawakami, K. and Nakahara, T. (1994) Importance of solute partitioning in biphasic oxidation of benzyl alcohol by free and immobilized whole cells of *Pichia pastoris*. *Biotechnol. Bioeng.*, **43**, 918–924.

Khmelnitsky, Y.L., Belova, A.B., Levashov, A.V. and Mozhaev, V.V. (1991a) Relationship between surface hydrophilicity of a protein and its stability against denaturation by organic solvents. *FEBS Lett.*, **284**, 267–269.

Khmelnitsky, Y.L., Mozhaev, V.V., Belova, A.B., Sergeeva, M.V. and Martinek, K. (1991b) Denaturation capacity: a new quantitative criterion for selection of organic solvents as reaction media in biocatalysis. *Eur. J. Biochem.*, **198**, 31–41.

Kim, J. and Dordick, J.S. (1993) Pressure affects enzyme function in organic media. *Biotechnol. Bioeng.*, **42**, 772–776.

Klibanov, A.M. (1989) Enzymatic catalysis in anhydrous organic solvents. *Trends Biochem. Sci.*, **14**, 141–144.

Kuhl, P. and Halling, P.J. (1991) Salt hydrates buffer water activity during chymotrypsin-catalysed peptide synthesis. *Biochim. Biophys. Acta*, **1078**, 326–328.

Kuhl, P., Halling, P.J. and Jakubke, H.-D. (1990) Chymotrypsin suspended in organic solvents with salt hydrates is a good catalyst for peptide synthesis from mainly undissolved reactants. *Tetrahedron Lett.*, **31**, 5213–5216.

Kvittingen, L., Sjursnes, B., Anthonsen, T. and Halling, P.J. (1992) Salt hydrates to buffer optimal water level during lipase catalysed synthesis in organic media: a practical procedure for organic chemists. *Tetrahedron*, **48**, 2793–2802.

Laane, C., Boeren, S., Vos, K. and Veeger, C. (1987) Rules for optimization of biocatalysis in organic solvents. *Biotechnol. Bioeng.*, **30**, 81–87.

Liu, W.R., Langer, R.L. and Klibanov, A.M. (1991) Moisture-induced aggregation of lyophilized proteins in the solid state. *Biotechnol. Bioeng.*, **37**, 177–184.

Martins, J.F., de Sampaio, T.C., de Carvalho, I.B. and Barreiros, S. (1994) Lipase catalyzed esterification of glycidol in nonaqueous solvents: solvent effects on enzymatic activity. *Biotechnol. Bioeng.*, **44**, 119–124.

McMinn, J.H., Sowa, M.J., Charnick, S.B. and Paulaitis, M.E. (1993) The hydration of proteins in nearly anhydrous organic solvent suspensions. *Biopolymers*, **33**, 1213–1224.

Mozhaev, V.V., Khmelnitsky, Y.L., Sergeeva, M.V., Belova, A.B., Klyachko, N.L., Levashov, A.V. and Martinek, K. (1989) Catalytic activity and denaturation of enzymes in water/organic cosolvent mixtures. α-chymotrypsin and laccase in mixed water/alcohol, water/glycol and water/formamide solvents. *Eur. J. Biochem.*, **184**, 597–602.

Murcus, Y. (1985) *Ion solvation*. John Wiley & Sons Ltd, Chichester.

Narayan, V.S. and Klibanov, A.M. (1993) Are water-immiscibility and apolarity of the solvent relevant to enzyme efficiency? *Biotechnol. Bioeng.*, **41**, 390–393.

Ottolina, G., Gianinetti, F., Riva, S. and Carrea, G. (1994) Solvent configuration influences enzyme activity in organic media. *J. Chem. Soc., Chem. Commun.*, 535–536.

Reimann, A., Robb, D.A. and Halling, P.J. (1994) Solvation of CBZ-amino acid nitrophenyl esters in organic media and the kinetics of their transesterification by subtilisin. *Biotechnol. Bioeng.*, **43**, 1081–1086.

Reslow, M., Adlercreutz, P. and Mattiasson, B. (1987) Organic solvents for bioorganic synthesis 1. Optimization of parameters for a chymotrypsin catalyzed process. *Appl. Microbiol. Biotechnol.*, **26**, 1–8.

Rupley, J.A., Gratton, E. and Careri, G. (1983) Water and globular proteins. *Trends Biochem. Sci.*, **8**, 18–22.

Russell, A.J. and Klibanov, A.M. (1988) Inhibitor-induced enzyme activation in organic solvents. *J. Biol. Chem.*, **263**, 11624–11626.

Russell, A.J., Chatterjee, S., Rapanovich, I. and Goodwin, J.G. (1992) Mechanistic enzymology in anhydrous organic solvents. In *Biomolecules in Organic Solvents* (eds A. Gómez-Puyou, A. Darszon and M.T. Gómez-Puyou), CRC Press, London, pp. 91–111.

Ryu, K. and Dordick, J.S. (1989) Free energy relationships of substrate and solvent hydrophobicities with enzymatic catalysis in organic media. *J. Amer. Chem. Soc.*, **111**, 8026–8027.

Ryu, K. and Dordick, J.S. (1992a) How do organic solvents affect peroxidase structure and function? *Biochemistry*, **31**, 2588–2598.

Ryu, K. and Dordick, J.S. (1992b) Quantitative and predictive correlations for peroxidase catalysis in organic media. *Biotechnol. Techniques*, **6**, 277–282.

Schneider, L.V. (1991) A three-dimensional solubility parameter approach to nonaqueous enzymology. *Biotechnol. Bioeng.*, **37**, 627–638.

Secundo, F., Riva, S. and Carrea, G. (1992) Effects of medium and of reaction conditions on the enantioselectivity of lipases in organic solvents and possible rationales. *Tetrahedron: Asymmetry*, **3**, 267–280.

Ståhl, M., Jeppsson-Wistrand, U., Månsson, M.-O. and Mosbach, K. (1991) Induced stereo-selectivity and substrate selectivity of bio-imprinted α-chymotrypsin in anhydrous organic media. *J. Amer. Chem. Soc.*, **113**, 9366–9368.

Stevenson, D.E. and Storer, A.C. (1991) Papain in organic solvents: determination of conditions suitable for biocatalysis and the effect on substrate specificity and inhibition. *Biotechnol. Bioeng.*, **37**, 519–527.

Svensson, I., Wehtje, E., Adlercreutz, P. and Mattiasson, B. (1994) Effects of water activity on reaction rates and equilibrium positions in enzymatic esterifications. *Biotechnol. Bioeng.*, **44**, 549–556.

Ulbrich-Hofmann, R. and Selisko, B. (1993) Soluble and immobilized enzymes in water-miscible organic solvents: glucoamylase and invertase. *Enzyme Microb. Technol.*, **15**, 33–41.

Valivety, R.H., Johnston, G.A., Suckling, C.J. and Halling, P.J. (1991) Solvent effects on bio-catalysis in organic systems: equilibrium position and rates of lipase catalyzed esterification. *Biotechnol. Bioeng.*, **38**, 1137–1143.

Valivety, R.H., Halling, P.J. and Macrae, A.R. (1992a) *Rhizomucor miehei* lipase remains highly active at water activity below 0.0001. *FEBS Lett.*, **301**, 258–260.

Valivety, R.H., Halling, P.J., Peilow, A.D. and Macrae, A.R. (1992b) Lipases from different sources vary widely in dependence of catalytic activity on water activity. *Biochim. Biophys. Acta*, **1122**, 143–146.

Valivety, R.H., Halling, P.J. and Macrae, A.R. (1992c) Reaction rate with suspended lipase catalyst shows similar dependence on water activity in different organic solvents. *Biochim. Biophys. Acta*, **1118**, 218–222.

Vazquez-Duhalt, R., Fedorak, P.M. and Westlake, D.W.S. (1992) Role of enzyme hydrophobicity in biocatalysis in organic solvents. *Enzyme Microb. Technol.*, **45**, 837–841.

van Erp, S.H.M., Kamenskaya, E.O. and Khmelnitsky, Y.L. (1991) The effect of water content and nature of organic solvent on enzyme activity in low-water media. A quantitative description. *Eur. J. Biochem.*, **202**, 379–384.

Volkin, D.B., Staubli, A., Langer, R. and Klibanov, A.M. (1991) Enzyme thermoinactivation in anhydrous organic solvents. *Biotechnol. Bioeng.*, **37**, 843–853.

Vulfson, E.N., Sarney, D.B. and Law, B.A. (1991) Enhancement of subtilisin-catalysed inter-esterification in organic solvents by ultrasound irradiation. *Enzyme Microb. Technol.*, **13**, 123–126.

Wangikar, P.P., Graycar, T.P., Estell, D.A., Clark, D.S. and Dordick, J.S. (1993) Protein and solvent engineering of subtilisin BPN′ in nearly anhydrous organic media. *J. Amer. Chem. Soc.*, **115**, 12231–12237.

Wescott, C.R. and Klibanov, A.M. (1994) The solvent dependence of enzyme specificity. *Biochim. Biophys. Acta*, **1206**, 1–9.

West, J.B., Hennen, W.J., Lalonde, J.L., Bibbs, J., Zhong, Z., Meyer, E.F. and Wong, C.-H. (1990) Enzymes as synthetic catalysts: mechanistic and active-site considerations of natural and modified chymotrypsin. *J. Amer. Chem. Soc.*, **112**, 5313–5320.

Wheeler, C.J. and Croteau, R. (1986) Terpene cyclase catalysis in organic solvent/minimal water media: demonstration and optimization of (+)-α-pinene cyclase activity. *Arch. Biochem. Biophys.*, **248**, 429–434.

Wu, J. and Gorenstein, D.G. (1993) Structure and dynamics of cytochrome c in nonaqueous solvents by 2D NH-exchange NMR spectroscopy. *J. Amer. Chem. Soc.*, **115**, 6843–6850.

Xu, Z.-F., Affleck, R., Wangikar, P., Suzawa, V., Dordick, J.S. and Clark, D.S. (1994) Transition state stabilization of subtilisins in organic media. *Biotechnol. Bioeng.*, **43**, 515–520.

Yang, Z. and Robb, D.A. (1991) Enzyme activity in organic media. *Biochem. Soc. Trans.*, **20**, 13S.

Yang, Z. and Robb, D.A. (1993) Comparison of tyrosinase activity and stability in aqueous and nearly nonaqueous environments. *Enzyme Microb. Technol.*, **15**, 1030–1036.

Yang, Z. and Robb, D.A. (1994) Partition coefficients of substrates and products and solvent selection for biocatalysis under nearly anhydrous conditions. *Biotechnol. Bioeng.*, **43**, 365–370.

Yang, Z., Robb, D.A. and Halling, P.J. (1992) Variation of tyrosinase activity with solvent at a constant water activity, in *Biocatalysis in nonconventional media* (eds J. Tramper *et al.*). Elsevier, Amsterdam, pp. 585–592.

Yang, Z., Zacherl, D. and Russell, A.J. (1993) pH dependence of subtilisin dispersed in organic solvents. *J. Amer. Chem. Soc.*, **115**, 12251–12257.

Yennawar, H.P., Yennawar, N.H. and Farber, G.K. (1994) X-ray crystal structure of γ-chymotrypsin in hexane. *Biochemistry*, **33**, 7326–7336.

Yennawar, H.P., Yennawar, N.H. and Farber, G.K. (1995) A structural explanation for enzyme memory in nonaqueous solvents. *J. Amer. Chem. Soc.*, **117**, 577–585.

Zaks, A. and Klibanov, A.M. (1984) Enzymatic catalysis in organic media at 100°C. *Science*, **224**, 1249–1251.

Zaks, A. and Klibanov, A.M. (1985) Enzyme-catalyzed processes in organic solvents. *Proc. Natl. Acad. Sci. USA*, **82**, 3192–3196.

Zaks, A. and Klibanov, A.M. (1988a) Enzymatic catalysis in nonaqueous solvents. *J. Biol. Chem.*, **263**, 3194–3201.

Zaks, A. and Klibanov, A.M. (1988b) The effect of water on enzyme action in organic media. *J. Biol. Chem.*, **263**, 8017–8021.

4 New enzymatic properties in organic media

A. ZAKS

4.1 Introduction

There has been a remarkable progress in our understanding of properties of enzymes in organic media in the last decade. Although the catalysis by enzymes suspended in organic solvents was first observed almost 30 years ago (Dastoli *et al.*, 1966) the research in this area had progressed to only a limited extent till the early 1980s. The first reports covered mostly hydrolytic enzymes and dealt primarily with applied aspects of biocatalysis in non-aqueous media without focusing on the mechanistic aspects of this phenomenon (Oyama *et al.*, 1981; Yokozeki *et al.*, 1982). Not until a fundamental study of a large number of biocatalysts had been undertaken (see Klibanov, 1986, for a review) was enzymatic catalysis in non-aqueous media accepted and recognized as a general phenomenon.

It is now believed that enzymes are catalytically active in organic solvents because they remain trapped in the native conformation. The protein's inability to unfold in non-aqueous media stems in part from electrostatic interactions which are enhanced in organic solvents due to the low dielectric of most solvents, and also from an increased intramolecular hydrogen-bond network (Hartsough and Merz, 1993). The structural integrity of proteins in non-aqueous environments has been supported by a number of experiments including ^{15}N NMR (Burke *et al.*, 1989), solid-state ^{13}C NMR (Burke *et al.*, 1992) and X-ray crystallography (Fitzpatrick *et al.*, 1993; Yennawar *et al.*, 1994).

It is intuitive that protein stability in non-aqueous environments should be radically different from that in water. Since the water molecules that surround an enzyme in an aqueous solution contribute to all major forces that stabilize the folded conformation, including van der Waals interactions, salt bridges, and hydrogen bonds, removal of the water around a protein should alter protein stability. Similarly, since water plays a crucial role in enzyme–substrate interactions, one should expect that substrate specificity of enzymes in non-aqueous environments would be altered.

Although the mere existence of the solvent effect on enzyme specificity and stability is well accepted, the mechanism of protein stabilization in organic solvents and the dependence of the catalytic parameters on various solvent characteristics have not been fully elucidated.

The goal of this chapter is to present the most recent viewpoint on the effect of media on the specificity and stability of enzymes in non-aqueous media.

4.2 Specificity of enzymes in non-aqueous media

4.2.1 Substrate specificity

It is well recognized that the binding is the driving force of molecular catalysis (Fersht, 1985; Menger, 1993). Since the binding energy of an enzyme with a substrate is determined by the difference between the energy of the enzyme–substrate complex and the energy of an enzyme and a substrate in solution interacting with the solvent molecules (Fersht, 1985), the binding is always influenced by the solvent. Zaks and Klibanov (1986) reported that the substrate specificity of α-chymotrypsin, subtilisin, and esterase changed dramatically upon the replacement of an aqueous reaction medium with an organic solvent. In fact, the substrate specificity of α-chymotrypsin in octane was reversed compared to that in water. Since hydrophobic interactions contribute significantly to substrate binding to chymotrypsin in water (Dorovska et al., 1972), it was suggested that the reversal of substrate specificity in organic solvents was due to the lack of hydrophobic interactions in non-aqueous media.

Qualitatively similar results were obtained with polyethyleneglycol-modified chymotrypsin, trypsin, and subtilisin in benzene (Gaertner and Puigserver, 1989). While k_{cat}/K_M of chymotrypsin in water varied between aromatic and basic amino acids by more than three orders of magnitude, in benzene k_{cat}/K_M values were similar. Furthermore, N-Bz-L-Lys-OMe, which was hydrolyzed in water by chymotrypsin at the rate of about 1/15 of that of N-Bz-L-Tyr-OEt, became three times more reactive substrate in benzene. The authors showed that the specificity of chymotrypsin towards active-site specific inhibitors in organic solvents was also reversed. Phenylmethylsulfonyl fluoride, which reacts with the active-site serine and quickly inactivates the enzyme in water, had little effect on chymotrypsin in benzene. By contrast, more hydrophilic (4-amidinophenyl)-methanesulfonyl fluoride completely inactivated chymotrypsin within 3 h in a non-aqueous medium.

Clapés and Adlercreutz (1991) studied the catalytic parameters of α-chymotrypsin-catalyzed esterification of N-acetyl amino acids in organic media. They found that while the specificity of the enzyme for the amino acid side chain in ethyl acetate and acetonitrile was approximately the same as in water, the specificity towards N-protecting groups was reversed, i.e. a much higher rate was observed with N-acetyl than with N-benzyloxycarbonyl derivatives. The authors suggested that more favorable hydrophobic interactions were responsible for the higher rate of hydrolysis of the substrates with the large non-polar protecting groups in water. In organic solvents, where the hydrophobic interactions play no role, the less sterically hindered substrates with smaller N-protecting groups, such as acetyl, become more favorable. This explanation was also consistent with the fact that in organic solvents log (k_{cat}/K_M) correlated better with molar refractivity of the amino acid side chain, which is a measure of the bulk of the substituent, than with its hydrophobicity.

The lack of hydrophobic interactions in non-aqueous media was also found to be responsible for the change in the specificity of proteases towards nucleophiles in organic media (Gololobov et al., 1992). The authors demonstrated that the reactivity of the unprotected amino group of amino acid amides in an acyl transfer reaction in water was mainly determined by the partitioning of the nucleophile from the medium to the binding subsite of the enzyme, and correlated well the nucleophile hydrophobicity. Not surprisingly, the specificity of α-chymotrypsin towards amino acid amides in organic solvents did not depend on their hydrophobicity. Instead, it correlated with the normalized van der Waals volume of the amino acid, which is determined by the size of the side chain.

A thermodynamic model that predicted the substrate specificity of subtilisin Carlsberg in non-aqueous media on the basis of specificity of the enzyme in water and physicochemical characteristics of the solvent was developed by Wescott and Klibanov (1993a). The authors determined k_{cat}/K_M for the trans-esterification of N-acetyl-L-phenylalanine and N-acetyl-L-serine with propanol in 20 anhydrous solvents, and found that the $(k_{cat}/K_M)_{Ser}/(k_{cat}/K_M)_{Phe}$ varied 68-fold. They rationalized the results on the basis of the thermodynamic cycle shown in Scheme 4.1. Since:

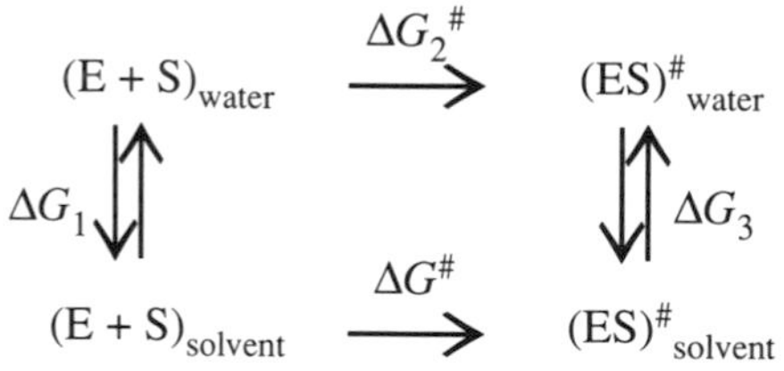

$$(E + S)_{water} \xrightarrow{\Delta G_2^{\#}} (ES)^{\#}_{water}$$

$$\Delta G_1 \Updownarrow \qquad \qquad \Updownarrow \Delta G_3$$

$$(E + S)_{solvent} \xrightarrow{\Delta G^{\#}} (ES)^{\#}_{solvent}$$

Scheme 4.1

$$\Delta G^{\#} = \Delta G_1 + \Delta G^{\#}_2 + \Delta G_3 \qquad (4.1)$$

$$\Delta G^{\#} = -RT \ln [(k_{cat}/K_M)_{solvent}(h/kT)] \qquad (4.2)$$

$$\Delta G_2^{\#} = -RT \ln [(k_{cat}/K_M)_{water}(h/kT)] \qquad (4.3)$$

$$\Delta G_1 = RT \ln P + RT \ln P_E, \qquad (4.4)$$

where h, k, P, and P_E are the Planck constant, the Boltzman constant, and the solvent to water partition coefficients for the substrate and the enzyme respectively, then:

$$\ln [(k_{cat}/K_M)_{solvent}(h/kT)] = -\ln P - \ln P_E + \ln [(k_{cat}/K_M)_{water}(h/kT)] - \Delta G_3/RT$$

$$(4.5)$$

Subtracting equation (4.5) written for N-Ac-L-Phe from that for N-Ac-L-Ser and assuming that $\Delta G_{3,\text{Ser}} = \Delta G_{3,\text{Phe}}$ the authors obtained equation (4.6):

$$\log \left[(k_{\text{cat}}/K_M)_{\text{Ser}}/(k_{\text{cat}}/K_M)_{\text{Phe}} \right]_{\text{solvent}} =$$

$$\log \left(P_{\text{Phe}}/P_{\text{Ser}} \right) + \log \left[(k_{\text{cat}}/K_M)_{\text{Ser}}/(k_{\text{cat}}/K_M)_{\text{Phe}} \right]_{\text{water}} \qquad (4.6)$$

Equation (4.6) allows one to correlate the ratio of the specificity parameters for two substrates in any organic solvents with that in water, provided that the partition coefficients for the substrates are known. Since the model is based on thermodynamic analysis, which is independent of the enzyme, it should be applicable to any enzyme–substrate pair.

Although the above model had a good correlation coefficient it had a minor drawback. It required the direct measurement of partition coefficients, which is impossible for water-miscible solvents. The application of this model was greatly improved when a new approach was developed that allowed the P ratios to be calculated (Westcott and Klibanov, 1993b). The approach was based on the following correlation:

$$P_{\text{Phe}}/P_{\text{Ser}} = (\gamma_{\text{Ser}}/\gamma_{\text{Phe}})_{\text{solvent}} \times (\gamma_{\text{Phe}}/\gamma_{\text{Ser}})_{\text{water}} \qquad (4.7)$$

where γ is the thermodynamic activity coefficient of the substrate, which can be calculated for a given molecule on the basis of van der Waals volumes and surface areas of the groups. The dependence of the substrate specificity of subtilisin Carlsberg on the ratio of calculated $P_{\text{Phe}}/P_{\text{Ser}}$ solvent to water partition coefficients was found to have a correlation coefficient of 0.96 (Figure 4.1). This unexpectedly high correlation implies that the change of substrate specificity of enzymes in organic solvents stems to a large extent from the energy of desolvation of the substrate.

Janssen and Halling (1994) used a similar method to relate enzyme specificity in a non-aqueous medium to a chosen standard state. The authors experimentally measured the rate of *Candida rugosa* lipase-catalyzed esterification of sulcatol with fatty acids in toluene, and showed that the enzyme exhibited a preference for C4, C8, C10 and C12 fatty acids. However, when the specificity was corrected to the gas phase or dilute aqueous standard states, a strong preference for longer chain fatty acids was observed.

The use of genetic engineering in combination with the catalysis in non-aqueous media was found to be a powerful tool for altering the substrate specificity of enzymes. It was shown that the effect of site-specific mutations on the substrate specificity of subtilisin BNP′ depends on the nature of the organic medium (Wangikar *et al.*, 1993). While a polar active site mutation (Gly166Asn) reduced k_{cat}/K_M of the enzyme-catalyzed transesterification of N-Ac-L-Ser-OEt with n-propanol in acetone 30-fold, it activated the enzyme by a factor of 4.5 in hexane. A similar effect was observed with N-Ac-L-Phe-OEt. The authors proposed that the effect was due to significant transition-state stabilization by a polar mutation in a non-polar solvent. Thus, some mutations that have a

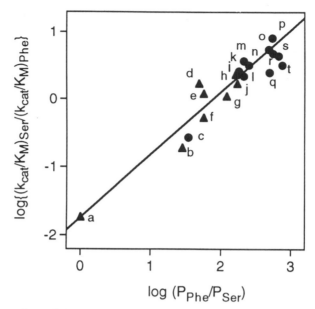

Figure 4.1 Dependence of the substrate specificity of subtilisin Carlsberg in water-miscible (▲) and water-immiscible (●) solvents on the ratio of calculated solvent to water partition coefficients of N-Ac-L-Phe-OEt and N-Ac-L-Ser-OEt. Solvents: (a) water, (b) *tert*-butyl alcohol, (c) *tert*-amyl alcohol, (d) acetonitrile, (e) dioxane, (f) pyridine, (g) acetone, (h) 2-butanone, (i) methyl acetate, (j) tetrahydrofuran, (k) ethyl acetate, (l) isopropyl acetate, (m) *tert*-butyl acetate, (n) diethyl ether, (o) chloroform, (p) dichloromethane, (q) octane, (r) toluene, (s) benzene, (t) carbon tetrachloride (adopted from Wescott and Klibanov, 1993b).

deleterious effect on activity in aqueous solutions improve activity in non-aqueous media.

Although the effect of media on specificity has been demonstrated predominantly on lipases and proteases, a number of studies indicate that the phenomenon is not limited to hydrolytic enzymes. For example, the dependence of V_{max}/K_M of horse-radish peroxidase on hydrophobicity of para-substituted phenols in organic solvents was found to be different from that in water (Ryu and Dordick, 1989, 1992). While in an aqueous medium the values of V_{max}/K_M for methoxyphenol and *tert*-butyl phenol were virtually the same, they varied by more than ten-fold in butyl acetate.

While the nature of the solvent has a tangible effect on catalysis, other methods exist that can alter the substrate specificity of an enzyme in a given solvent. Lyophilization of chymotrypsin and subtilisin from aqueous solutions containing competitive inhibitors (Russell and Klibanov, 1988; Zaks and Klibanov, 1988) not only increased their enzymatic activity up to 100-fold, but also changed the substrate specificity. For example, k_{cat}/K_M values of subtilisin lyophilized from 10 mM phosphate buffer, pH 7.8, in octanol in the trans-esterification of N-Ac-L-Phe-OMe and N-Ac-L-Ala-OMe with propanol were similar. However, the enzyme recovered from the solution containing a competitive inhibitor, N-Ac-L-Tyr-NH$_2$, reacted ten times more efficiently with the latter substrate than with the former one. Similar effects were observed with α-chymotrypsin (Ståhl *et al.*, 1991). It was hypothesized that when a ligand binds to the enzyme in an aqueous medium, it induces conformational changes (analogous to regulation by allosteric receptors) that remain unchanged during lyophilization and subsequently in a solvent when the ligand is washed away.

In a follow-up study it was found that the ligands not only induce the changes in the substrate specificity of enzymes but also protect the enzymes against reversible denaturation during lyophilization (Burke *et al.*, 1992; Dabulis and Klibanov, 1993).

4.2.2 Stereoselectivity

Enantioselectivity, which is perhaps the most attractive feature of enzyme-catalyzed synthesis, stems from the unique ability of enzymes to discriminate between enantiomers of a racemic pair. It is brought about by the free-energy difference of the diastereomeric transition states for two enantiomers. Few methods exist that are known to affect enantioselectivity of enzymatic reactions. They include site-directed mutagenesis (Weinhold *et al.*, 1991; Ozaki and Ortiz de Montellano, 1994), and use of enantioselective inhibitors (Guo and Sih, 1989), coenzyme analogs (Zheng and Phillips, 1992), bile salts (Zahalka *et al.*, 1991), temperature (Phillips, 1992), and water miscible co-solvents (Lam *et al.*, 1986). The use of non-aqueous solvents was found to be, perhaps, the most versatile tool in changing the stereoselectivity of enzymes (see Klibanov, 1990).

Margolin *et al.* (1987) noticed that enantioselectivity of subtilisin in organic solvents was different from that in water, a feature that allowed the authors to develop facile incorporation of D-amino acids under non-aqueous conditions. A detailed mechanistic investigation of this phenomenon that followed later revealed that enantioselectivity of subtilisins Carlsberg and BPN', elastase, α-lytic protease, α-chymotrypsin, and trypsin were lower in non-aqueous media than in water (Sakurai *et al.*, 1988). Moreover, an inverse correlation between enantioselectivity of subtilisin in transesterification of N-acetylalanine chloroethyl ester and propanol and hydrophobicity of the solvent was observed, i.e. enantioselectivity of the enzyme was lower in the solvents with higher hydrophobicity. The findings were rationalized on the basis of the notion that binding of a substrate to subtilisin is accompanied by the release of water molecules from the enzyme (Dewar and Storch, 1985). The productive binding of D-amino acid does not involve an incorporation of the amino acid side chain into the hydrophobic binding pocket of the enzyme and therefore is accompanied by the release of fewer water molecules compared to the binding of the L-isomer. The authors concluded that, since the release of water molecules into a hydrophobic solvent is thermodynamically unfavorable, the reactivity of the L-enantiomer in these solvents is diminished to a greater extent than that of the

D-enantiomer, and that explains the drop in enantioselectivity. The model also predicted that a difference between enantioselectivity in water and in an organic solvent should be higher for the substrates with bulky side chains. This was found to be the case experimentally.

A similar inverse correlation between selectivity and solvent hydrophobicity was found in *Candida cylindracea* lipase-catalyzed esterification of 2-chloro-propionic acid with *n*-butanol (Gubicza and Kelemen-Horvàth, 1993) and *Pseudomonas* sp. lipase-catalyzed hydrolysis of compound **1** (Terradas *et al.*, 1993).

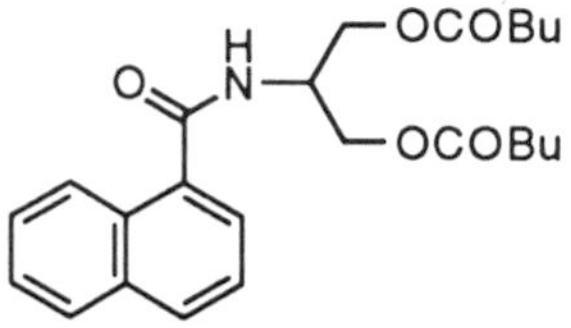

1

In the last-named case the enzyme's prochiral selectivity varied from >30 in acetonitrile to 2.6 in carbon tetrachloride. The authors hypothesized that, in a stereoselective mode of binding, the hydrophobic binding pocket of the lipase is occupied by the naphthyl moiety. In non-stereoselective mode, the naphthyl moiety is partitioned into the solvent and therefore the diester substrate may be oriented in a *pro-S* or *pro-R* configuration with equal probability. Since it is thermodynamically advantageous for the hydrophobic naphthyl moiety to partition into the hydrophobic binding pocket of the enzyme in hydrophilic solvents (Smithrud and Diederich, 1990), the stereoselective mode of binding prevails. The validity of the model was also supported by the finding that 1 M 1-naphthol, which competes with **1** for the binding pocket, reduces the enzyme's selectivity ten-fold.

A physicochemical rational behind the effect of media on enantioselectivity of enzymes was expanded further by Fitzpatrick and Klibanov (1991). The authors found that despite a good correlation between solvent hydrophobicity and the enantioselectivity of transesterification of *N*-acetyl amino acids catalyzed by subtilisin Carlsberg (Sakurai *et al.*, 1988), the same type of correlation for subtilisin-catalyzed transesterification of vinyl butyrate with the *sec*-phenethyl alcohol **2** was poor.

Instead, the subtilisin's enantioselectivity factor $(k_{cat}/K_M)_S/(k_{cat}/K_M)_R$ corre-

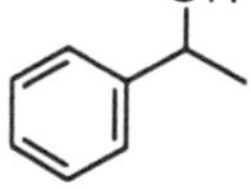

2

lated well with the dipole moment and dielectric constant of the medium. This disparity was attributed to the ester–substrate and the alcohol–substrate binding to distinctly different enzyme binding sites, and therefore a new model that explained the solvent effect on alcohol–substrate selectivity was introduced. The model inferred the presence of both a large and a small pocket in the binding site of subtilisin. The authors suggested that the binding of the slower-reacting enantiomer of **2** created high steric hindrances between the phenyl group and the small binding pocket of the enzyme. Therefore, any decrease in the protein rigidity that diminishes this steric hindrance should increase the reactivity of the slower-reacting enantiomer to a greater extent than that of the faster-reacting enantiomer, because the latter has not experienced the steric hindrances in the first place. Since the protein's rigidity stems mainly from electrostatic interactions and intramolecular hydrogen bonding, it is higher in the solvents with low dielectric (dioxane) than in those with high dielectric (acetonitrile). Therefore, enantioselectivity of subtilisin in dioxane is higher.

The validity of the above model was later supported by computer modeling experiments that revealed that the phenyl group of the S-enantiomer of **2** occupied the spacious opening while the methyl group was cramped in a small cavity of the active site (Fitzpatrick *et al.*, 1992). The binding of the R-enantiomer was just the opposite. Also consistent with the model was a significant jump in the reactivity of a slower-reacting R-enantiomer upon the transition from dioxane to acetonitrile, and a drop of enantioselectivity upon the addition of water and water mimics such as formamide which are known to increase the protein flexibility (Fitzpatrick and Klibanov, 1991).

Although the above models do explain the effect of the media on the enantioselectivity of subtilisin Carlsberg, they are not universal. For example, solvent hydrophobicity had only a very minor effect on enantioselectivity of porcine pancreatic lipase-catalyzed transesterifications of trifluoroethyl butyrate with 2-octanol and 1-phenylethanol (Kanerva *et al.*, 1990). The enantioselectivity factor in transesterification of the 2,2,2-trifluoroethyl ester of Z-norvaline with methanol catalyzed by *Pseudomonas cepacia* lipase varied from 1 in dimethylformamide (DMF) to 45 in benzene. However, no correlation between solvent hydrophobicity and the lipase selectivity was found (Miyazawa *et al.*, 1992). Similarly, the enantioselectivity of *Candida cylindracea* lipase-catalyzed esterification of a number of 2-hydroxy acids with primary alcohols did not have a clear correlation with the log P of the medium (Parida and Dordick, 1991, 1993).

An unexpected bell-shaped correlation between enantioselectivity of *Pseudomonas* sp. lipase and solvent hydrophobicity was observed by Nakamura *et al.* (1991, 1994) when the organic solvents were separated into two groups: cyclic and acyclic. To explain the different behavior of cyclic and acyclic solvents the authors suggested that site-specific enzyme–solvent interactions affected the enantioselectivity of the lipase, and since, on average, cyclic molecules are bulkier than acyclic, the solvent molar volume (V_m) should be taken into account. Indeed, a good correlation between enantioselectivity and a combination of Kirkwood parameter and V_m was obtained (Nakamura *et al.*, 1994).

Bovara *et al.* (1991) and Secundo *et al.* (1992) reported that the enantiomeric ratio of lipase PS in transesterification of (±)-*trans*-sobrerol (**3**), (±)-sulcatol (**4**) and (±)-3-bromo-5-hydroxymethyl isoxazoline (**5**) with activated esters varied from more than 500 in *tert*-amyl alcohol to 69 in tetrahydrofuran. No correlation between selectivity and hydrophobicity (log P) or dielectric constant of the media was observed.

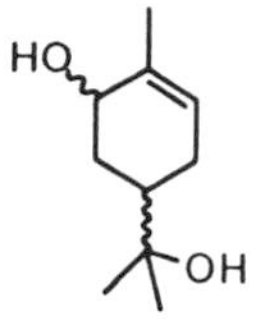

3

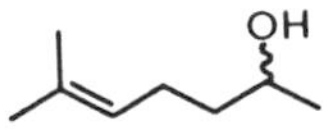

4

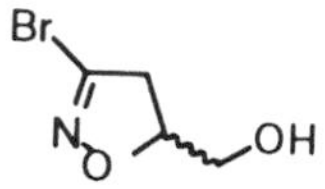

5

Enantioselectivity of porcine pancreatic lipase (PPL) in the acylation of **4** was only slightly influenced by the solvent and by the nature of the acylating agent. Interestingly, while V_{max} for the acylation of **4** was similar in all the solvents tested, the K_M values increased dramatically with increasing solvent polarity (Secundo *et al.*, 1992). Unfortunately, no explanation of this unexpected phenomenon was given. The authors acknowledged that although enantio-selectivity of both PPL and lipase PS was influenced by the solvent, stereo-chemical outcome of the reaction could not be predicted on the basis of physico-chemical properties of the solvent only: the trend varied for a given substrate as a function of the enzyme and for a given enzyme as a function of the substrate. Similar to the hypothesis of Nakamura *et al.* (1994), the authors suggested that the enzyme's selectivity is controlled, at least in part, by local non-specific inter-actions between the solvent and the enzyme that affect the reactivity of one enantiomer to a greater extent than another.

In order to evaluate the effect of solvent geometry on the activity, enantio- and regioselectivity of enzymes, several lipases and proteases were tested in two enantiomerically pure solvents, (*R*)- and (*S*)-carvone (Ottolina *et al.*, 1994a, b). Although the activity of all the enzymes tested depended on the solvent con-figuration, no significant variation in the enzyme's regio- and enantio-selectivity was observed.

The full magnitude of the effect of the reaction medium on enantioselectivity of enzymes was revealed by Tawaki and Klibanov (1992), who demonstrated that enantiomeric preference of protease from *Aspergillus oryzae* could be inverted by the reaction medium. While the transesterification of *N*-acetyl-phenylalanine chloroethyl ester with propanol in hydrophilic solvents such as acetonitrile, DMF, and pyridine was faster for the L-enantiomer, the opposite was true in hydrophobic toluene, octane, and tetrachloromethane. The authors explained this phenomenon on the basis of two alternative modes of binding for the L- and the D-enantiomer: in a productive mode the L-enantiomer binds to the enzyme in such a way that the benzyl group is partitioned into the hydrophobic binding pocket of the enzyme. Contrariwise, the D-isomer, in order to have a scissile bond in a proper arrangement, has the hydrophobic side chain partitioned into the solvent. The preferential partitioning of the side chain into the hydrophobic solvents favors the productive binding and consequently faster hydrolysis of the D-enantiomer.

An inversion of enantioselectivity was also observed in the case of *Candida cylindracea* lipase-catalyzed esterification of derivatives of 2-phenoxypropionic acid (Wu *et al.*, 1991; Ueji *et al.*, 1992). While the product in the *R*-configuration (ee 84–86%) was obtained in hydrophobic carbon tetrachloride and hexane, acetone and acetonitrile favored the formation of the esters in the *S*-configuration (ee 16–23%). The authors hypothesized that changes in the enzyme conformation caused by direct interactions of the catalyst with the solvents were responsible for the inversion of enantioselectivity.

A remarkable solvent-induced reversal of stereoselectivity of *Pseudomonas* sp. lipase was observed in the hydrolysis of 1,4-dihydropyridine derivatives, **7** (Hirose *et al.*, 1992).

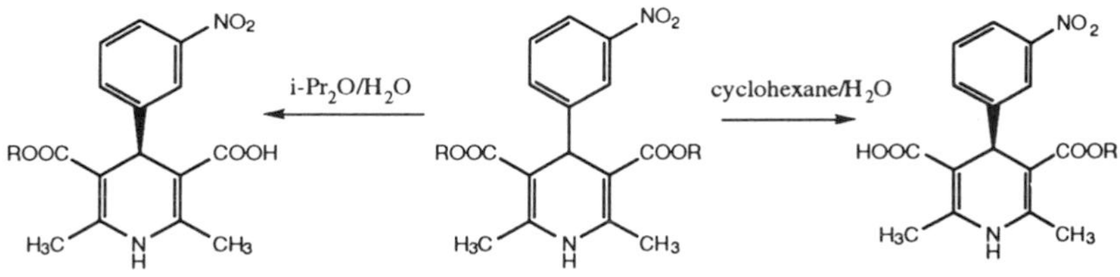

When isopropyl ether was used as a reaction medium the product in the *S*-configuration (**6**) was obtained in 87% yield and >99% ee. However, when the hydrolysis was carried out in cyclohexane saturated with water, the *R*-isomer (**8**) was obtained in 88% yield and 89% ee. Since log *P* values of the solvents were similar, it seemed unlikely that the inversion of enantioselectivity was brought about by a difference in the partitioning of the substrate between the enzyme and the media. The authors attributed the change in enantioselectivity to direct interaction between the solvent and the enzyme.

The majority of reports on the effect of water and water-mimicking solvents on selectivity of enzymes in organic media indicate that enantioselectivity of enzymes increases with the increase of water content of the media (Bodnár *et al.*, 1990; Kitaguchi *et al.*, 1990; Gubicza and Kelemen-Horvàth, 1993; Högberg *et al.*, 1993; Arroyo and Sinisterra, 1994). Although no explanation of this phenomenon was given, one may argue that since in most cases the increase of enantioselectivity was accompanied by the increase of the reaction rate, the addition of water alleviates unfavorable enzyme–solvent interactions and restores the enzyme's inherently more selective native conformation.

4.2.3 Regioselectivity

There have been only few studies of the effect of media on regioselectivity of enzymes (Rubio et al., 1991; Ottolina et al., 1994a). Rubio et al. (1991) reported that the ratio of the reaction rates (v_1/v_2) of P. cepacia lipase-catalyzed transesterification of 9 with butanol in organic solvents differed significantly.

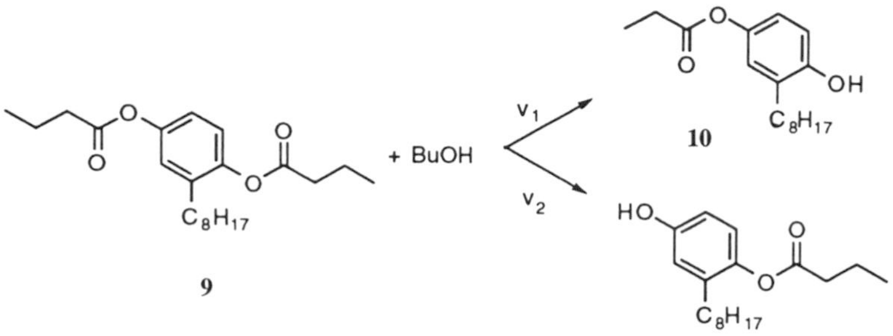

Moreover, a good correlation was found between the regioselectivity factor, $(k_{cat}/K_M)_1/(k_{cat}/K_M)_2$, and solvent hydrophobicity parameter, log P. To explain this finding, the authors suggested that the lipase has a hydrophobic cleft in the vicinity of the catalytic site. If the binding of 9 occurs in such a way that the octyl moiety occupies a putative hydrophobic cleft, the proximal butyryl group is placed in the catalytic site, leading to the formation of 10. An alternative binding places a distal butyryl group in the vicinity of the catalytic site, leading to the formation of 11. Since in hydrophobic solvents there is no thermo-dynamic advantage for the octyl moiety to partition into the hydrophobic site of the enzyme, the formation of 11 is preferred. In hydrophilic solvents the favorable partitioning of the octyl moiety into the binding site of the enzyme leads to the formation of 10.

Regioselectivity of lipoprotein lipase and subtilisin was not affected to any significant extent when enantiomerically pure solvents were used as reaction medium (Ottolina et al., 1994a).

4.2.4 Chemoselectivity

Chemoselectivity, i.e. the ability to discriminate between chemically distinct functional groups, is yet another property of enzymes that can be affected by the medium. Chinsky et al. (1989) reported that Aspergillus niger lipase-catalyzed acylation of 6-amino-1-hexanol proceeded with an overwhelming preference for the hydroxyl group. This unexpected selectivity, which was opposite to that of the chemical reaction, allowed the authors to produce a number of monoesters of amino alcohols in good yield without requiring any protecting groups. Tawaki and Klibanov (1993) extended the study and reported that chemoselectivity of enzymes can be greatly affected by the reaction medium. Butyrylation of a number of amino alcohols in a variety of organic solvents revealed that the ratio of the rate of *O*- to *N*-acylation was markedly dependent on the solvent. For example, the chemoselectivity of *Pseudomonas* sp. lipase in the acylation of *N*-α-benzoyl-L-lysinol (**12**) with trifluoroethyl butyrate varied from 1.1 in *tert*-butyl alcohol to 21 in 1,2-dichloroethane, with the hydroxyl group being more reactive than the amino group in all the solvents tested.

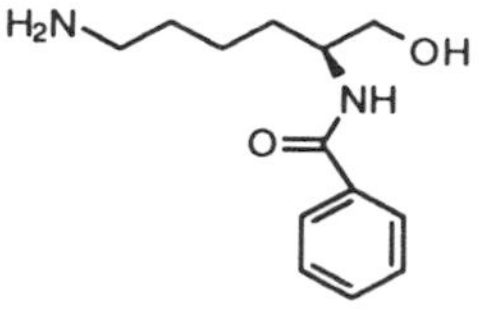

12

Surprisingly, in the same solvents *Mucor meihei* lipase exhibited the preference for the amino group, with chemoselectivity varying 18-fold. In this case, a correlation between chemoselectivity of the lipase and the hydrogen-bonding parameter was observed: while the enzyme catalyzed the acylation of the hydroxyl group of amino alcohols in the solvents with low propensity to form hydrogen bonds, it exhibited a strong preference for the amino groups in the solvents with high propensity to form hydrogen bonds. The authors reasoned that in order to carry out a nucleophilic attack on the acyl enzyme intermediate the nucleophilic groups should be free from any hydrogen bonding. Since the hydroxyl group has a higher tendency to form hydrogen bonds, it is deactivated to a higher extent than the amino group in solvents that permit hydrogen bonding. Therefore, in these solvents the acylation of the amino group is preferred.

4.3 Thermal stability of enzymes in non-aqueous media

4.3.1 Mechanistic considerations

The issue of thermal stability of enzymes in organic solvents encompasses a number of subjects that include the effect of hydration on protein structure and dynamics, protein–solvent interactions, and a solvent contribution to thermodynamics of protein folding.

It is well accepted that water molecules which surround the proteins in aqueous solutions make a major contribution to protein stability. Water plays a crucial role in hydrophobic interactions, which are the major force of protein stabilization in aqueous solutions (Kauzmann, 1959; Tanford, 1962, 1978; Creighton, 1984); it also contributes to van der Waals interactions, salt bridges, and hydrogen bonds (Dill, 1990). Moreover, by interacting with both folded and unfolded states of a protein water determines the entire energetics of the folding–unfolding process. It is not surprising, therefore, that manipulation of the nature of the medium and the amount of water surrounding a protein has a profound effect on its stability.

Although the quantitative evaluation of the contribution of water to protein stability is very complex, the effect of the removal of water from a protein can be estimated qualitatively. For example, since the hydrophobic interactions are brought about by reorganization of the hydrogen-bond network of water molecules (Kauzmann, 1959; Tanford, 1980), the removal of water from the surface of the protein should undoubtedly diminish the hydrophobic effect and therefore destabilize the protein.

Conversely, dehydration of a protein enhances the contribution of hydrogen bonding to protein stability. Similar to all other forces, the energy of hydrogen bonding represents the difference between energy of molecules in solutions interacting with water and molecules in a complex interacting with each other. The removal of water from the system is accompanied by a reduction of the total number of intermolecular hydrogen bonds between water and a protein molecule. This results in stabilization of intramolecular hydrogen bonding and, as a consequence, in higher protein rigidity. Indeed, the dramatically increased hydrogen-bond network of a protein in an organic solution was confirmed by molecular dynamic simulations (Hartsough and Merz, 1993). Moreover, since the dominant component of a hydrogen bond is electrostatic, the strength of hydrogen bonding should increase in solvents with lower dielectric.

The consequences of the removal of water on electrostatic interactions are more complex. It is generally agreed that electrostatic interactions within a protein are enhanced by both the lower dielectric of the protein interior and by an entropic factor. The entropic contribution is attributed to water molecules that form a less ordered and more energetically favorable state when charged groups come together and expel water to solution during the folding process (Kauzmann, 1959). The removal of water from around the protein reduces the contribution of the entropic factor to the overall stabilization of the folded state. Moreover, since about 20% of buried charged groups do not form ion pairs and are solvated (Rashin and Hong, 1984), the removal of this water should expose the charges and destabilize the folded conformation even more. However, the replacement of water by a medium with significantly lower dielectric should strengthen the ion pairing, so that the detrimental effect of the entropic factor may be negated and the overall effect of dehydration on electrostatics may be positive. In fact, the increase of rigidity of horse liver alcohol dehydrogenase (Guinn et al., 1991) and α-chymotrypsin (Affleck et al., 1992a) in solvents with low dielectric indicate that this may be the case.

LIVERPOOL
JOHN MOORES UNIVERSITY
AVRIL ROBARTS LRC
TEL. 0151 231 4022

It is well accepted that irreversible thermo-inactivation of enzymes in aqueous solutions is usually preceded by a reversible unfolding (Lumry and Eyring, 1954), a process which requires conformational mobility. In this regard, the effect of hydration on protein dynamics becomes of primary importance. The ability of water to enhance protein flexibility stems from its high propensity to form multiple hydrogen bonds with a protein and from the dielectric screening of electrostatic interactions by protein-bound water molecules. It is not surprising, therefore, that when most water is removed or replaced by an organic solvent, protein flexibility is decreased significantly (Rupley *et al.*, 1980, 1983; Bone and Pethig, 1985; Schinkel *et al.*, 1985; Burke *et al.*, 1993; Hartsough and Merz, 1993).

It is important to emphasize that although in a dry state proteins are much more rigid than in water, surprisingly very little water is needed to induce the mobility. For example, less than a monolayer of water on the surface of lysozyme was enough to increase internal protein motions up to the level in a true solution (Schinkel *et al.*, 1985). Similarly, the presence of only 1% water in THF was shown to enhance greatly the dynamic flexibility of cytochrome c (Wu and Gorenstein, 1993) and subtilisin Carlsberg (Affleck *et al.*, 1992b).

4.3.2 Hydration and thermal stability

When proteins are exposed to elevated temperatures for a prolonged period of time they inevitably undergo a process of thermal unfolding which exposes the reactive groups and buried hydrophobic areas. This leads to irreversible chemical changes, monomolecular scrambling and aggregation. These chemical processes which are highly accelerated at elevated temperatures and often lead to inactivation include deamidation of asparagine and glutamine residues, hydrolysis of peptide bonds at aspartic acid residues, thiol–disulfide interchange, destruction of disulfide bonds, oxidation of cysteines, isomerization of prolines, glycation of amino groups, and others (Volkin and Klibanov, 1989; Volkin and Middaugh, 1992; Volkin *et al.*, 1995). Most of these processes involve water, and therefore do not take place in a water-free environment such as anhydrous organic solvents. Furthermore, increased protein rigidity in dry solvents, mentioned earlier, hinders any reversible unfolding process, which also contributes to thermal stability.

In fact, it was shown that a number of enzymes suspended in dry organic solvents exhibited thermostability far superior to that in aqueous solutions (Table 4.1). For example, porcine pancreatic lipase (PPL) had a remarkably high stability in dry organic media, with a half-life of about 12 h at 100°C (Zaks and Klibanov, 1984). This half-life represented more than a 1000-fold stabilization over that in an aqueous buffer, where the enzyme inactivated almost instantaneously at that temperature. Not surprisingly, thermal stability of PPL strongly depended on the concentration of water (Figure 4.2): in the presence of only 1%

LIVERPOOL
JOHN MOORES UNIVERSITY
AVRIL ROBARTS LRC
TEL. 0151 231 4022

Table 4.1 Stability of enzymes in non-aqueous vs. aqueous media

Enzyme	Conditions	Thermal property	References
PPL	tributyrin	$t_{1/2} < 26$ h	Zaks and Klibanov (1984)
	aqueous, pH 7.0	$t_{1/2} < 2$ min	
Candida lipase	tributyrin/heptanol	$t_{1/2}$ 1.5 h	Zaks and Klibanov (1984)
	aqueous pH 7.0	$t_{1/2} < 2$ min	
Chymotrypsin	octane, 100°C	$t_{1/2}$ 80 min	Zaks and Klibanov (1988)
	aqueous, pH 8.0, 55°C	$t_{1/2}$ 15 min	Martinek *et al.* (1977)
Subtilisin	octane, 110°C	$t_{1/2}$ 80 min	Russell and Klibanov (1988)
Lysozyme	cyclohexane, 110°C	$t_{1/2}$ 140 h	Ahern and Klibanov (1986)
	aqueous	$t_{1/2} < 10$ min	
Ribonuclease	nonane, 110°C, 6 h	95% activity remains	Volkin and Klibanov (1990)
	aqueous, pH 8.0, 90°C	$t_{1/2} < 10$ min	
F_1-ATPase	toluene, 70°C	$t_{1/2} > 24$ h	Garza-Ramos *et al.* (1989)
	aqueous, 70°C	$t_{1/2} < 10$ min	
Alcohol dehydrogenase	heptane, 55°C	$t_{1/2} > 50$ days	Kaul and Mattiasson (1993)
Hind III	heptane, 55°C, 30 days	no loss of activity	Kaul and Mattiasson (1993)
Lipoprotein lipase	toluene, 90°C, 400 h	40% activity remains	Ottoline *et al.* (1992)
β-Glucosidase	2-propanol, 50°C, 30 h	80% activity remains	Tsitsimpikou *et al.* (1994)
Tyrosinase	chloroform, 50°C	$t_{1/2}$ 90 min	Yang and Robb (1993)
	aqueous solution, 50°C	$t_{1/2}$ 10 min	
Acid phosphatase	hexadecane, 80°C	$t_{1/2}$ 8 min	Toscano et al. (1990
	aqueous, 70°C	$t_{1/2}$ 1 min	
Cytochrome oxidase	toluene, 0.3% water	$t_{1/2}$ 4.0 h	Ayala *et al.* (1986)
	toluene, 1.3% water	$t_{1/2}$ 1.7 min	

water in the organic medium, the stability of the lipase approached that in an aqueous solution. A similar trend was observed with the stability of horse liver alcohol dehydrogenase (Deetz and Rozzell, 1988) and F_1-ATPase (Garza-Ramos *et al.*, 1989). The latter enzyme had a half-life in toluene at 70°C of more than 24 h, whereas at this temperature in all-aqueous medium the enzyme denatured in a matter of minutes.

A number of other enzymes exhibited improved thermostability in non-aqueous environments. For example, the half-life of lysozyme in cyclohexane at 100°C was 140 h — almost 1000-fold higher than in water (Ahern and Klibanov, 1986). Similarly, the half-life of chymotrypsin at 100°C (Zaks and Klibanov, 1988) and that of subtilisin at 110°C in dry octane (Russell and Klibanov, 1988) were found to be 270 and 80 min respectively. For comparison, the half-life of chymotrypsin in an aqueous solution of pH 8.0 at 55°C is only about 15 min (Martinek *et al.*, 1977). Ribonuclease, which lost 50% activity in less than 10 min at 90°C at pH 8.0, retained over 95% activity after incubation in anhydrous nonane at 110°C for 6 h (Volkin *et al.*, 1991).

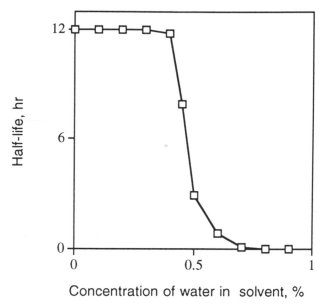

Figure 4.2 The time course of thermal inactivation at 100°C of porcine pancreatic lipase in a 2 M solution of heptanol in tributyrin at different concentrations of water (adapted from Zaks and Klibanov, 1984).

4.3.3 Mechanism of protein inactivation in organic solvents

It is accepted that increased structural rigidity and a slower rate of chemical deterioration are responsible for the high thermal stability of many proteins in dry organic solvents. However, the prolonged exposure of proteins to elevated temperatures even under apparently anhydrous conditions leads to their inactivation.

A number of processes have been found to be responsible for protein inactivation in non-aqueous solvents. It was suggested that the chemical processes that cause inactivation of dry powders in air, which include Mailard-type condensation of ε-amino groups of lysines with carbonyls of reducing sugars (Carpenter and Morgan, 1962), transamidation between the ε-amino groups of lysine and asparagines and glutamines, and loss of cysteines (Bjarnason and Carpenter, 1970), should also contribute to protein inactivation in organic solvents. Indeed, it was reported that the thermo-inactivation of ribonuclease, chymotrypsin and lysozyme in anhydrous nonane was caused by a combination of physical association and chemical crosslinking (Volkin *et al.*, 1991). The latter process was the result of intermolecular disulfide interchange and transamidation.

Since hydrophobic solvents are essentially inert with respect to interactions with the proteins, the thermostability of proteins suspended in hydrophobic media is usually higher than that in hydrophilic media (Zaks and Klibanov, 1984,

1988; Wheeler and Croteau, 1986; Laane *et al.*, 1987; Reslow *et al.*, 1987; Ayala *et al.*, 1986; Volkin *et al.*, 1991; Tsitsimpikou *et al.*, 1994). Moreover, in hydrophobic solvents enzymes exhibit greatly enhanced thermal denaturation temperatures, which increase with decreasing water content of the protein (Volkin *et al.*, 1991).

4.3.4 Stabilization of enzymes in organic solvents

It has been reported that thermal stability of α-chymotrypsin in octane strongly depends on the pH of the aqueous solution from which the protein was lyophilized (Zaks and Klibanov, 1988). It was shown later that not only the pH but also the presence of substrate analogs (Russell and Klibanov, 1988) and simple salts (Schulze and Klibanov, 1991) in aqueous solutions of proteins before lyophilization affected their stability in organic solvents. The magnitude of the effect of pH on stability depended on experimental conditions and varied for different proteins, but in some cases it was very significant. As depicted in Figure 4.3, the half-life of subtilisin BNP′ recovered from a solution of pH 7.9 in dimethylformamide at 45°C was 25-fold higher than that of the enzyme lyophilized from the same solution at pH 6.0 (Schulze and Klibanov, 1991).

It was also observed in this study that the time course of inactivation of subtilisin BPN′ in anhydrous *tert*-amyl alcohol, based on activity in the solvent itself, differed from the inactivation based on activity measured in an aqueous

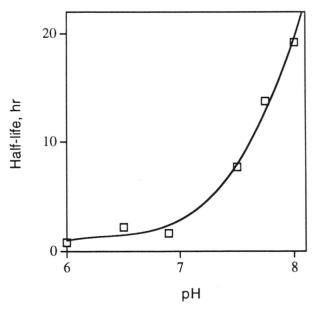

Figure 4.3 The dependence of the half-life of subtilisin BPN′ in anhydrous dimethylformamide at 45°C on the pH of the aqueous solution from which the enzyme was lyophilized (reproduced with permission from Schulze and Klibanov, 1991).

solution. The mechanistic analysis of this phenomenon revealed that in the organic solvent the subtilisin underwent a reversible inactivation, and that the major contributor to this process was dehydration, i.e. partitioning of the essential water from the enzyme into the bulk solvent. When the enzyme was assayed under aqueous conditions, the inactivation was not detected because the presence of the essential water was restored. Analogously, the authors argued, the addition of some exogenous water to the organic solvent should prevent the partitioning of the essential water and stabilize the enzyme against the reversible inactivation. Indeed, the addition of 2% water to *tert*-amyl alcohol led to a significant stabilization of the subtilisin. As expected, higher amounts of water in the solvent (4%) destabilized the subtilisin.

Ester substrates also increased the stability of subtilisins Carlsberg and BNP' in organic solvents, presumably via formation of acyl enzymes (Schulze and Klibanov, 1991).

Chemical cross-linking of enzyme crystals has been found to increase thermal stability of a number of proteins. Whereas the activity of cross-linked thermolysin after an 18-day incubation in ethyl acetate at 55°C remained virtually unchanged, free thermolysin was inactivated completely after 4 days (St Clair and Navia, 1992).

A number of reports describe the use of genetic engineering to increase stability and activity of enzymes in organic solvents and aqueous-organic mixtures (Arnold, 1988, 1990; Wong *et al.*, 1990; Chen *et al.*, 1991). It was assumed that the incorporation of amino acids that can provide internal hydrogen bonding and increase the number of salt bridges and the degree of internal crosslinking should produce more solvent-resistant enzymes (Arnold, 1988). Consequently, a mutant subtilisin E that had the N218S substitution (known to improve hydrogen bonding in a related subtilisin BNP' in an aqueous solution) was designed. The half-life of the mutant in 40% DMF at 50°C was doubled compared to that of the wild-type enzyme (Chen *et al.*, 1991).

To investigate the relationship between the charge on a protein surface and stability in non-aqueous solvents, Martinez and Arnold (1991) substituted four charged residues on the surface of α-lytic protease by a number of amino acids. They found that seven substitutions at two out of the four positions improved the stability of the protein in 84% dimethylformamide. Interestingly, six of these stabilizing substitutions were large, hydrophobic amino acids. The authors concluded that the replacement of a charged amino acid with a hydrophobic one removed the unfavorable interactions of the protein with the solvent and thus led to increased protein stability.

Random mutagenesis was also found to have some utility in generating more active and solvent-resistant enzymes (Chen and Arnold, 1993). A variant of subtilisin E was found which hydrolyzed a peptide substrate in a DMF/water mixture 256-fold more efficiently than the wild type. The effect of mutagenesis on the stability was significantly lower: the most stable mutant was only two times more stable in 40% DMF at 50°C than the wild-type enzyme.

Not surprisingly, mutations that improved stability of a protein in aqueous media had a positive effect on its stability in organic solvents. One subtilisin mutant which contained six stabilizing mutations was found to have an effective half-life in anhydrous DMF 50 times higher than that of the native enzyme (Wong *et al.*, 1990).

4.4 Conclusions

It is apparent that non-aqueous media affect the catalytic behavior and stability of enzymes in a great number of ways. Although the structure of some enzymes, subtilisin, for example, in non-aqueous media is indistinguishable from that in water (Fitzpatrick *et al.*, 1993), the structures of cytochrome c and chymotrypsin change slightly, especially in the vicinity of side chains that directly interact with the solvent (Wu and Gorenstein, 1994; Yennawar *et al.*, 1994). It is intuitive that these structural changes, as well as the direct binding of the solvent to the protein surface near the active site and the removal of the essential water, may have a major effect on the catalytic properties of enzymes in non-aqueous environments. Remarkably, in many cases the structural changes contribute very little to the specificity of enzymes in non-aqueous solvents, and the major contributor to the dramatic change in specificity turns out to be the energy of desolvation of the substrate.

Despite that seemingly simple mechanism, the specificity of enzymes in organic solvents cannot be accurately predicted yet. The substrate specificity as well as stereo-, regio-, and chemoselectivity of some enzymes have been correlated with a number of physicochemical parameters of the media including polarity, dielectric constant, propensity to form hydrogen bonds, molar volume, etc. Although the correlations were sufficiently reliable for some enzymes in specific cases, they failed to explain the behavior of other enzymes. It becomes apparent from a large body of data on enzyme specificity that has already accumulated (Wescott and Klibanov, 1994) that more than one physical property of a solvent can affect the substrate binding, and that the prevailing property has to be determined on an individual basis.

The question of thermostability of enzymes in non-aqueous media is more settled. The overwhelming majority of enzymes are much more stable in dry organic solvents than in water. The removal of water from the protein increases the strength of intramolecular hydrogen bonds and salt bridges that stabilize proteins in their native conformation and creates a high kinetic barrier between the native and the unfolded state. Moreover, the elimination of water from the system prevents the occurrence of a number of detrimental chemical reactions that lead to protein inactivation in aqueous solutions. The thermostability of enzymes is usually higher in hydrophobic organic solvents that are essentially inert towards the charged amino acid side chains found on a protein surface. The lyophilization of proteins from aqueous solutions at the optimal pH, and the use of lyoprotectants, increases the thermostability of enzymes in non-aqueous media even further.

References

Affleck, R., Haynes, C.A. and Clark, D.S. (1992a) Solvent dielectric effects on protein dynamics. *Proc. Natl. Acad. Sci. USA*, **89**, 5167–5170.

Affleck, R., Xu, Z.-F., Suzawa, V. *et al.* (1992b) Enzymatic catalysis and dynamics in low-water environments. *Proc. Natl. Acad. Sci. USA*, **89**, 1100–1104.

Ahern, T.J. and Klibanov, A.M. (1986) Why do enzymes irreversibly inactivate at high temperature? In *Protein Structure, Folding and Design* (Oxender, D.L., ed.), Alan R. Liss, New York, pp. 283–289.

Arnold, F.H. (1988) Protein design for non-aqueous solvents. *Protein Engineering*, **2**(1), 21–25.

Arnold, F.H. (1990) Enzyme engineering for non-aqueous solvents. *Tibtech*, **8**, 244–255.

Arroyo, M. and Sinisterra, J.V. (1994) High enantioselective esterification of 2-arylpropionic acids catalyzed by immobilized lipase from *Candida antarctica*: a mechanistic approach. *J. Org. Chem.*, **59**, 4410–4417.

Ayala, G., Tuena de Gomez-Puyou, M.T., Gomez-Puyou, A. and Darszon, A. (1986) Thermostability of membrane enzymes in organic solvents. *FEBS Lett.* **203**(1), 41–43.

Bjarnason, J. and Carpenter, K.J. (1970) Mechanisms of heat damage in proteins. *Brit. J. Nutr.*, **24**, 313–329.

Bodnár, J., Gubicza, L. and Szabó, L.-P. (1990) Enantiomeric separation of 2-chloropropionic acid by enzymatic esterification in organic solvents. *J. Molec. Catal.*, **61**, 353–361.

Bone, S. and Pethig, R. (1982) Dielectric studies of the binding of water to lysozyme. *J. Mol. Biol.*, **157**, 571–575.

Bone, S. and Pethig, R. (1985) Dielectric studies of protein hydration and hydration-induced flexibility. *J. Mol. Biol.*, **181**, 323–326.

Bovara, R., Carrea, G., Ferrara, L. and Riva, S. (1991) Resolution of (±)-*trans*-sobrerol by lipase PS-catalyzed transesterification and effect of organic solvents on enantioselectivity. *Tetrahedron: Asymmetry*, **2**(9), 931–938.

Burke, P.A., Smith, S.O., Bachovchin, W.W. and Klibanov, A.M. (1989) Demonstration of structural integrity of an enzyme in organic solvents by solid-state NMR. *J. Amer. Chem. Soc.*, **111**, 8290–8291.

Burke, P.A., Griffin, R.G. and Klibanov, A.M. (1992) Solid-state NMR assessment of enzyme active center structure under non-aqueous conditions. *J. Biol. Chem.*, **267**(28), 20057–20064.

Carpenter, K.J. and Morgan, C.B. (1962) Chemical and nutritional changes in stored herring meal. *Brit. J. Nutr.*, **16**, 451–465.

Chen, K. and Arnold, F.H. (1993) Tuning the activity of an enzyme for unusual environments: sequential random mutagenesis of subtilisin E for catalysis in dimethylformamide. *Proc. Natl. Acad. Sci. USA*, **90**, 5618–5622.

Chen, K., Robinson, A.C., Van Dam, M.E., Martinez, P. *et al.* (1991) Enzyme engineering for non-aqueous solvents. II. Additive effects of mutations on the stability and activity of subtilisin E in polar organic media. *Biotechnol. Prog.*, **7**, 125–129.

Chinsky, N., Margolin, A.L. and Klibanov, A.M. (1989) Chemoselective enzymatic monoacylation of bifunctional compounds. *J. Amer. Chem. Soc.*, **111**, 386–388.

Clapés, P. and Adlercreutz, P. (1991) Substrate specificity of α-chymotrypsin-catalyzed esterification in organic media. *Biochim. Biophys. Acta*, **1118**, 70–76.

Creighton, T.E. (1984) *Proteins: Structure and Molecular Properties*, Freeman, New York.

Dabulis, K. and Klibanov, A.M. (1993) Dramatic enhancement of enzymatic activity in organic solvents by lyoprotectants. *Biotechnol. Bioeng.*, **41**, 566–571.

Dastoli, F.R., Musto, N.A. and Price, S. (1966) Reactivity of active sites of chymotrypsin suspended in organic medium. *Arch. Biochem. Biophys.*, **115**, 44–47.

Deetz, J.S. and Rozzell, J.D. (1988) Enzyme-catalyzed reactions in non-aqueous media. *Trends Biotechnol.*, **6**, 15–19.

Dewar, M.J.S. and Storch, D.M. (1985) Alternative view of enzyme reactions. *Proc. Natl. Acad. Sci. USA*, **82**, 2225–2229.

Dill, K.A. (1990) Dominant forces in protein folding. *Biochemistry*, **29**(31), 7133–7155.

Dorovska, V.N., Varfolomeyev, S.D., Kazanskaya, N.F. *et al.* (1972) The influence of the geometric properties of the active center on the specificity of α-chymotrypsin catalyst. *FEBS Lett.*, **23**(1), 122–124.

Fersht, A. (1985) *Enzyme Structure and Mechanism*, 2nd edn. W.H. Freeman and Co., New York.

Fitzpatrick, P.A. and Klibanov, A.M. (1991) How can the solvent affect enzyme enantioselectivity? *J. Amer. Chem. Soc.*, **113**, 3166–3171.

Fitzpatrick, P.A., Ringe, D. and Klibanov, A.M. (1992) Computer-assisted modeling of subtilisin enantioselectivity in organic solvents. *Biotechnol. Bioeng.*, **40**, 735–742.

Fitzpatrick, P.A., Steinmetz, A.C.U., Ringe, D. and Klibanov, A.M. (1993) Enzyme crystal structure in neat organic solvents. *Proc. Natl. Acad. Sci. USA*, **90**, 8653–8657.

Gaertner, H. and Puigserver, A. (1989) Kinetics and specificity of serine proteases in peptide synthesis catalyzed in organic solvents. *Eur. J. Biochem.*, **181**, 207–213.

Garza-Ramos, G., Darszon, A., Tuena de Gomez-Puyou, M. and Gomez-Puyou, A. (1989) Catalysis and thermostability of mitochondrial F_1-ATPase in toluene–phospholipid–low water systems. *Biochemistry*, **28**, 3177–3182.

Garza-Ramos, G., Darszon, A., Tuena de Gomez-Puyou, M. *et al.* (1990) Enzyme catalysis in organic solvents with low water content at high temperatures. The adenosinetriphosphatase of submitochondrial particles. *Biochemistry*, **29**, 751–757.

Gololobov, M.Y., Voyushina, T.L., Stepanov, V.M. and Adlercreutz, P. (1992) Organic solvent changes the chymotrypsin specificity with respect to nucleophiles. *FEBS Lett.*, **307**(3), 309–312.

Gubicza, L. and Kelemen-Horvàth, I. (1993) Effect of water-mimicking additives on the synthetic activity and enantioselectivity in organic solvents of a *Candida cylindracea* lipase. *J. Mol. Catalysis*, **84**, L27–L32.

Guinn, R.M., Skerker, P.S., Kavanaugh, P. and Clark, D.S. (1991) Activity and flexibility of alcohol dehydrogenase in organic solvents. *Biotechnol. Bioeng.*, **37**, 303–308.

Guo, Z.-W. and Sih, C.J. (1989). Enantioselective inhibition: a strategy for improving enantioselectivity of biocatalytic systems. *J. Amer. Chem. Soc.*, **111**, 6836–6841.

Hartsough, D.S. and Merz, Jv.K.M. (1993) Protein dynamics and solvation in aqueous and nonaqueous environments. *J. Amer. Chem. Soc.*, **115**, 6529–6537.

Hirose, Y., Kariya, K., Sasaki, I. *et al.* (1992) Drastic solvent effect on lipase-catalyzed enantioselective hydrolysis of prochiral 1,4-dihydropyridines. *Tetrahedron Lett.*, **33**(47), 7157–7160.

Högberg, H.-E., Edlund, H., Berglund, P. and Hedenström, E. (1993) Water activity influences enantioselectivity in a lipase-catalyzed resolution by esterification in an organic solvent. *Tetrahedron: Asymmetry*, **4**(10), 2123–2126.

Janssen, A.M. and Halling, P.J. (1994) Specificities of enzymes 'corrected for solvation' depend on the choice of the standard state. *J. Amer. Chem. Soc.*, **116**, 9827–9830.

Kanerva, L.T., Vihanto, J., Halme, M.H. *et al.* (1990) Solvent effects in lipase-catalyzed transesterification reactions. *Acta Chem. Scand.*, **44**, 1032–1035.

Kaul, R. and Mattiasson, B. (1993) Improving the shelf life of enzymes by storage under anhydrous organic solvents. *Biotechnol. Lett.*, **7**(8), 585–590.

Kauzmann, W. (1959) Some factors in the interpretation of protein denaturation. *Adv. Protein Chem.*, **14**, 1–63.

Kauzmann, W. (1987) Thermodynamics of unfolding. *Nature*, **325**, 763–764.

Kitaguchi, H., Itoh, I. and Ono, M. (1990) Effects of water-mimicking solvents on the lipase-catalyzed esterification in an apolar solvent. *Chem. Lett.*, 1203–1206.

Klibanov, A.M. (1986) Enzymes work in organic solvents. *Chemtech*, **16**, 354–359.

Klibanov, A.M. (1990) Asymmetric transformations catalyzed by enzymes in organic solvents. *Acc. Chem. Res.*, **23**, 114–120.

Laane, C., Boeren, S.,. Vos, K. and Veeger, C. (1987) Rules for optimisation of biocatalysis in organic solvents. *Biotechnol. Bioeng.*, **30**, 81–87.

Lam, L.K.P., Hui, R.A.H.F. and Jones, J.B. (1986) Enzymes in organic synthesis. 35. Stereoselective pig liver esterase catalyzed hydrolysis of 3-substituted glutarate diesters. Optimization of enantiomeric excess via reaction conditions control. *J. Org. Chem.*, **51**, 2047–2050.

Lumry, R. and Eyring, H. (1954) Conformational changes of proteins. *J. Phys. Chem.*, **58**, 110–120.

Margolin, A.L., Tai, D.-F. and Klibanov, A.M. (1987) Incorporation of D-amino acids into peptides via enzymatic condensation in organic solvents. *J. Amer. Chem. Soc.*, **109**, 7885–7887.

Martinek, K., Klibanov, A.M., Goldmacher, V.S. and Berezin, I.V. (1977) The principles of enzyme stabilization. I. Increase in thermostability of enzymes covalently bound to a complementary surface of a polymer support in a multipoint fashion. *Biochim. Biophys. Acta*, **485**, 1–12.

Martinez, P. and Arnold, F.H. (1991) Surface charge substitutions increase the stability of α-lytic protease in organic solvents. *J. Amer. Chem. Soc.*, **113**, 6336–6337.

Menger, F.M. (1993) Enzyme reactivity from an organic perspective. *Acc. Chem. Res.*, **26**, 206–212.

Miyazawa, T., Mio, M., Watanabe, Y. *et al.* (1992) Lipase-catalyzed transesterification procedure for the resolution of non-protein amino acids. *Biotechnol. Lett.*, **14**(9), 789–794.

Nakamura, K., Takebe, Y., Kitayama, T. and Ohno, A. (1991) Effect of solvent structure on enantioselectivity of lipase-catalyzed transesterification. *Tetrahedron Lett.*, **32**(37), 4941–4944.

Nakamura, K., Kinoshita, M. and Ohno, A. (1994) Effect of solvent on lipase-catalyzed trans-esterification in organic media. *Tetrahedron*, **50**(16), 4681–4690.

Ottolina, G., Carrea, G., Riva, S. *et al.* (1992) Effect of enzyme form on the activity, stability and enantioselectivity of lipoprotein lipase in toluene. *Biotechnol. Lett.*, **14**(10), 947–952.

Ottolina, G., Bovara, R., Riva, S. *et al.* (1994a) Activity and selectivity on some hydrolases in enantiomeric solvents. *Biotechnol. Lett.*, **16**(9), 923–928.

Ottolina, G., Gianinetti, f., Riva, S. and Carrea, G. (1994b) Solvent configuration influences enzyme activity in organic media. *J. Chem. Soc., Chem. Commun.*, 535–536.

Oyama, K., Nishimura, S., Nonaka, Y. *et al.* (1981) Synthesis of an aspartame precursor by immobilized thermolysin in an organic solvent. *J. Org. Chem.*, **46**, 5241–5244.

Ozaki, S.-I. and Ortiz de Montellano, P.R. (1994) Molecular engineering of horseradish peroxidase. Highly enantioselective sulfoxidation on aryl aklyl sulfides by the Phe-41→Leu mutant. *J. Amer. Chem. Soc.*, **116**, 4487–4488.

Parida, S. and Dordick, J.S. (1991) Substrate structure and solvent hydrophobicity control lipase catalysis and enantioselectivity in organic media. *J. Amer. Chem. Soc.*, **113**, 2253–2259.

Parida, S. and Dordick, J.S. (1993) Tailoring lipase specificity by solvent and substrate chemistries. *J. Org. Chem.*, **58**, 3238–3244.

Phillips, R.S. (1992) Temperature effects on stereochemistry of enzymatic reactions. *Enzyme Microb. Technol.*, **14**, 417–419.

Rashin, A.A. and Hong, B. (1984) On the environment of ionizable groups in globular proteins. *J. Miol. Biol.*, **173**, 515–521.

Reslow, M., Adlercreutz, P. and Mattiasson, B. (1987) Organic solvents for bioorganic synthesis. *Appl. Microbiol. Biotechnol.*, **26**, 1–8.

Rubio, E., Fernandez-Mayorales, A. and Klibanov, A.M. (1991) Effect of the solvent on enzyme regioselectivity. *J. Amer. Chem. Soc.*, **113**, 695–696.

Rupley, J.A., Yang, P.-H. and Tollin, G. (1980) Thermodynamic and related studies of water interacting with proteins. In *Water in Polymers*, ACS Symposium Series Vol. 127, (ed. Rowland, S.P.). American Chemical Society, Washington, DC, pp. 111–132.

Rupley, J.A., Gratton, E. and Careri, G. (1983) Water and globular proteins. *Trends Biochem. Sci.*, **8**, 18–22.

Russell, A.J. and Klibanov, A.M. (1988) Inhibitor induced enzyme activation in organic solvents. *J. Biol. Chem.*, **263**, 11624–11626.

Ryu, K. and Dordick, J.S. (1989) Free energy relationships of substrate and solvent hydrophobicities with enzymatic catalysis in organic media. *J. Amer. Chem. Soc.*, **111**, 8026–8027.

Ryu, K. and Dordick, J.S. (1992) How do organic solvents affect peroxidase structure and function. *Biochemistry*, **31**, 2588–2598.

Sakurai, T., Margolin, A.L., Russell, A.J. and Klibanov, A.M. (1988) Control of enzyme enantioselectivity by the reaction medium. *J. Amer. Chem. Soc.*, **110**, 7236–7237.

Schinkel, J.E., Downer, N.W. and Rupley, J.A. (1985) Hydrogen exchange of lysozyme powders. Hydration dependence of initial motions. *Biochemistry*, **24**, 352–366.

Schulze, B. and Klibanov, A.M. (1991) Inactivation and stabilization of subtilisins in neat organic solvents. *Biotechnol. Bioeng.*, **38**, 1001–1006.

Secundo, F., Riva, S. and Carrea, G. (1992) Effects of medium and of reaction conditions on the enantioselectivity of lipases in organic solvents and possible rationales. *Tetrahedron: Asymmetry*, **3**(2), 267–280.

Smithrud, D.B. and Diederich, F. (1990) Strength of molecular complexation of apolar solutes in water and in organic solvents is predictable by linear free energy relationships: a general model for solvation effects on apolar binding. *J. Amer. Chem. Soc.*, **112**, 339–343.

St Clair, N.L. and Navia, M.A. (1992) Cross-linked enzyme crystals as robust biocatalysts. *J. Amer. Chem. Soc.*, **114**, 7314–7316.

Ståhl, M., Jeppsson-Wistrand, U., Månsson, M.-O. and Mosbach, K. (1991) Induced stereo-selectivity and substrate selectivity of bio-imprinted α-chymotrypsin in anhydrous organic media. *J. Amer. Chem. Soc.*, **113**, 9366–9368.

Tanford, C. (1962) Contribution of hydrophobic interactions to the stability of the globular conformation of proteins. *J. Amer. Chem. Soc.*, **84**, 4240–4247.

Tanford, C. (1978) The hydrophobic effect and the organization of living matter. *Science*, **200**, 1012–1018.

Tanford, C. (1980) *The Hydrophobic Effect*, Wiley, New York.

Tawaki, S. and Klibanov, A.M. (1992) Inversion of enzyme enantioselectivity mediated by the solvent. *J. Amer. Chem. Soc.*, **114**, 1882–1884.

Tawaki, S. and Klibanov, A.M. (1993) Chemoselectivity of enzymes in anhydrous media is strongly solvent-dependent. *Biocatalysis*, **8**, 3–19.

Terradas, F., Teston-Henry, M., Fitzpatrick, P.A. and Klibanov, A.M. (1993) Marked dependence of enzyme prochiral selectivity on the solvent. *J. Amer. Chem. Soc.*, **115**, 390–396.

Toscano, G., Pirozzi, D. and Greco, G. Jr. (1990) Enzyme stability in the presence of organic solvents. *Biotechnol. Lett.*, **12**(11), 821–824.

Tsitsimpikou, C., Voutou, D., Christakopoulos, P. *et al.* (1994) Studies of the effect of organic solvent on the stability of β-glucosidase from *Fusarium oxysporum. Biotechnol. Lett.*, **16**(1), 57–62.

Ueji, S., Fujino, R., Okubo *et al.* (1992) Solvent-induced inversion of enantioselectivity in lipase-cat-alyzed esterification of 2-phenoxypropionic acids. *Biotechnol. Lett.*, **14**(3), 163–168.

Volkin, D.B. and Klibanov, A.M. (1989) Minimizing protein inactivation. In *Protein Function, A Practical Approach.* (ed. T.E. Creighton), IRL Press, Oxford, pp. 1–24.

Volkin, D.B. and Middaugh, C.R. (1992) The effect of temperature on protein structure. In *Stability of Protein Pharmaceuticals, Part A: Chemical and Physical Pathways of Protein Degradation.* Chapter 8 (eds T.J. Ahern and M.C. Manning), Plenum Press, New York, pp. 215–247.

Volkin, D.B., Staubli, A., Langer, R. and Klibanov, A.M. (1991) Enzyme thermoinactivation in anhydrous organic solvents. *Biotechnol. Bioeng.*, **37**, 843–853.

Volkin, D.B., March, H. and Middaugh, C.R. (1995) Degradative covalent reactions important to protein stability. In *Methods in Molecular Biology: Protein Stability and Folding Protocols*, vol. 40, (ed. B.A. Shirley), Humana Press Inc., Totowa, NJ.

Wangikar, P.P., Graycar, T.P., Estell, D.A. *et al.* (1993) Protein and solvent engineering of subtilisin BPN′ in nearly anhydrous organic media. *J. Amer. Chem. Soc.*, **115**, 12231–12237.

Weinhold, E.G., Glasfield, A., Ellington, A.D. and Benner, S.A. (1991) Structural determinants of stereospecificity in yeast alcohol dehydrogenase. *Proc. Natl. Acad. Sci. USA*, **88**, 8420–8424.

Wescott, C.R. and Klibanov, A.M. (1993a) Solvent variation inverts substrate specificity of an enzyme. *J. Amer. Chem. Soc.*, **115**, 1629–1631.

Wescott, C.R. and Klibanov, A.M. (1993b) Predicting the solvent dependence of enzymatic substrate specificity using semiempirical thermodynamic calculations. *J. Amer. Chem. Soc.*, **115**, 10362–10363.

Wescott, C.R. and Klibanov, A.M. (1994) The solvent dependence of enzyme specificity. *Biochim. Biophys. Acta*, **1206**, 1–9.

Wheeler, C.J. and Croteau, R. (1986) Terpene cyclase catalysis in organic solvent/minimal water media: Demonstration and optimization of (+)-α-pinene cyclase activity. *Arch. Biochem. Biophys.*, **248**, 429–434.

Wong, C.-H., Chen, S.-T., Hennen, W.J. *et al.* (1990) Enzymes in organic synthesis: Use of subtilisin and a highly stable mutant derived from multiple site-specific mutations. *J. Amer. Chem. Soc.*, **112**, 945–953.

Wu, J. and Gorenstein, D.G. (1993) Structure and dynamics of cytochrome c in nonaqueous solvents by 2D NH-exchange NMR spectroscopy. *J. Amer. Chem. Soc.*, **115**, 6843–6850.

Wu, S.-H., Chu, F.-Y. and Wang, K.-T. (1991) Reversible enantioselectivity of enzymatic reactions by media. *Bioorg. Med. Chem. Lett.*, **1**(7), 339–342.

Yang, Z. and Robb, D.A. (1993) Comparison of tyrosinase activity and stability in aqueous and nearly nonaqueous environments. *Enzyme Microb. Technol.*, **15**, 1030–1036.

Yennawar, N.H., Yennawar, H.P. and Farber, G.K. (1994) X-ray crystal structure of γ-chymotrypsin in hexane. *Biochemistry*, **33**, 7326–7336.

Yokozeki, K., Yamanaka, S., Takinami, K. *et al.* (1982) Application of immobilized lipase to regiospecific interesterification of triglyceride in organic solvent. *Eur. J. Appl. Microbiol. Biotechnol.*, **14**, 1–4.

Zahalka, H.A., Dutton, P.J., O'Doherty, B. *et al.* (1991) Bile salt modulated stereoselection in the cholesterol esterase catalyzed hydrolysis of α-tocopheryl acetates. *J. Amer. Chem. Soc.*, **113**, 2797–2799.

Zaks, A. and Klibanov, A.M. (1984) Enzymatic catalysis in organic media at 100°C. *Science*, **224**, 1249–1251.

Zaks, A. and Klibanov, A.M. (1986) Substrate specificity of enzymes in organic solvents vs. water is reversed. *J. Amer. Chem. Soc.*, **108**, 2767–2768.

Zaks, A. and Klibanov, A.M. (1988) Enzymatic catalysis in non-aqueous solvents. *J. Biol. Chem.*, **263**, 3194–3201.

Zheng, C. and Phillips, R.S. (1992) Effect of coenzyme analogs on enantioselectivity of alcohol dehydrogenase. *J. Chem. Soc., Perkin Trans. 1*, 1083–1084.

5 Enzymatic resolutions of alcohols, esters, and nitrogen-containing compounds

C.J. SIH, G. GIRDAUKAS, C.-S. CHEN and J.C. SIH

5.1 Introduction

The notion that enzymes are catalytically active in organic solvents was known as early as the turn of the century. However, it was not until the mid-1980s that Klibanov and coworkers advocated the full exploitation of the potential use of enzymes in organic solvents. This is reflected in the publication profile for the period between 1967 and 1991 (Figure 5.1). It is readily evident that since 1984 there has been a continued rapid increase in the number of papers, reaching over 180 in 1991.

Consequently, in a decade, preparative enzymatic synthesis has gained added versatility. The deployment of enzymes in monophasic organic solvents at low water activity has several advantages: enzymes may be more stable in organic solvents than in water and enzyme enantioselectivity may be enhanced in organic media; water-insoluble organic substrates are sometimes transformed at faster rates in organic solvents than in an aqueous medium; undesired side reactions involving water are repressed. However, the most important advantage is the possibility to shift the thermodynamic equilibrium to favor synthesis over hydrolysis. Thus, hydrolytic enzymes such as lipases and proteases have been extensively used for preparative enantioselective syntheses via transacylation processes in organic media. The objective of this chapter is to discuss some general principles in conducting biocatalytic reactions using hydrolytic enzymes in organic solvents, and to highlight selective enantioselective transformations to underscore the usefulness of this nonaqueous methodology.

5.2 Quantitative analysis of enantioselective biocatalytic reactions

Since enzymatic reactions in organic solvents also obey Michaelis–Menten kinetics, the same equations that have been developed for aqueous systems can also be applied to non-aqueous systems. Thus, for the most commonly encountered competing reactions (equations 5.1 and 5.2), the rates for the formation of P and Q are $v = (V_{max}/K_m)R$ and $v = V_{max}'/K_m')S$ respectively, and the reaction is pseudo-first order with respect to the enantiomers, R and S where

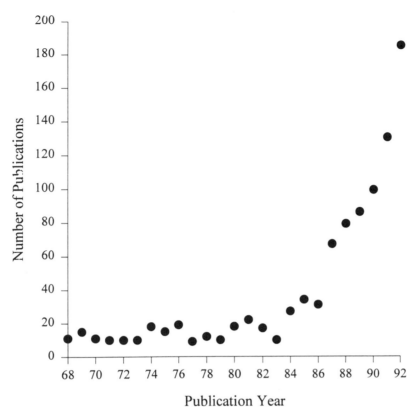

Figure 5.1 Number of publications on enzymes in organic solvents (data obtained from *CAS Online* search performed on February 15, 1995).

Enz is the biocatalyst. The enantiomer ratio, E (1), is simply the ratio of the pseudo-first order rate constants, $E = (V_{max}/K_m)/(V_{max}'/K_m')$. Implicit in this derivation is the assumption that R is the fast reacting enantiomer and the system is irreversible and devoid of product inhibition.

$$R + \text{Enz} \rightleftarrows R - \text{Enz} \rightarrow \text{Enz} + \text{P} \qquad (5.1)$$

$$S + \text{Enz} \rightleftarrows S - \text{Enz} \rightarrow \text{Enz} + \text{Q} \qquad (5.2)$$

The E value is a measure of the enantioselective behavior of the biocatalytic system and allows one to make quantitative predictions relating the extent of conversion of a racemate and the enantiomeric composition of the recovered substrate (R,S) and product (P,Q), as well as a comparison of the stereo-selectivity of different biocatalytic systems.

The E value may be readily calculated from equation (5.3) or (5.4) by experi-mentally determining the extent of conversion (c) and either the enantiomeric excess of the remaining substrate (ee_s) or that of the product (ee_p). Alternatively, one can also obtain E directly from knowledge of ee_s and ee_p using equation (5.5), since $c = ee_s/(ee_s + ee_p)$.

$$E = \frac{\ln[(1 - c)(1 - ee_s)]}{\ln[(1 - c)(1 - ee_s)]} \tag{5.3}$$

$$E = \frac{\ln[(1 - c)(1 - ee_p)]}{\ln[(1 - c)(1 - ee_p)]} \tag{5.4}$$

$$E = \frac{\ln\left[\dfrac{1 - ee_s}{1 - ee_s/ee_p}\right]}{\ln\left[\dfrac{1 - ee_s}{1 + ee_s/ee_p}\right]} \tag{5.5}$$

If the conformation of the enzyme is not altered significantly in organic solvents or if the reaction proceeds via similar transition states, one would expect the stereochemical preference and the E value not to change appreciably from those observed in aqueous solutions (2).

Enantioselective transesterifications of racemic (R and S) alcohols and amines in organic solvents are generally conducted using achiral acyl donors (A) such as vinyl or isopropenyl acetates (IPA), or anhydrides to avoid the problem of reversibility of the reaction (3,4). While the competing reactions (equations 5.6 and 5.7) are second order (first order with respect to R or S and first order with respect to A), the relationship between the enantiomer ratio E, ee_s and c is still governed by equation (5.3). This homocompetitive equation is valid for all kinetic resolutions where the reaction is first order with respect to R or S and any order with respect to A.

$$R + A \xrightarrow{E_{NZ}} P + B \tag{5.6}$$

$$S + A \xrightarrow{E_{NZ}} Q + B \tag{5.7}$$

When the reaction is reversible, the treatment becomes more complex (2). However, for preparative purposes one normally uses experimental conditions favoring the reaction to go to completion. When an excess amount of achiral acyl acceptor or donor is used in the kinetic resolutions, the mechanistic treatment is simplified, and equation (5.3) may be used reliably for the calculation of the E value.

On the other hand, when the reaction is second order with respect to R or S but not with respect to R and S (equations 5.9 and 5.10), it can be readily shown that the enantiomer ratio, E is related to c and ee_s by equation (5.8) (5, 6).

$$E = \frac{(1 + ee_s)}{(1 - ee_s)} \times \frac{1 - (1 - c)(1 - ee_s)}{1 - (1 - c)(1 + ee_s)} \tag{5.8}$$

While kinetic resolutions involving second-order kinetics are uncommon, one could envisage the asymmetric dimerization of chiral hydroxy acids when the consumption of the monomer is second order as an example. A published case is the enantioselective polymerization of racemic bis(2,2,2-trichloroethyl) *trans*-3,4-epoxyadipate with 1,4-butanediol using porcine pancreatic lipase as the catalyst (7).

$$R + R \xrightarrow{k_1} R - R \quad \text{(fast)} \tag{5.9}$$

$$S + S \xrightarrow{k_2} S - S \quad \text{(slow)} \tag{5.10}$$

The derivation of equation (5.8) is based on the assumption that no meso dimer (R–S or S–R) is formed. When $c \rightarrow 1$, the limit of ee_s is $(E - 1)/(E + 1)$. In contrast to first-order kinetic resolutions, with second-order cases it is not possible to reach 100% ee_s for high conversions since $(E - 1)/(E + 1)$ remains

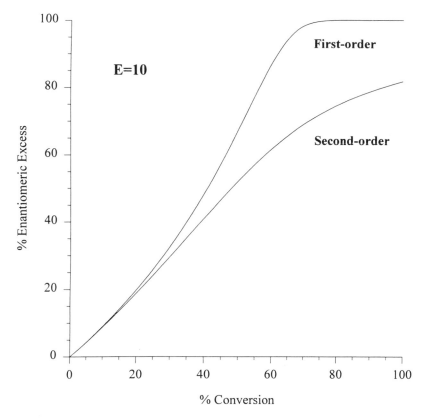

Figure 5.2 Comparison of the computed curves for first-order and second-order kinetic resolutions ($E = 10$).

less than 1 unless $E \to \infty$. However, for systems with E values of 200 and $c \to 1$, ee_s could reach 99%. Figure 5.2 shows the computed curves ($E = 10$) for the remaining substrate fractions for first-order and second-order kinetic resolutions.

In cases where a meso dimer (R–S) also is formed, an amplification of enantioselectivity could be achieved even when $E = 1$ provided that one starts out with a partially resolved substrate (8). Such enhancement of enantio-selectivity becomes especially pronounced when $k_{RS} \gg k_R$ or k_S. This phenomenon is sometimes referred to as the meso effect and occasionally may be encountered during the column-chromatographic separation of partially-enriched racemates (9). That is, the enantiomer in an enriched mixture which may interact non-covalently to form homo (chiral) and hetero (meso) dimers, which are resolved on the column, resulting in an enhancement of the optical purity of the homochiral dimer fraction.

$$R + R \overset{k_R}{\to} R - R \tag{5.11}$$

$$S + S \overset{k_S}{\to} S - S \tag{5.12}$$

$$R + S \overset{k_{RS}}{\to} R - S \quad \text{(fast)} \tag{5.13}$$

5.3 Stereochemical recognition of lipases

With the natural substrate of triacylglycerols, lipases exhibit regiospecificity (Table 5.1) but lack significant enantiotropic discrimination.

This deficiency in prochiral discrimination, however, does not truly reflect the remarkable ability of lipases as enantioselective catalysts toward a wide spectrum of substrates. The discrepancy in the stereochemical behavior raises

Table 5.1 Positional specificity of lipases (9)

Positional specificity	Lipase origin
1,3-specific	porcine pancreas, *Rhizopus arrhizus, Rhizopus delemar* *Rhizopus japonicus, Rhizopus javanicus, Aspergillus niger* *Saccharomycopsis lipolytica, Penicillium camembertii* *Chromobacterium viscosum, Mucor javanicus, Mucor miehei*
Intermediary	*Phycomyces nitens, Pseudomonas fluorescens, Alcaligenes* sp.
Non-specific	*Candida cylindracea, Geotrichum candidum, Achromobacter* sp.

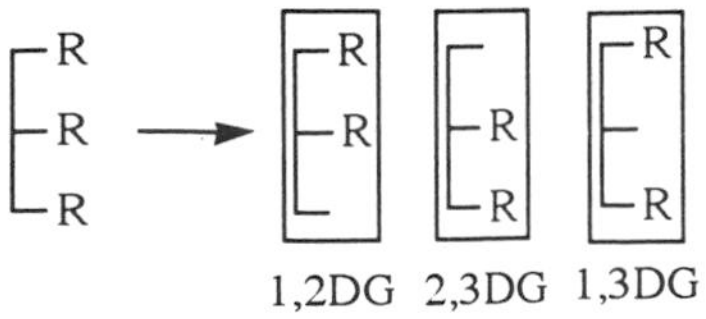

positional specificity constant (PSI)
= [1,2(2,3)DG – 1,3DG × 2]
× 100/[1,2(2,3)DG + 1,3DG × 2]
PSI = 100: 1,3-specific
70 < PSI < 80: Intermediary
–20 < PSI < 30: Nonspecific

1,2DG 2,3DG 1,3DG

an interesting question with regard to the origin of enantioselection in lipase catalysis. Here, we propose a simplified kinetic model to account for this phenomenon. It should be noted that since the modes of differentiation between enantiotopes and between optical isomers are virtually identical, the basic kinetic principle can be applied to either case.

Inherent in interfacial catalysis is the presence of two successive equilibria, i.e. reversible binding of the lipase to the interface and the subsequent interaction with the monomeric substrate. When two enantiotropic triglycerides or two

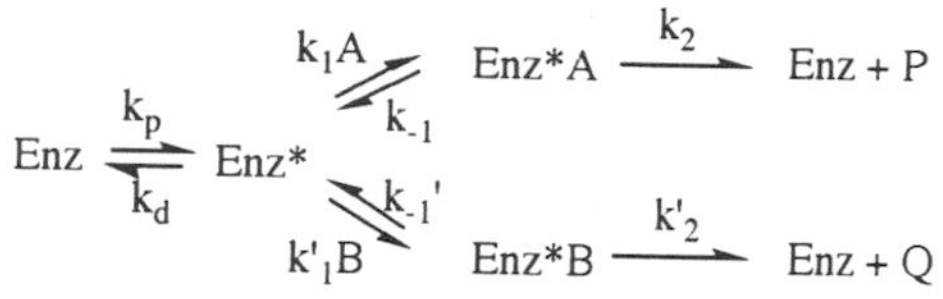

$$\text{Enz} \underset{k_d}{\overset{k_p}{\rightleftharpoons}} \text{Enz*} \underset{k'_1 B}{\overset{k_1 A}{\rightleftharpoons}} \begin{matrix} \text{Enz*A} \xrightarrow{k_2} \text{Enz} + \text{P} \\ \\ \text{Enz*B} \xrightarrow{k'_2} \text{Enz} + \text{Q} \end{matrix}$$

Graph 5.1

competing enantiomers (A and B) compete for the enzyme, the kinetic scheme may be envisaged as shown in Graph 5.1, where: k_p and k_d denote the rate constants for the adsorption and desorption of the lipase onto and from interface, respectively; k_1, k_{-1}, k_2, k'_1, k'_{-1}, and k'_2 are apparent rate constants for the binding and subsequent catalytic steps. It is apparent that the specificity for the competing substrates can be expressed by the ratio of apparent specificity constants, $(k_{cat}/K_m)_A{}^{app}$ and $(k_{cat}/K_m)_B{}^{app}$ (equation 5.14).

$$E = (k_{cat}/K_m)_A/(k_{cat}/K_m)_B \tag{5.14}$$

Representation of the specificity constants in equation (5.14) with individual kinetic constants gives equation (5.15).

$$E = \{k_1/[1 + (k_{-1}/k_2)]\}/\{k'_1/[1 + (k'_{-1}/k'_2)]\} \tag{5.15}$$

Recent structural evidence indicates that the binding domains of lipases consist of a large hydrophobic surface that is exposed upon interfacial activation by substrate (11). As this hydrophobic pocket has an intrinsic affinity for triacylglycerols, the resulting complex is expected to be highly stable. Consequently, the dissociation constant (k_{-1}) will be relatively smaller than that of the subsequent catalytic step (k_2), i.e. $k_{-1} \ll k_2$ and $k'_{-1} \ll k'_2$. Equation (5.15) can thus be reduced to equation (5.16), which indicates a partial or complete loss of specificity.

$$E = k_1/k'_1 \tag{5.16}$$

Accordingly, this hypothesis may also in part account for the low stereoselectivity encountered with the cleavage of the primary ester bond as compared to that of the secondary ester bond due to the higher catalytic constant. In principle, high enantioselection is achieved only when the monomeric substrate is in rapid equilibrium between the free and enzyme-bound forms before catalysis takes place (i.e. $k_{-1} \gg k_2$ and $k'_{-1} \gg k'_2$) (equation 5.17).

$$E = (k_1 k_2/k_{-1})/(k'_1 k'_2/k'_{-1}) \tag{5.17}$$

This premise is supported by the lipase-mediated acylation of 2-substituted glycerols in organic media (12) (Table 5.2). This reverse catalysis of triglyceride hydrolysis gave products of high enantiomeric purity, which is in line with the hydrophilic nature of the substrates.

However, it should be kept in mind that the stereochemical behavior of lipases is complex and the degree of stereoselectivity cannot be intuitively predicted by examining structures. For example, in the preparation of optically active hydroxyalkanoic acids, the use of a non-hydrolyzable t-butyl ester, instead of the free acid, at the carboxyl terminus significantly improved the enantio-selectivity in the lipase-mediated hydrolysis of the acyloxy ester (Table 5.3) (13). Especially worthy of note was that the enantioselectivity was found to be independent of the relative distance between the t-butyl ester and the acetoxy function, suggesting that the enhancement was not solely due to the steric effect of the t-butyl ester function.

Bhalerao and coworkers (14) reported a similar finding on the enantioselective hydrolysis of racemic $(\omega - 2)$-acetoxy-ω-bromoalkenoic esters to afford the corresponding R acid by $C.$ $rugosa$ lipase. The use of a n-butyl ester instead of a methyl ester greatly enhanced the enantioselectivity in a similar fashion. Again, the enhancement was independent of the chain length of the substrate (Graph 5.2).

In addition, the enantioselectivity may also be affected by changes in the micro-environments surrounding the catalytic domain. Deveer and coworkers (15) reported the dependence of stereoselectivity of $Pseudomonas$ $glumae$ lipase catalysis on surface pressure. With pseudoglycerides as substrates, it was found that increasing the surface pressure from 10 to 30 mN/m resulted in a twofold

Table 5.2 Lipase-catalyzed enantioselective esterification of 2-O-substituted glycerols

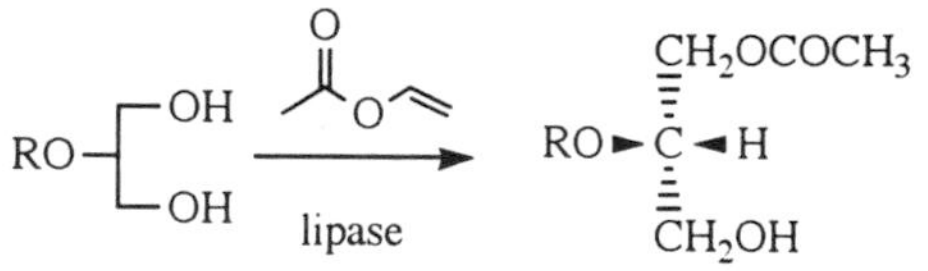

R	Lipase	Conversion (%)	Enantiomeric excess (%)
$PhCH_2-$	P	100	94
	LP	12	84
CH_3-	P	100	70
	LP	100	92
CH_3CH_2-	P	100	90
	LP	100	89

P: $Pseudomonas$ $cepacia$ lipase
LP: $Chromobacterium$ $viscosum$ lipase

Table 5.3 Lipase-catalyzed hydrolysis of racemic acetoxyalkanoic acids and their *t*-butyl esters

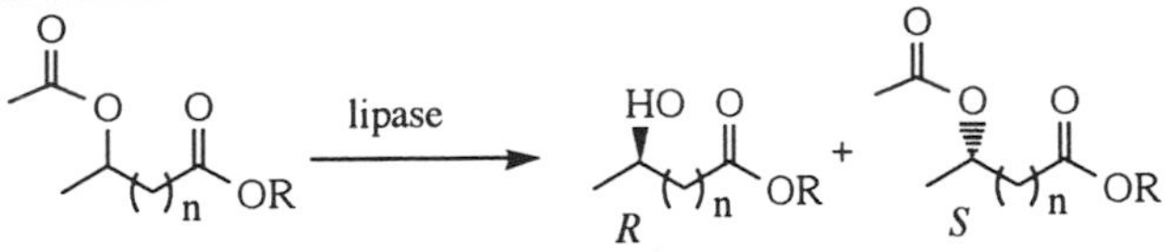

				Enantiomeric ratio (E)			
R =		H				C(CH$_3$)$_3$	
n =	1	5	10	1	5	10
Lipase AK	1	6	2	>100	>100	>100
Lipase K-10	3	3	3	>100	>100	>100

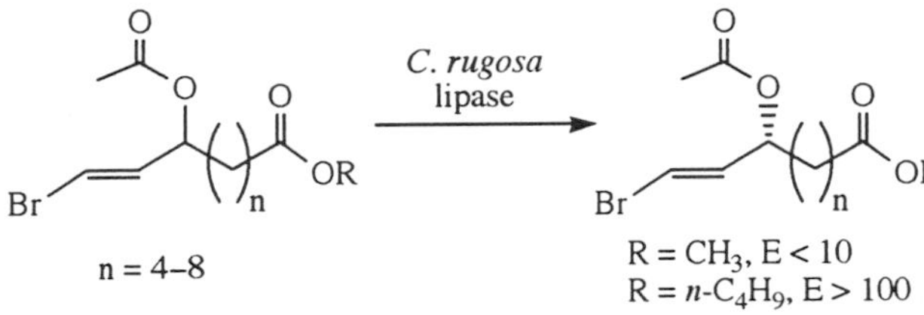

Graph 5.2

increase in the stereoselectivity index (v_R/v_S), of which the underlying mechanism remained unclear. This pressure control of enantioselectivity appears to be common among different types of enzymes. Russell and coworkers (16) reported that, in supercritical fluoroform, both subtilisin and *Aspergillus* protease became more enantioselective as pressure increased. These authors attributed such a selectivity enhancement to the effect of pressure on the dielectric constant of the solvent, i.e. fluoroform became more hydrophilic as pressure increased, which could account for the increase in enantioselectivity.

5.4 Working models for predicting stereoselectivity

The X-ray crystallographic structures of a number of lipases have been reported, which include human pancreatic lipase (17), *Rhizomucor miehei* lipase (18), *Geotrichum candidum* lipase (19), and *Candida rugosa* (formerly *cylindracea*, CCL) lipase (11, 20). In all of them, the active sites consist of Ser-His-Asp/Glu catalytic triads that are occluded from media by a polypeptide flap (lid). The topological location of the flaps varied among these lipases, and their length and complexity increased with the size of the molecule. Evidence suggested that a water–lipid interface facilitated conformational changes that opened access to the active site serine and exposed a number of hydrophobic residues. This flexibility allowed the creation of a large hydrophobic area for interfacial interactions with substrate molecules. Comparison of the three-dimensional structures among different lipases showed similarities in essential structural and functional features. For example, Cygler and coworkers (20) reported that the crystal structures of *C. rugosa* lipase–inhibitor and *R. miehei* lipase–inhibitor complexes were superimposible using the strand-turn-helix motif around the catalytic Ser as an anchor. This finding, in combination with other structural information, led to the postulation that in all lipases that share the α/β-hydrolase fold, such as those of *C. rugosa* and *G. candidum*, the positioning of the ester bond and the stereochemistry of hydrolysis should be alike (20).

The three-dimensional structures could provide useful insights into the stereochemical recognition at the catalytic domain. However, the crystal structures of many lipases have yet to be established. An alternative approach to envisage the mode of enantioselective discrimination is to examine the enantioselectivity data (*E* values) of a wide spectrum of substrates. Accordingly, a working model can be constructed to correlate the relationships between selectivity and substrate structures. For example, a 'two-site' model analogous to the Prelog's rule was put forward by two independent groups to predict the stereopreferences for the lipases of *Candida rugosa*, *Pseudomonas cepacia* (P-30; Amano), and *Pseudomonas* sp. SAM-II and AK (21, 22). For these lipases, esters of secondary carbinols of the configuration shown in Figure 5.3 will be preferentially hydrolyzed (L and M refer to large and medium groups, respectively).

Figure 5.3 Two-site binding model.

Although this working model is inadequate to predict the stereochemical outcome for many substrates, it underscores the structural homology among the catalytic domains of these lipases. A more sophisticated cubic lattice binding model was postulated for *Pseudomonas cepacia* lipase by Sih and coworkers (23) (Figure 5.4). The substrate binding domain is divided into four quadrants, and the enantiomer with the spatial arrangement shown in Figure 5.4 will be preferentially reacted.

This model has been applied to the prediction of many artificial substrates with a high degree of accuracy. Compared to lipase P-30, the mode of enantio-

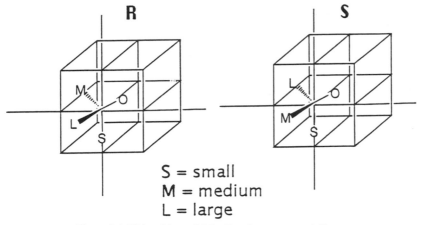

S = small
M = medium
L = large

Figure 5.4 Sih's cubic model for *Pseudomonas cepacia* lipase.

selection for *C. rugosa* lipase, especially toward acyclic substrates, appears to be more difficult to predict, which may be in part due to the heterogeneity of the lipase (see the next section). A model for *C. rugosa* lipase was advanced by Faber and coworkers, but was limited to bicyclic alcohols with the *endo-* configuration (24). Moreover, some lipase preparations such as porcine pancreatic lipase (PPL) contain contaminating hydrolases in substantial quantities. For these lipases, attempts to construct a rational model with high confidence may prove to be futile (25).

5.5 Molecular and submolecular heterogeneity of *Candida rugosa* lipase

Among numerous lipases that have been used for chiral kinetic resolutions, *C. rugosa* lipase displays a unique capability for enantioselectivity enhancement through a variety of strategies (Table 5.4) (26–29). Moreover, one may observe varying degrees of enantioselectivity with *C. rugosa* lipase preparations from different commercial sources.

Another interesting phenomenon associated with *C. rugosa* lipase is that chemical modification of the lysine residues leads to an enhanced activity in esterification accompanied by a reduced hydrolytic activity (30).

These apparent changes in the stereochemical and catalytic behaviors may in part be explained by the presence of multiple isoforms in the crude preparations, which is common among many extracellular microbial lipases (31). Alberghina and coworkers (32) reported the cloning and sequencing of two lipase genes from *C. rugosa*. These two genes coded for 543-amino acid proteins with 79% sequence homology, of which the predicted isoelectric points were 4.5 and 4.9 (32). Despite the high degree of structural similarity among several isoforms,

Table 5.4 Enhancement of enantioselectivity in *C. rugosa* lipase catalysis

Selectivity-enhancing factors	Examples

1. Enantioselective inhibition (25)

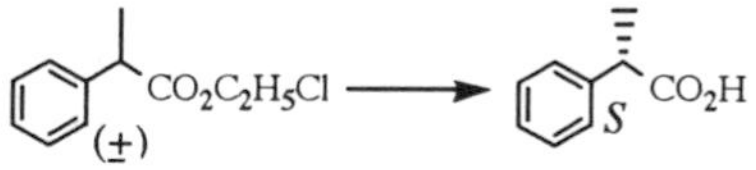

+ inhibitor	E
None	10
Dextromethophan	>100

2. Bile salt/organic solvent treatment (26)

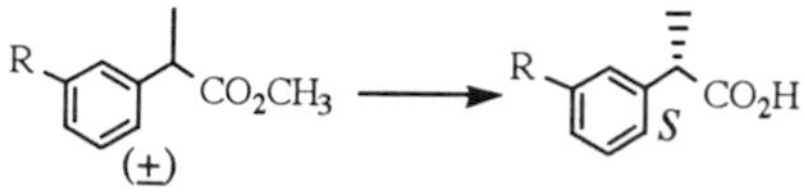

R	crude lipase	treated lipase
H	10	>100
PhCO	4	>100

3. Solvent hydrophobicity (27)

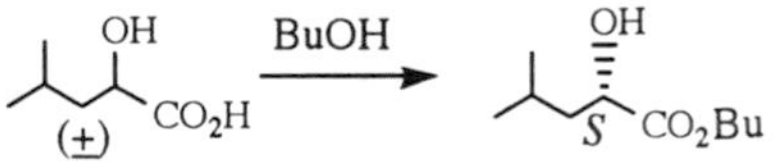

organic medium   log P   E^a

R	crude lipase	treated lipase
cyclohexane	3.2	5.2
toluene	2.5	110
diethyl ether	0.85	7.2
THF	0.49	4.2
dioxane	−1.1	4.4

4. Enzyme immobilization (28)

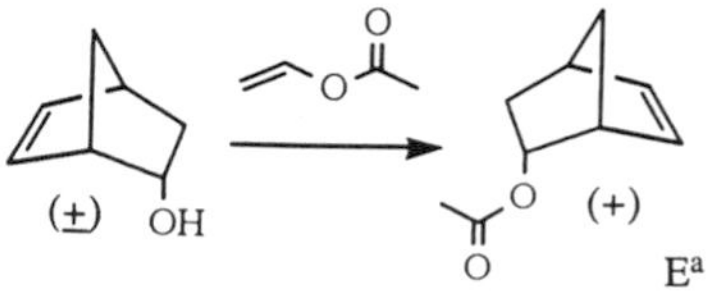

E^a

native lipase	10
lipase immobilized onto an epoxy-activated polymer	>100

^{a} The E values were calculated from the reported (V_{max}/K_m)s their enzymatic properties appeared to be different. Shaw and coworkers (33) reported the isolation of two distinct forms of *C. rugosa* lipase by ion-exchange chromatography, which displayed different optimal pH values (7.0 and 5.0, respectively) and fatty acid specificity. Also, separation of the crude lipase on hydrophobic interaction chromatography gave several active fractions displaying distinct substrate specificities (34). Conceivably, as these isoforms may exhibit different stereochemical behaviors, one may optimize stereoselectivity by selectively inhibiting the less specific enzyme(s).

Furthermore, recent data suggested that the increase in stereoselectivity may be due to conformational changes of the lipase. Sih and coworkers (27) noted that treatment of *C. rugosa* lipase with deoxycholate, followed by solvent precipitation with a 1:1 mixture of ether and ethanol, resulted in a change of the chromatographic behavior of the protein accompanied by significantly higher enantioselectivity (Entry 2, Table 5.4). This finding is in accord with reports by Cygler's group, showing that *C. rugosa* lipase existed in two conformational states, open and closed, in reference to the orientation of the flap (11, 20). According to the crystal structure, the hydrophobic portion of the flap, which is buried in the closed conformation, becomes exposed and forms one wall of the binding cleft in the open conformation. Moreover, the 'open' and 'closed' crystal forms of the enzyme could be induced by different crystallization conditions [2-methyl-2,4-pentandiol (11) and PEG 8000 (35), respectively]. These findings led to the notion that the flap movement might not be essential for lipase activation, and that *C. rugosa* lipase existed in different conformational states, i.e. open, intermediate, and closed, all of which may be catalytically active.

Presumably, the stripping of bound lipids from the crude *C. rugosa* lipase by detergent/solvent treatments may facilitate the conversion from the closed to the open form, resulting in an enhancement of enantioselectivity. Although the underlying mechanism and factors affecting this conformational transition remain to be determined, this simple manipulation is of practical use to produce a more enantioselective enzyme for industrial applications. It is noteworthy that such conformational heterogeneity may be a common phenomenon among different lipases, since a flap motif has been found in all of the lipase structures reported to date. These conformational changes on chiral recognition are thus of great interest to both enzymologists and organic chemists, and warrant further investigation.

5.6 Sequential biocatalytic kinetic resolutions

Substrates with dual functionalities undergo consecutive enantioselective catalyses, either hydrolysis or esterification, resulting in higher optical yield than systems possessing a single catalytic step. The concept of synergistic coupling, first introduced by Sih (36), has evolved into several strategies for the resolution of different substrates, the principles of which are as follows.

5.6.1 Asymmetric catalysis of meso/prochiral diesters (or diols)

This concept is illustrated in Graph 5.3. Suppose that S is an achiral diester (or diol) with a plane of symmetry, which is transformed by an enzyme to yield two enantiomeric monoesters P (fast forming) and Q (slow forming); in turn, they undergo further concurrent catalysis by the same enzyme to furnish the diol (or diester) R. In principle, if the same stereochemical preference is maintained, the relative rate of catalysis follows the order $k_1 > k_3$ and $k_2 > k_4$. Consequently, the inherent consecutive resolution step enhances the optical purity of the chiral intermediate, P, by preferentially removing the minor enantiomer Q. The principle of this approach is illustrated by the examples shown in Table 5.5.

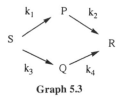

Graph 5.3

It is noteworthy that despite the low selectivity of the initial enantioselective catalytic step (k_1/k_3), a highly optically enriched monoester intermediate may still be obtained, albeit at lower chemical yield. For example, even with an initial selectivity of 3.6 for substrate 3, the monoester of 95% ee was recovered in 31% yield. This strategy therefore extends the usefulness of enzymes of low enantioselectivity for asymmetric synthesis.

Table 5.5 Optical purity enhancement in enzymatic asymmetric catalysis

substrate	mode of catalysis	stereo-preference	k_1	k_2	k_3	k_4	Ref.
			(arbitrary units)				
AcO ⟍⤴⤴⟋ OAc **1**	hydrolysis/ PPL	pro-*S*	15.6	0.66	1	3.0	(35)
HOⅢ⟨ ⟩ⅢOH **2**	acetylation/ P-30	pro-*S*	4	0.76	1	7.2	(36)
AcO ⟍⤴⟋ OAc **3**	hydrolysis/ PPL	pro-*R*	3.6	0.18	1	0.69	(37)

5.6.2 Consecutive kinetic resolution of racemic axially disymmetric substrate

This approach is shown in Graph 5.4. During the enantioselective catalysis of racemic axially disymmetric diesters/diols, two consecutive kinetic resolution steps are operating in tandem. Provided that the stereochemical preference remains the same, the second step will likewise improve the enantioselectivity of the first resolution step, resulting in high optical yield of the final product (R + S, fraction). This strategy has been successfully applied to the preparation of chiral auxiliaries such as binaphthol ((±)-**4**) (39) and *trans*-2,4-pentanediol ((±)-**5**) (40) with optical purities greater than 98% (Table 5.6).

$$A \xrightarrow{k_1} P \xrightarrow{k_2} R$$

$$B \xrightarrow[k_3]{} Q \xrightarrow[k_4]{} S$$

Graph 5.4

5.6.3 Coupled kinetic resolution of racemic monofunctional alcohols in organic media

In organic media, the resolution of racemic alcohols can be conducted by enantioselective esterification or by deacylation of their corresponding esters. However, in many cases, either method fails to achieve a satisfactory optical yield, and the products have to be recycled to increase the enantiomeric purity. In principle, if the rate of deacylation is comparable to that of the subsequent acylation and an achiral acyl donor (e.g. vinyl acetate) is added to the system

Table 5.6 Consecutive kinetic resolution of axially disymmetric molecules

			kinetic constants			
substrate	mode of catalysis	stereo- preference	k_1	k_2	k_3	k_4
			(arbitrary units)			
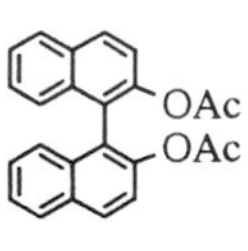 (±) ![OAc OAc binaphthyl structure] (±)-**4**	hydrolysis/ microbial cells (*Absidia glauca*)	*S*	12.2	205	1	3.7
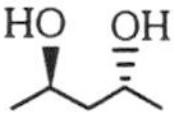 (±) HO OH ![pentanediol structure] (±)-**5**	acetylation with hexanoic acid/ Lipase AK	*2R,4R*	100	1.1	1	0.02

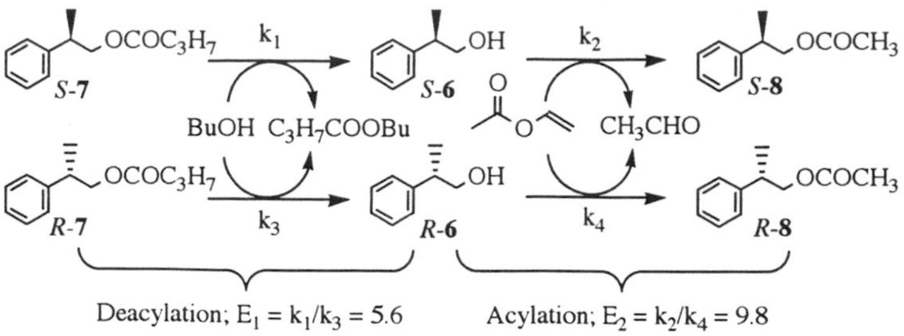

Deacylation; $E_1 = k_1/k_3 = 5.6$ Acylation; $E_2 = k_2/k_4 = 9.8$

Scheme 5.1

midway through the reaction, the two kinetic resolution steps can be coupled in tandem. The basis of this approach parallels that of product recycling without the separation of the alcohol intermediate. This strategy is illustrated by the resolution of racemic 2-phenyl-1-propanol (±)-**6** by PPL (41) as shown in Scheme 5.1.

The observed E values for the deacylation of (±)-**7** and the acylation of (±)-**6** were 5.6 and 9.8, respectively, with S stereopreference, giving products with low to moderate optical purity. The coupling was achieved by adding excess amounts of vinyl acetate to the deacylation reaction at approximately 50% conversion. The resulting S-**8** could thus be obtained with optical purity greater than 90% in up to 30% chemical yield. In contrast, the ee values attainable at 30% conversion for a single catalytic step with E values of 5.6 and 9.8 were 0.62 and 0.75, respectively.

This strategy was further extended to the use of two enzymes by Rakels and coworkers (42). The reported system consisted of the hydrolysis of racemic methyl 2-chloropropionate ((±)-**9**) by carboxylesterase NP ($E = 6.5$), followed by the dehalogenation of the resulting (S)-2-chloropropionic acid (S-**10**) by DL-dehalogenase to afford the corresponding lactic acid (S-**11**) ($E = 6.8$)(Scheme 5.2). In line with the aforementioned example, significant enhancement in the optical purity of the final product was noted.

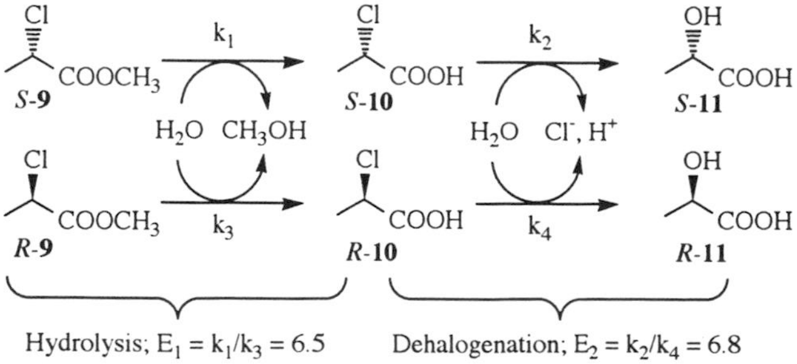

Hydrolysis; $E_1 = k_1/k_3 = 6.5$ Dehalogenation; $E_2 = k_2/k_4 = 6.8$

Scheme 5.2

However, in the two-enzyme system, the optical purity of the final product depends on the relative rates of the consecutive catalyses. Moreover, the usefulness of this strategy is limited by the availability of two 'compatible' biocatalysts.

5.7 Acyl donors and acceptors

Enzymatic acyl transfer reactions in organic solvents have been widely used for the preparation of enantiomerically pure compounds (43, 44). One important application of this methodology is in the kinetic resolution of chiral esters, alcohols and amines. The catalytic mechanism involves the reaction of the acyl donor with the enzyme to form the acyl-enzyme, which is then deacylated by an acyl acceptor. In the kinetic resolution of (a) chiral racemic esters (acyl donor), the chirality-determining step resides in the acylation step, whereas in the kinetic resolution of (b) racemic alcohols and amines, it is the deacylation of the acyl-enzyme that is the enantioselective step, as shown in Graph 5.5. Depending

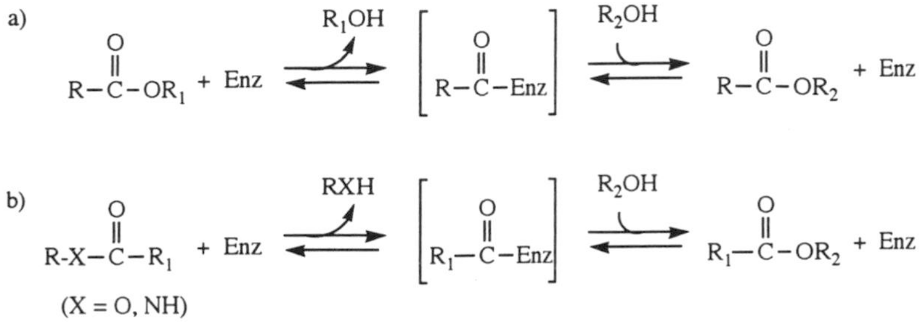

Graph 5.5

on the properties and the concentrations of the acyl donor and acceptor as well as the design of reaction conditions, the thermodynamic equilibrium may be shifted to favor product formation. Since no net amount of water is formed in acyl transfer, the loss of enantioselectivity due to reversibility of the reaction may be avoided. While several different approaches have been utilized for the displacement of the equilibrium in enantioselective transesterifications, the most widely used approach is the use of enol esters as irreversible acyl donors (Graph 5.6) (3, 4, 43, 44).

Enol esters are equivalent to activated esters, for they are cleaved at rates $10-10^3$ times faster than simple alkyl esters (4). Vinyl and isopropenyl acetates are the most commonly used acyl donors and enol esters with varying acyl

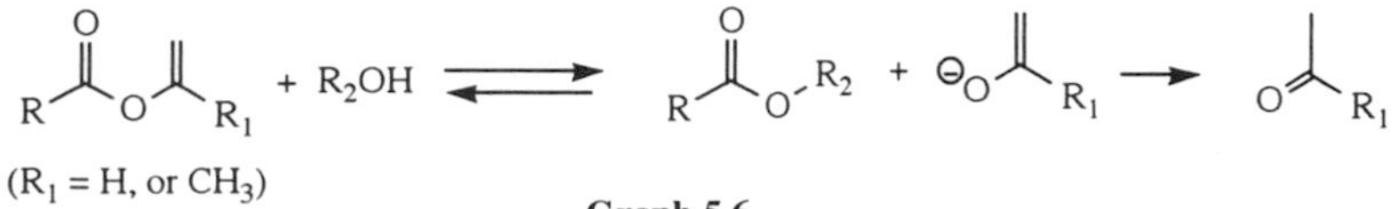

($R_1 = H$, or CH_3)

Graph 5.6

groups may be easily prepared. Since the enolate liberated in the reaction rapidly tautomerizes to acetaldehyde or acetone, the reaction is irreversible and product inhibition is not a concern. Although the reaction rate is faster with vinyl acetate as compared to isopropenyl acetate, the acetaldehyde liberated may inactivate the enzyme due to Schiff base formation with the amino group of lysine (43, 44) especially when the reaction is conducted in a neat solution of the enol ester. Hence, the reactions are normally carried out in a suitable non-polar organic solvent such as *tert*-butyl methyl ether using a few equivalents of enol ester.

As far as the kinetic resolution of chiral esters are concerned, the ester (acyl donor) is attacked by a nucleophile (serine) on the enzyme to form an acyl-enzyme intermediate with the liberation of R_1–OH; the acyl group is then transferred on to R_2–OH with the regeneration of free enzyme. In contrast to hydrolytic reactions, where water is always in excess, the concentration of the nucleophile R_2–OH in acyl transfer reactions is always limited. As a consequence, transesterification reactions are generally reversible and the reversibility of the reaction is determined by the relative nucleophilicity of the nucleophile and the leaving group, both of which are able to compete for the acyl-enzyme intermediate in the forward and reverse directions. The nucleophilicity of the leaving group may be decreased by the introduction of electron-withdrawing substituents to shift the reaction towards completion. For example, the use of activated esters, such as 2-haloethyl, cyanomethyl, trichloroethyl and trifluoroethyl esters, not only enhances the reaction rate but also reduces the nucleophilicity of the leaving group; they are commonly used on a laboratory scale (45).

5.8 Solvent and enzyme enantioselectivity

The enantioselectivity of biocatalysts is governed by the enantiomer ratio, E, the ratio of the specificity constants, k_{cat}/K_m of the reactants. An enzyme can assume various conformations in different organic media, which is manifested in changes in the kinetic parameters, k_{cat} and K_m, of the biocatalytic system. It is therefore apparent that enzyme stereoselectivity may be altered by the nature of the organic solvent. Although numerous attempts have been made to predict the selectivity enhancement mediated by medium-engineering, no rationale of general validity has emerged from these studies. Since it is not yet possible to predict enzyme conformational changes in different organic media, it is not surprising that there is a lack of correlation between the physicochemical properties of the solvent and the E values, an index of enzyme stereoselectivity.

Many empirical studies have shown that E values of biocatalytic kinetic resolution systems are improved markedly by organic solvents (46, 47). As many of the studies were conducted using impure preparations contaminated with enzymes of varying degrees of stereoselectivity, and/or with opposite stereo-chemical preferences, the interpretation of such results should be made with considerable caution. It is likely that the observed enhancement in stereoselectivity is not solely due to an alteration of the intrinsic enantioselective properties of the pure enzyme, but rather it is the cumulation of changes in the k_{cat}/K_m values of all the enzymes competing in the organic reaction medium.

One remarkable example is the report by Hirose and coworkers (48) on the enantiotropically-selective hydrolysis of prochiral dihydropyridine dicarboxy-lates catalyzed by *Pseudomonas* sp. lipase (AH). In water-saturated cyclo-hexane, the *R*-isomers were preferentially formed with *ee* values ranging from 88 to 91%, whereas in water-saturated di-isopropyl ether, the *S*-isomers were formed with *ee* values as high as 99%. On the other hand, with lipase of *Pseudomonas cepacia*, no inversion of stereochemical preference was observed, and only the *R* monoesters of high optical purities were obtained. These results strongly suggest that the lipase preparation (AH) derived from *Pseudomonas* sp. consisted of enzymes of opposite stereochemical preference, with the pro-*R* selective enzyme dominating in cyclohexane and the pro-*S* selective enzyme dominating in di-isopropyl ether.

In comparing the stereoselectivity of hydrolysis in water and transesterification in organic solvent, it is difficult to make accurate predictions. In some cases, poor enantioselectivity is observed in water whereas the reverse reaction in organic solvent may yield high enantioselectivity, or vice versa. The outcome would depend on conformational changes of the enzyme in different media. Klibanov and coworkers (49) found that the enantioselectivity of aspergillo-peptidase B changed with the nature of the solvent. Further, they showed that the hydrophobicity of the solvent correlated with the preference for the D or L enantiomer. They suggested that there are two distinct modes in which the substrate can bind to the enzyme and still be correctly poised for catalysis: that is, one to bind the L-enantiomer in water and a second to bind the D-enantiomer in an organic solvent.

Several groups have shown that when an enzyme is lyophilized from a solution containing a molecule with a particular stereochemistry and is then suspended in organic solvent, the activity of the enzyme is much higher towards the substrate with the same stereochemistry (50, 51). The explanation for this behavior is that the enzyme assumes a particular conformation when it binds to the substrate. This conformation is frozen in place when the enzyme is moved into an organic solvent, since mobility is lower in organic solvents than in water.

To examine the effects of organic solvents on enzyme catalysis, Farber and coworkers (52) selected γ-chymotrypsin, which is catalytically active in hexane. Also, other workers have shown that the specificities at P1 and P1′ sites change (53, 54) when the enzyme is suspended in different organic media, and that chymotrypsin can be imprinted by using D-amino acid as a substrate in organic solvents. It was shown that if chymotrypsin was lyophilized in the presence of *N*-acetyl-D-tryptophan and then resuspended in an organic solvent, it was able to accept D-amino acids as substrates in the synthetic direction. Enzyme which was lyophilized either with no ligand or with N-acetyl-L-tryptophan showed greatly reduced activity toward D-amino acid substrates. However, the crystal structures of γ-chymotrypsin in the presence of N-acetyl-D-tryptophan in hexane and water showed only minor differences. Comparison of these structures suggested that the altered enantioselectivity of enzymes in organic solvents was due to changes in binding affinity for N-acetyl-D-tryptophan rather than the presence of a second binding pocket in the enzyme. Chymotrypsin bound both D- and L-amino acids in the same pocket in either water or hexane, but the binding affinity changed between the two solvents. The authors concluded that N-acetyl-D-tryptophan bound more tightly to the active site when the protein was in hexane. The lack of enzyme activity when chymotrypsin was lyophilized with N-acetyl-L-tryptophan could simply be due to tight binding of this inhibitor in the organic solvent. The low enzymatic activity found when chymotrypsin was lyophilized with no ligand suggested that N-acetyl-D-tryptophan acted as a lyoprotectant. The observed changes in V_{max} which were ascribed to imprinting were really due to changes in the active enzyme concentration rather than true changes in k_{cat}.

5.9 General behavior of enzymes in organic solvents

In general, while the thermostability of biocatalysts is raised when it is transferred from an aqueous to an organic medium, the catalytic activities of enzymes are often decreased by several orders of magnitude. The Dutch group (55) first introduced a useful and convenient measure of the compatibility of organic solvents with enzymatic activity by using the logarithm of their partition ratio in a biphasic water–octanol mixture, commonly designated as the log P value. Generally, lipophilic solvents with log P values greater than 4 support a high degree of activity retention whereas more hydrophilic solvents with log P values below 2 are generally unsuitable for biocatalysis. These findings are consistent with the view that hydrophilic solvents strip away the essential or core water from the enzyme which is required for catalysis (56). Suitable solvents may be selected from aromatics such as toluene and benzene, simple alicyclics or aliphatics such as cyclohexane or C5–C8 hydrocarbons, chlorinated lipophilic solvents such as chloroform and dichloromethane or ethers such as diethyl, di-isopropyl and preferably tert-butyl methyl ether.

Enzymes have a large number of side chains whose ionization state is controlled by pH. These groups are important when dried enzyme preparations suspended in low-water organic media are used as catalysts, because the pH of the aqueous solution from which the enzyme is dried affects its subsequent activity. A number of workers have reported what may be counterion effects on the activity of dried enzyme preparations in organic media. Reaction rates were significantly affected by the nature of the buffer species in the solution from which the enzyme was dried.

Recently, it was reported (57) that the transesterification activity of subtilisin Carlsberg in anhydrous hexane is strongly dependent on the KCl content of the lyophilized preparation, and increased sharply when the salt content approached 98%. The k_{cat}/K_m for transesterification of N-acetyl-L-phe-OEt with 1-propanol is over 3750 times higher than that for the salt-free enzyme. Activation of catalysis due to KCl was also observed in a variety of different organic solvents. The improvement of catalytic activity was primarily the result of a dramatic increase in k_{cat} rather than a decreased K_m. These authors proposed that the observed activation upon entrainment of the enzyme in a salt matrix is due to a protective effect afforded by the matrix against deactivation by direct contact with the organic solvent.

On the other hand, a recent study (58) using immobilized enzyme showed no significant effect by salt ions. This was the case even when salt ions were in considerable excess of that needed to balance protein charges. They proposed that the activity variations noted with freeze-dried powders were probably due to changes in the microscopic structure rather than to molecular-scale interactions.

Many reviews (59) discussing the advantages and disadvantages of replacing water with organic media and examples comparing the relative rates of reactions in aqueous and non-aqueous systems are now available. In general, the water activity is now considered the most concise way to characterize the effect of low-water systems. It should be emphasized that even in neat organic solvents, enzymes suspended in the form of lyophilized powders carry substantial amounts of adhered water. The amount of water required for catalytic activity varies with different enzymes.

5.10 Resolution of alcohols in organic solvents

An inherent problem encountered in the kinetic resolution of alcohols in organic solvents by lipase-catalyzed transesterification (ester formation) is the reverse reaction of the achiral alcohol component of the acyl transfer reagent with the optically active ester product. The reverse reaction leads to a decrease in optical purity. To circumvent this problem irreversible acyl transfer reagents have been developed. Enol acetates like vinyl acetate and isopropenyl acetate for acetylation of chiral alcohols are now the reagents of choice for carrying out lipase acylation of alcohols in organic solvents. The same goal has also been realized by employing oxime esters and acid anhydrides. Excellent reviews on the use of these reagents with variety of chiral alcohols have appeared in the literature (42). This section is not meant to provide a completely comprehensive review on the kinetic resolution of alcohols in organic solvents. Emphasis will be placed on selected substrates which are of importance in organic synthesis and are difficult to secure by chemical methods. For the most part the subject matter will be limited to the literature since 1993, because most of the earlier material has already appeared in review articles (43, 44, 59–61).

5.10.1 Resolution of diols and polyols

5.10.1.1 1,2-Diols. The lipase-catalyzed acylation of terminal 1,2-diols is highly regioselective (62, 63) but shows low enantioselectivity in the mono-acylation step (64). A method for the kinetic resolution of 3-(phenoxy)-1,2-propanediols and 3-benzyloxy-1,2-propanediols without additional protection–deprotection, using a lipase catalyzed sequential transesterification with lipase Amano PS and vinyl acetate, has been reported by Theil (65) and Herradon (66). In the first step of this one-pot procedure the racemic 1,2-diols

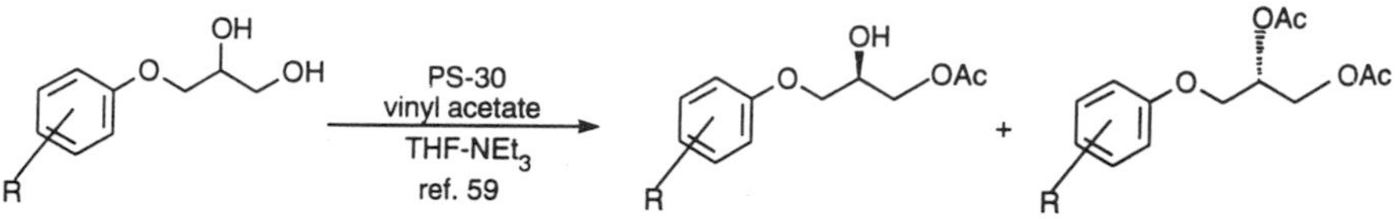

R	time (h)	% Yield (% ee)	% Yield (% ee)
H	92	48 (85)	49 (79)
4-Me	57	57 (66)	42 (93)
2-Me	96	45 (93)	45 (80)
4-OMe	28	48 (96)	52 (94)
2-OMe	44	58 (63)	42 (87)
4-Cl	25	48 (94)	49 (92)
2-Cl	24	59 (55)	38 (88)
4-*t*-Bu	50	50 (99)	50 (93)
2-*t*-Bu	78	82 (3)	17 (34)

were acylated regioselectively at the primary hydroxy group without enantio-selection. The subsequent acylation at the secondary hydroxy group of the primary monoacetate formed was responsible for high enantioselection. In general, the derivatives with substituents in the 4-position of the aromatic ring showed significantly higher enantioselectivities than the corresponding derivatives with substituents in the 2-position. A substituent in the 4-position, independent of its electronic properties, appeared to be a prerequisite for high enantioselectivity of the resolution procedure, because the unsubstituted derivative (R = H) showed an enantioselectivity comparable to the 2-substituted compound. The unsubstituted benzyloxy compound exhibited similar enantio-selectivity towards Pseudomonas fluorescens lipase, (PFL) and vinyl acetate (66). The replacement of the aryloxy by an alkyl substituent caused a dramatic decrease of enantioselectivity. In contrast to 1-phenyl-1,2-ethanediol, which was converted with moderate enantioselectivity into the monoacetate and the

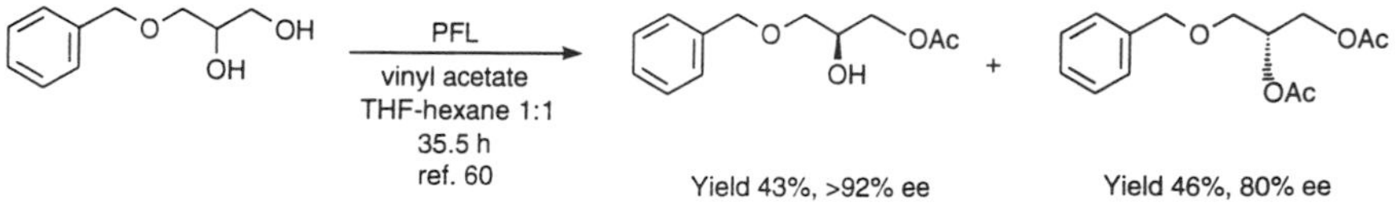

Yield 43%, >92% ee   Yield 46%, 80% ee

diacetate, PS-catalyzed esterification of the aliphatic substrates gave very low optical purities of secondary monoester, primary monoester, diester and unchanged diol (59).

R	% Yield (% ee)	% Yield (% ee)
Et	45 (61), S	38 (49), R
n-Pr	40 (19), S	52 (11), R
C(OH)Me$_2$	63 (17), S	34 (14), S
Ph	60 (66), R	40 (93), S

Several racemic 1,2-diols bearing a tertiary benzylic alcohol stereogenic center can be esterified regioselectively by vinyl acetate and lipases in organic media (67, 68). With the correct choice of lipase, and if the non-phenyl substituent at the tertiary center is unsaturated, E values of > 250 were obtained. This work illustrates that: (1) regio- and enantioselective recognition could be achieved at a reaction site removed from the stereogenic center in a substrate; (2) that chiral recognition can be determined only by the subtle effect of a double bond relatively far removed from the reaction site — polar or sterically demanding substituents were usually used to obtain such effects.

R^1	R^2	% c	% ee	% ee
Ph	Et	13	1	7
Ph	n-Pr	8	0	0
Ph	CH$_2$CHCH$_2$	49	76	78
Ph	MeCHCHCH$_2$	59	94	65
Ph	CH$_2$Ph	52	>99	93
Ph	H	50	>97	>97

PS-30, vinyl acetate in chloroform, ref. 7

Ph	Me	40–60	<10	<10
PhCH$_2$	Me	40–60	92–98	92–98
Me$_2$CCHCH$_2$	Me	40–60	92–98	92–98

5.10.1.2 Acyclic diols and their derivatives. Many 2-substituted-1,3-diols have been resolved in organic solvents via lipase catalyzed irreversible acylations (69–72). 2-Methyl-1,3-propanediol was resolved with 60% enantioselectivity using PFL and vinyl acetate as the solvent. The optically pure (S)-monoacetate was prepared in > 98% *ee* from the racemic diol if the diol was completely

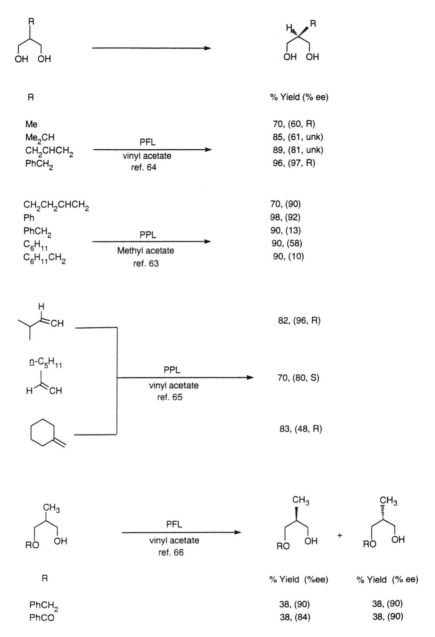

reacted to yield 60% of diacetate and 40% of monoacetate. The results with monoacylation were much more selective if the monoether and monobenzoate were used as the substrate with PFL, and vinyl acetate in chloroform. A kinetic resolution of 1,3-butanediol was accomplished using a continuous flow system by sequential esterification using immobilized lipase SP382 (*Candida* sp.) in butyl acetate to produce (R)-1,3-diacetoxybutane of 85.8% *ee* (73). When the esterification of the diol was carried out with SP382 using acetic and butyric anhydrides, the reaction to the desired (R)-1,3-diacetoxybutane proceeded rapidly, but the enzyme was inactivated.

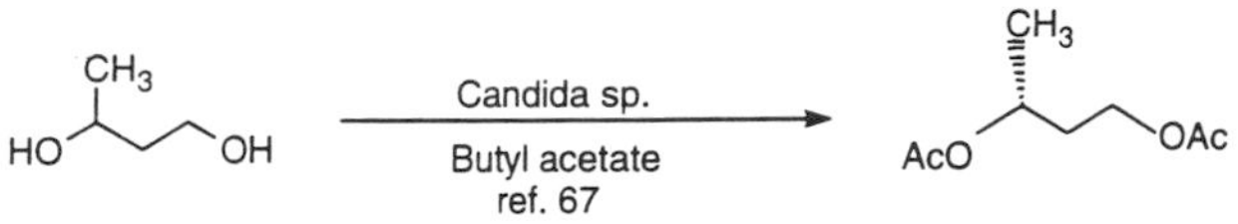

The irreversible transesterification of 2-methyl-1,4-butanediol and its derivatives was studied with PFL in chloroform and vinyl acetate (74, 75). With 2-methyl-1,4-butanediol at 66% conversion (4 h), the (R)-(+)-diol was isolated with 70% *ee*. At longer reaction time (24 h), the diol completely reacted. The resulting mixture of monoacetates was reduced with lithium aluminium hydride (LAH) and the (R)-(+)-diol was isolated with 20% *ee*. The resolution was much more successful using the 4-benzyl ether derivative. The PFL-catalyzed transacetylation in chloroform afforded the (R)-alcohol and the (S)-acetate (38–40% yield), both with a high *ee* (98% and 85%, respectively). The corresponding 1-benzyl ether regioisomer did not work as well indicating, that when

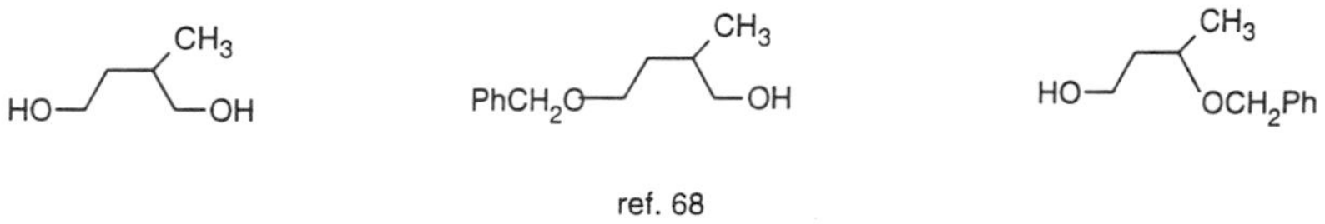

ref. 68

the stereogenic center was further removed from the reacting group, the stereochemical discrimination of the two enantiomers was less efficient. The regioselectivity and enantioselectivity of 2-methyl-1,4-butanediol were compared to the 2-methylene- and 2-epoxy-1,4-butanediols (75). Transesterification of

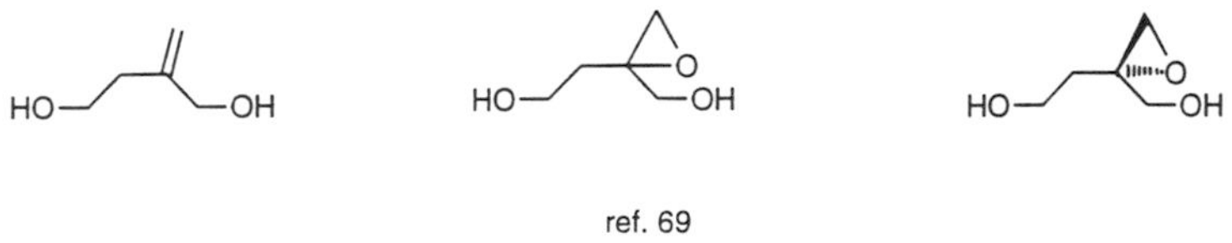

ref. 69

2-methyl-1,4-butanediol with PS-30 and vinyl acetate was monitored at different time intervals, and the ratio of 70:30 (in preference for acylation at the C-1 position) found at the beginning was maintained through the entire reaction. With the 2-methylene compound the ratio was 90:10 and with the epoxide 97:3. At 60% conversion the remaining (S)-epoxy diol was found to be 86% *ee* (determined after reduction with LAH and conversion of the triol to the known (S)-(–)-acetonide (69). 2,5-Hexanediol was found to be a good substrate for lipase catalyzed esterification (76, 77). An advantage of this method was the monoacetate derived from the meso diol in the starting material was convertible to the diacetate via inversion of configuration of the hydroxyl group by the Mitsunobu procedure.

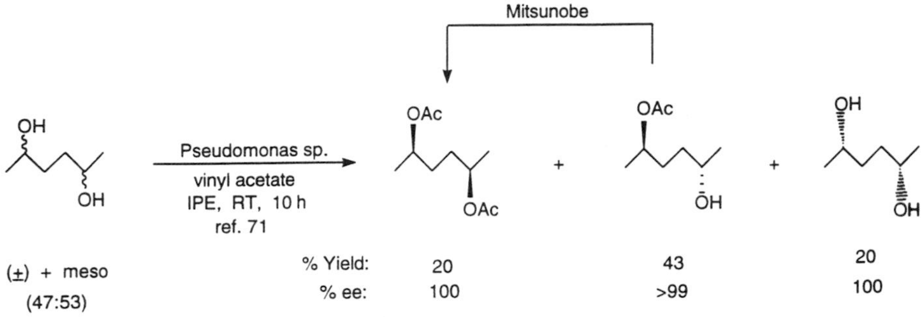

% Yield:	20	43	20
% ee:	100	>99	100

5.10.2 Lipase resolution of allylic and propargyl alcohols

(E)-4-Hydroxy-hex-2-enal displayed extremely low activity and very poor selectivity in lipase-catalyzed acylations in organic solvents. Better results were obtained in the preparation of the dimethyl acetal derivative (78). The same result was observed in the resolution of 4-hydroxyalk-2-ynals where conversion of the aldehyde group to the ethyl acetal derivative gave better results (79). It was noteworthy that there was little difference in reactivity and enantioselectivity between the allylic and propargyllic compounds.

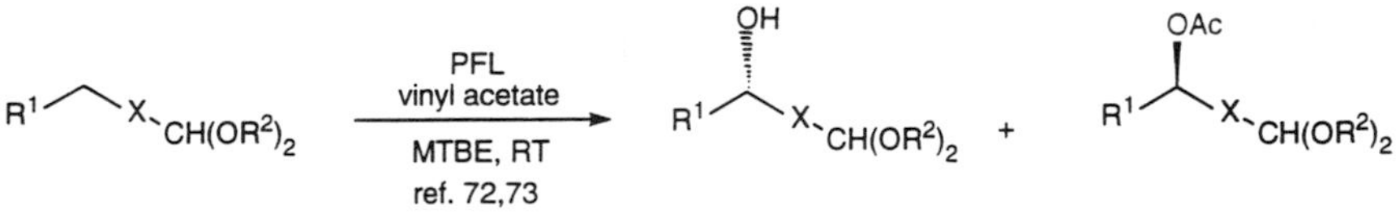

R^1	R^2	X	time, % c	% Yield, (% ee)	% Yield, (% ee)
C_2H_5	Me	CH=CH	1 d, 51	44, (>95)	48, (90)
C_5H_{11}	Me	CH=CH	5 d, 52	43, (>95)	46, (89)
C_7H_{15}	Me	CH=Ch	5 d, 52	43, (>95)	46, (88)
Me	OEt	C≡C	1 d, 50	44, (>95)	38, (>95)
C_5H_{11}	OEt	C≡C	3 d, 48	46, (88)	37, (95)
C_7H_{15}	OEt	C≡C	4 d, 50	43, (95)	37, (95)
$C_3H_7CH=CH$	OEt	C≡C	3 d, 46	49, (82)	35, (>95)
Ph	OEt	C≡C	7 d, 48	46, (88)	37, (95)
$(Me)_2CH$	OEt	C≡C	7 d, 46	47, (74)	36, (87)
$(Me)_2CHCH_2$	OEt	C≡C	9 d, 20	70, (22)	15, (90)

Boaz and Zimmerman (80) reported on the lipase-catalyzed esterification of 3-butene-1,2-diol monotosylate. These workers found that inclusion of triethylamine afforded rapid conversion to 50%, eliminated multiple enzyme feeds, and enhanced enantioselectivity. The amine neutralized the inhibitory effect of *p*-toluenesulfonic acid that was produced during storage and/or generated during the lipase reaction. This effect was not limited to the allylic alcohol. Several other 1,2-diol mono-tosylates were prepared and submitted to

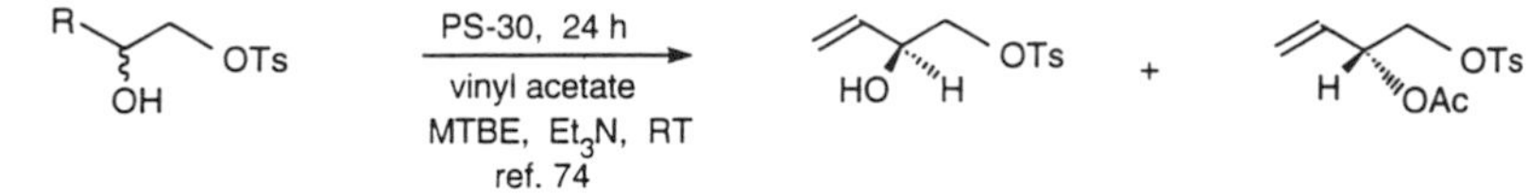

R = CH$_2$=CH, CH$_3$, CH$_2$Cl,
C$_2$H$_5$

98% ee   96% ee

the enzymatic esterification protocol with and without triethylamine. In all cases the amine accelerated the reaction and to different degrees, improved the enantioselectivity of the esterification reaction. A lipase-catalyzed resolution of *N,N*-dialyl-3-hydroxy-4-pentenamides was found to be successful, whereas the Katsuki–Sharpless kinetic resolution with the same substrates resulted only in recovery of the starting materials (81, 82).

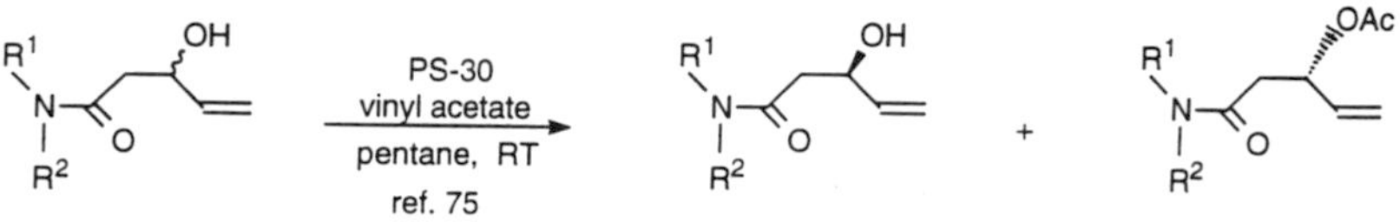

	time	% Yield, (% ee)	% Yield, (% ee)
R^1 = R^2 = PhCH$_2$	22 h	44, (99)	49, (98)
R^1 = Me, R^2 = PhCH$_2$	15 h	42, (>99)	45, (>99)

Lipase catalyzed acylation of (*E*)-methyl 4-hydroxyhept-2-eneoate with IPA (isopropyl acetate) proceeded at a convenient rate and the (*R*)-enantiomer of this substrate reacted significantly faster than its optical antipode (83). The enantio-selectivity of simple homologues (R = Me and Et) was similar. However, when the substituent R contained a branched chain (R = iso-Pr, iso-PrCH$_2$, cyclo-hexyl-CH$_2$), the reaction slowed down (6–7 days), the enantioselection was poor, and the *stereochemical preference of the enzyme was reversed.*

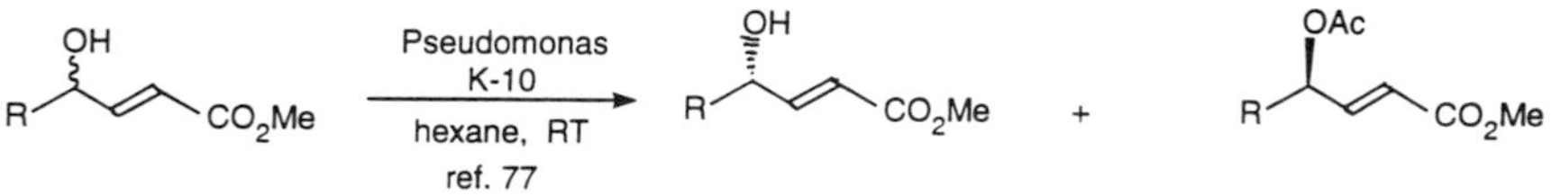

R	time, h	% Yield, (% ee)	% Yield, (% ee)
Me	49	37, (>95)	39, (91)
Et	50	44, (>95)	45, (>95)
n-Pr	400	42, (>95)	54, (74)

Enantioselective acylation of γ-hydroxy-α,β-unsaturated phenylsulfones can be resolved via lipase mediated acylation in organic solvent (84, 85). The enantioselectivity of the lipase-catalyzed acetylation of the allylic alcohols was almost independent of the length and size of the R group. The reaction time was 5–16 h for 50% conversion except when R = iso-Pr, (62 h) or $PhSO_2CH_2CH_2$ (162 h). The % ee for the remaining alcohol and monoacetate was very high, in the order of > 95–98%. The reaction failed for R = t-butyl (less than 5% conversion) and for the cis double-bond isomer (R = CH_3).

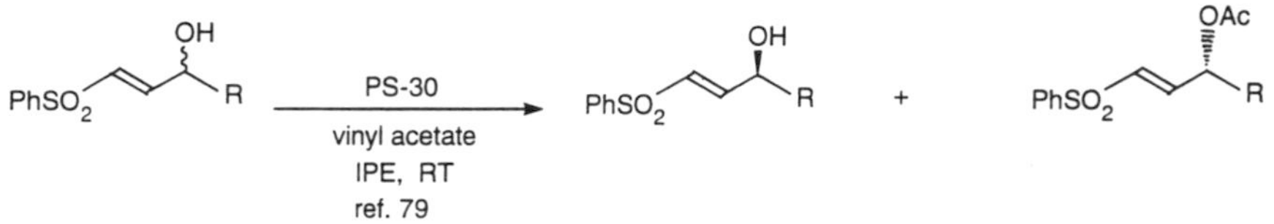

R	time h, % c	% Yield, (% ee)	% Yield, (% ee)
Me	5.5, 50	46, (>98)	47, (>95)
Et	3.5, 50	49, (>98)	49, (>95)
n-C_6H_{11}	10, 50	48, (>98)	47, (>95)
i-Pr	62, 50	46, (94)	48, (>95)
t-butyl	300, < 5		
$PhCH_2OCH_2CH_2$	16, 50	48, (88)	47, (>95)
$PhSO_2CH_2CH_2$	162, 50	45, (>98)	48, (>95)

(E)-Vinylsilanes have been resolved by Sparks and Panek (86). There was little difference in enantioselectivity with the various silyl groups examined. The reactivity of the dimethylphenylsilyl derivative in lipase acylation was much faster.

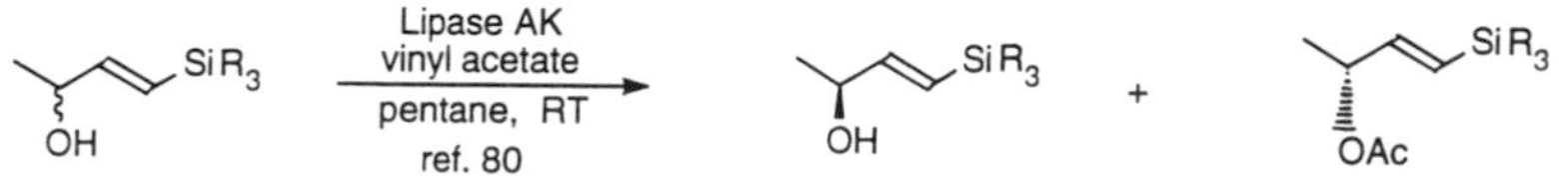

R	time h, % c	% Yield, (% ee)	% Yield, (% ee)
Me_3	24, 50	34.2, (>95)	34.5, (80)
t-Bu(Me)$_2$	13, 52	36, (>95)	39, (64)
Me_2Ph	2, 52	44, (>95)	47, (80)

Optically active 4-benzyloxy- and 4-(4-methoxyphenoxy)-3-hydroxy-1-butyne were resolved starting with racemic 4-benzyloxy-3-hydroxy-trimethylsily-1-butyne and 4-methoxyphenoxy-3-hydroxy-1-trimethylsilyl-1-butyne (87). It was interesting to note that, when the same lipase-catalyzed conditions were subjected to the terminal desilylated acetylenic substrates, the maximal optical yields of the products were decreased to impractical levels, although the chemical yield was satisfactory.

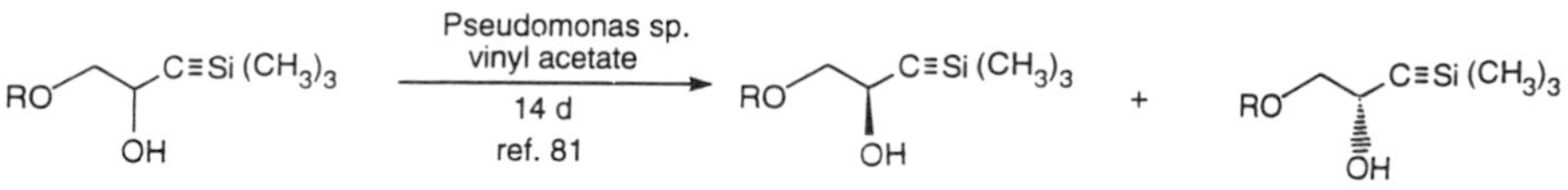

R	% Yield, (% ee)	% Yield, (% ee)
PhCH$_2$	60, (63)	39, (96)
p-MeOC$_6$H$_4$	61, (60)	36, (>99)

3-Hexyne-2,5-diol was subjected to lipase AK catalyzed acylation with vinyl acetate to afford unreacted diol (> 98% *ee*, 95% *de* [diastereomeric excess]), monoacetate (> 98% *ee*, 88% *de*), and diacetate (> 98% *ee*, 86% *de*) (88). The monoacetate and diacetate were subjected to a second round of AK-catalyzed reaction to amplify the diastereomeric excess. Finally, the hydrolysis of diacetate

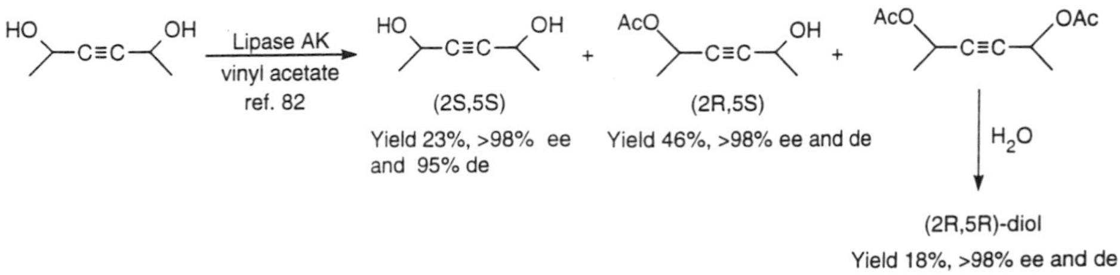

to (2*R*,5*R*)-diol completed the resolution process. It is noteworthy that the high enantiomeric and diastereomeric purities for the (2*S*,5*S*)-diol were achieved via the first round of enzyme-catalyzed reaction.

5.10.3 Lipase-mediated transesterification versus deacylation for resolution of racemic alcohols

The resolution of racemic alcohols can be effected by enantioselective acylation of the alcohol or deacylation (hydrolysis or alcoholysis) of the corresponding esters. The method of choice depends on many factors such as yield, purity, reaction rate, stability of starting materials and products and product isolation. In most cases, the two modes of reaction are enantiocomplementary. One major operational advantage of acylation is the avoidance of an additional step to prepare the ester derivative. In this section a few examples will be cited where acylation has been found to be advantageous.

1-Phenylethane-1,2-diol was studied by lipase hydrolysis and alcoholysis of the corresponding diacetate and the results compared with acylation in organic solvent (89). The rate of hydrolysis in PS-30 of the primary ester moiety was 2.6 times faster than that of the secondary ester. Both monoesters underwent further hydrolysis which resulted in low regioselectivity. The *S*-isomer of the racemic dipropionate ester was preferentially hydrolyzed, but with low optical purity.

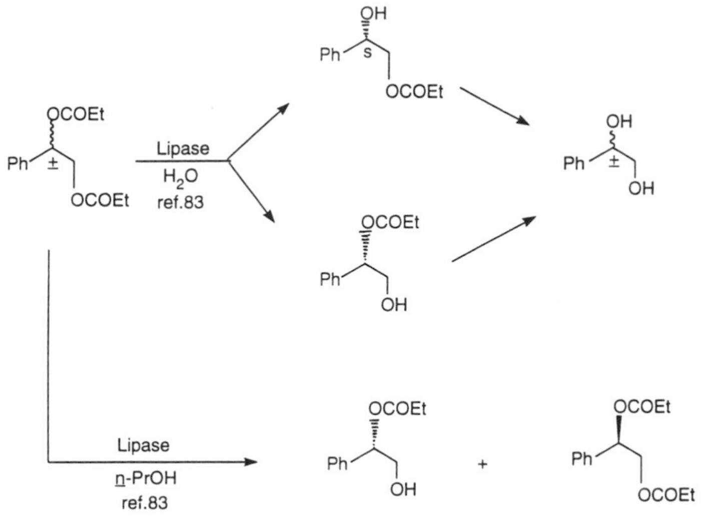

Alcoholysis with *n*-propanol (8 equivalents) and PS-30 supported on Celite in 2-methylbutan-2-ol as solvent terminated at 40% conversion and gave only the (*R*)-dipropionate ester and the secondary (*S*)-monopropionate ester. The optical purities were found to be 36% *ee* and 78% *ee*, respectively. Sequential acylation of 1-phenylethane-1,2-diol gave in high yields the (*S*)-dipropionate ester (93% *ee*) and primary (*R*)-propionate ester (77% *ee*) in high yields.

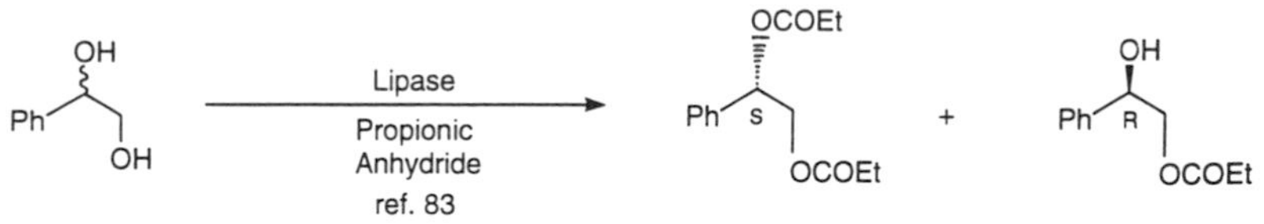

The kinetic resolution of 2-alkyl substituted carboxylic acids with lipases has not been successful. This was also true for the lipase-catalyzed hydrolysis of 2-methyl carboxylates (90, 91) as well as for the enantioselective esterification of racemic carboxylic acids (92) or the enantioselective alcoholysis of racemic 2-methyl alkanoates with alcohols (93). Hydrolysis of various esters of 2-ethyl-1-hexanol gave slow conversion rates and low optical yields (0–38% *ee*) (94). The lipase-catalyzed resolution of the racemic 2-alkyl substituted primary alcohols with vinyl acetate in methylene dichloride yielded the (*R*)-alcohols with optical purities between 96 and 99% *ee* by driving the extent of conversion to 60–80% (94). Similar results were obtained with the resolution of racemic 4-phenylthio- (95) and 4-phenylsulfonyl-2-methyl-1-butanol (95). Hydrolysis of their corresponding acetate and butanoate ester derivatives gave at best 20–30% *ee* of the product. Transesterification of these compounds with vinyl

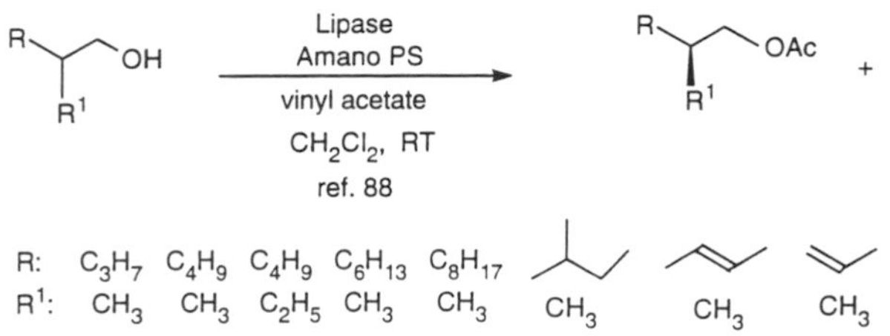

R: C_3H_7 C_4H_9 C_4H_9 C_6H_{13} C_8H_{17}
R^1: CH_3 CH_3 C_2H_5 CH_3 CH_3

acetate in chloroform in the presence of PFL proceeded with high enantio-selectivity in excellent yields. At 60% conversion quantitative yields of the alcohols were obtained; the *ee* of the alcohols was 98%.

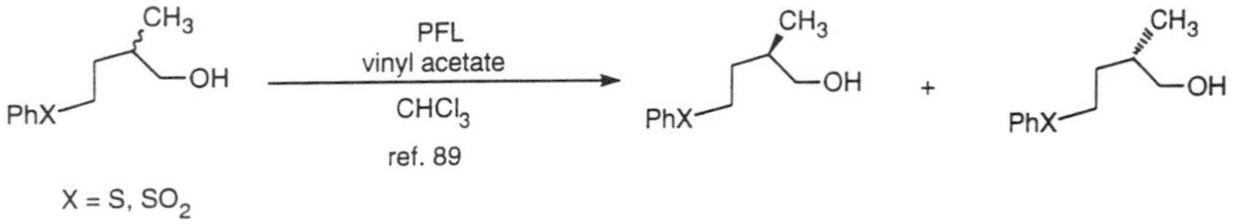

X = S, SO$_2$

The PPL lipase-catalyzed acylation of α-hydroxystannanes with 2,2,2-trifluoroethyl valerate afforded good yields and excellent enantiomeric purities (> 95% *ee*) of the (*S*)-valerate and (*R*)-α-hydroxystannane (96). Hydrolysis with CCL provided the product alcohol with respectable enantiomeric purity (90% *ee*); however, the isolated yields were < 10%, the conversions were low, and the reactions were very slow.

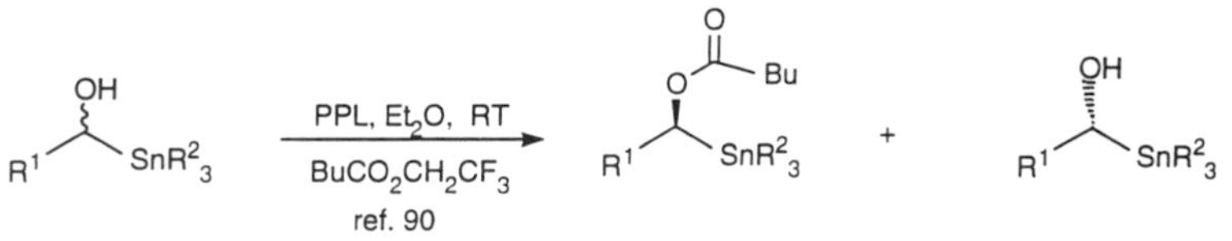

R^1	R^2	time (h), % c	% Yield, (% ee)	% Yield, (% ee)
Me	Me	48, 42.3	31, (97)	27, (71)
Et	Me	64, 36.1	36, (99)	36, (56)
Pr	Me	129, 6.7	7, (97)	68, (7)
Me	Et	84, 34.0	35, (99)	47, (51)
Et	Et	85, 12.6	14, (97)	57, (14)

The lipase-catalyzed resolution of 4-chloro-1-phenyl-1-butanol derivatives displayed greater enantioselectivity in organic solvent than in water (97, 98). At 50% conversion, the optical purities of the esters were 79% (R = H), > 95% (R = F) and > 95% (R = *t*-butyl), and > 95% *ee* in all three cases for the alcohol remaining. In the hydrolysis mode the optical purities were 80%, 78% and > 95% for the esters and 89%, 93% and > 95% for the remaining (*R*)-alcohol.

89.2% ee 80.1% ee

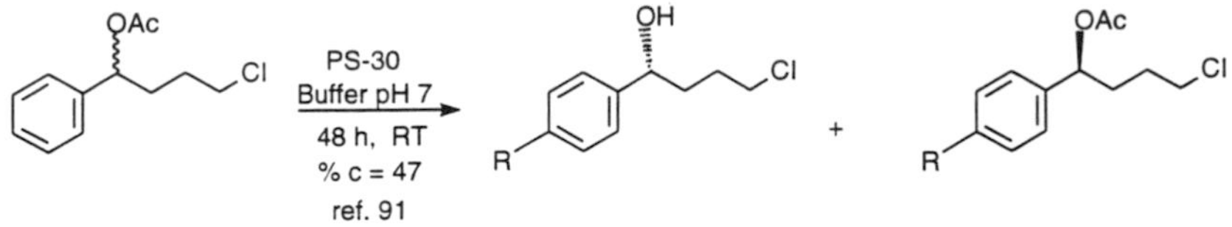

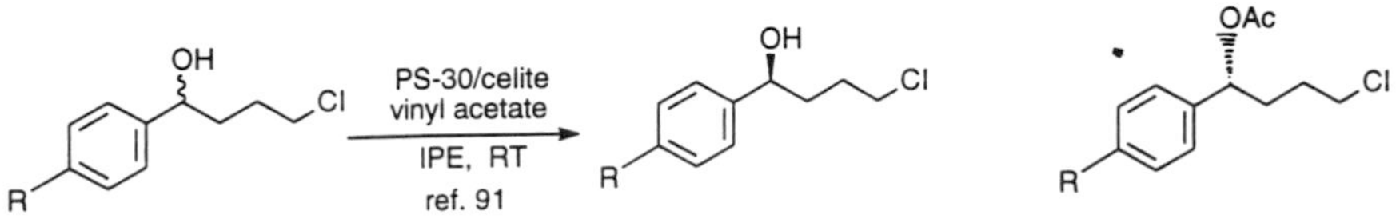

R	% c	% ee	% ee
H	55.8	>95	79.0
F	50.0	>95	>95
t-Butyl	50.0	>95	>95

The optically active tetrahydrofuranyl *cis*-monoacetate derivative was best obtained by acylation of the *cis*-diol with phenyl acetate in cyclohexane in the presence of lipase YS (99). The reaction proceeded more rapidly (1.5 h versus 96 h), better yield (53% versus 42%), and higher enantiomeric purity (71% *ee* versus 29% *ee*) than the hydrolysis of the *cis*-diacetate. Similarly, acylation of (±)-*trans*-diol in cyclohexane-THF with vinyl acetate afforded the optically active *trans*-diol, which could not be easily isolated from lipase hydrolysis of the (±)-*trans*-diacetate in buffer solution.

cis-diol

Yield 53%, 71% ee

(±)-trans

Yield 28%, 37% ee Yield 50%, 25% ee Yield 7%, 58% ee

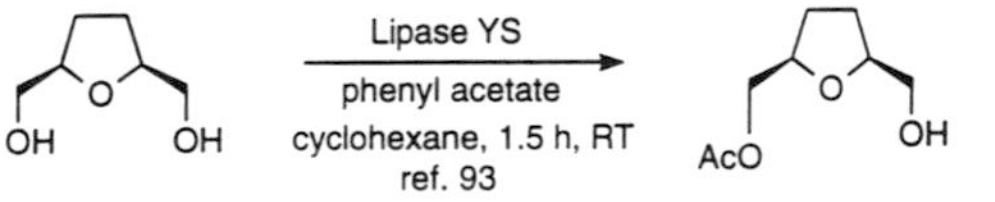

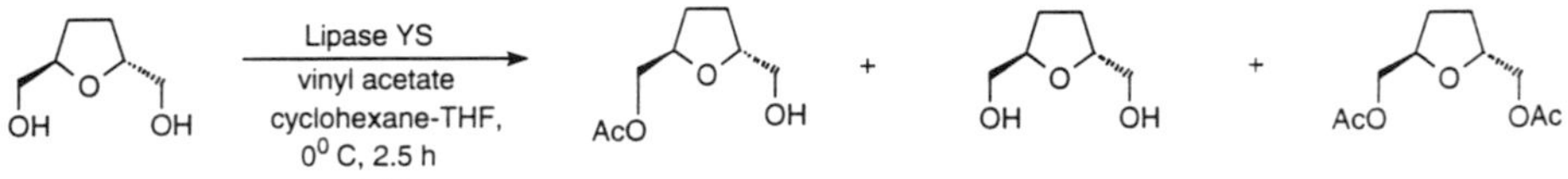

5.11 Resolution of acids and esters in organic solvents

The two primary methods for the resolution of racemic acids and esters in a non-aqueous environment are lipase-catalyzed alcoholysis and lipase-catalyzed esterification. Due to the wide scope of the subject matter and space constraints of the text, discussion will be confined to selected reports that have appeared in the literature since 1992.

(1) $R^1-\overset{\overset{\displaystyle R^2}{|}}{\underset{\pm}{CH}}- CO_2H$ + R^3-OH ⇌ $R^1-\overset{\overset{\displaystyle R^2}{|}}{\underset{+}{CH}}- CO_2R^3$ + $R^1-\overset{\overset{\displaystyle R^2}{|}}{\underset{-}{CH}}- CO_2H$ + H_2O

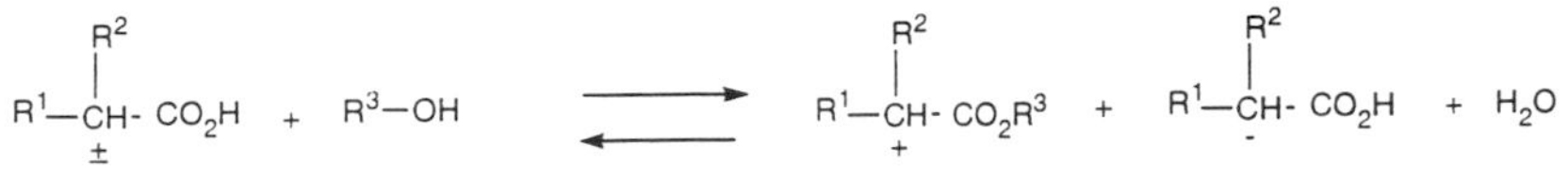

(2) $R^1-\overset{\overset{\displaystyle R^2}{|}}{\underset{\pm}{CH}}- CO_2R^3$ + R^4-OH ⇌ $R^1-\overset{\overset{\displaystyle R^2}{|}}{\underset{+}{CH}}- CO_2R^4$ + $R^1-\overset{\overset{\displaystyle R^2}{|}}{\underset{-}{CH}}- CO_2R^3$ + R^3-OH

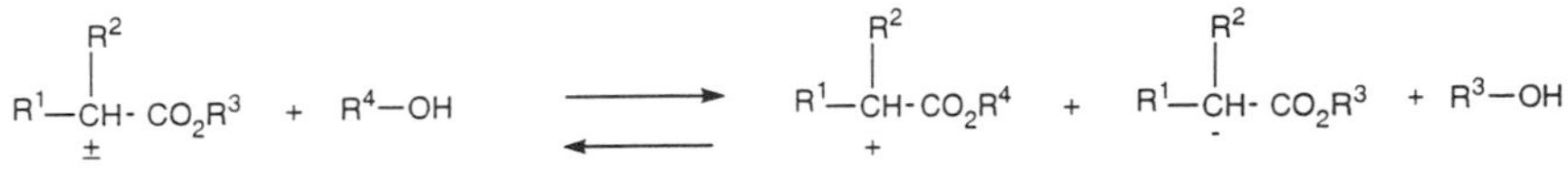

The resolution of α-substituted carboxylic acids based on an acyl transfer reaction between mixed carboxylic–carbonic anhydrides and alcohols was recently reported by the French workers (100). The mixed anhydrides were readily prepared in high yields by condensation of carboxylic acids with iso-propyl chlorocarbonate in the presence of N-methylmorpholine. The observed enantioselectivity at 50% conversion using the mixed anhydride method was modest, but the reaction rates were significantly higher, compared to other enzyme-catalyzed resolutions of acids or their corresponding esters.

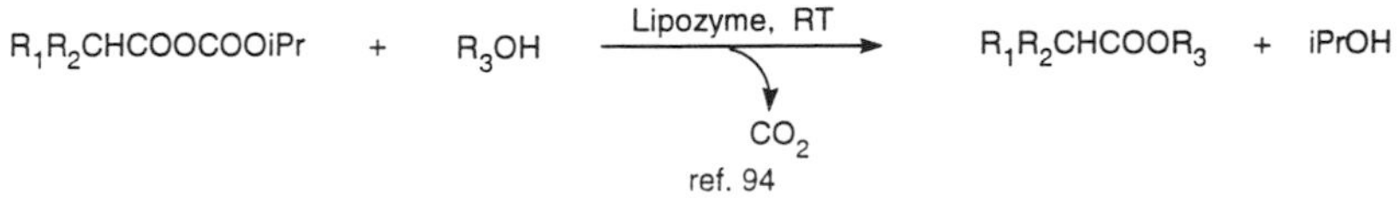

$$R_1R_2CHCOOCOOiPr \quad + \quad R_3OH \xrightarrow[\substack{\searrow \\ CO_2 \\ ref.\ 94}]{Lipozyme,\ RT} R_1R_2CHCOOR_3 \quad + \quad iPrOH$$

R_3 = n-propyl, n-butyl, n-C_8H_{17}

Ester	Alcohol R3	Solvent	time (min)	% c	% ee	E
1	n-propyl	ether	120	49	66	11
1	n-propyl	MTBE	60	55	90	20
1	n-octyl	MTBE	40	50	79	20
2	n-octyl	MTBE	60	47	74	24
3	n-butyl	MTBE	15	28	33	17
4	n-propyl	ether	60	29	67	7
4	n-octyl	ether	20	47	80	19

1; R_1 = phenyl, R_2 = methyl. 2; R_1 = p-butylphenyl, R_2 = methyl. 3; R_1 = phenyl, R_2 = methoxy. 4; R_1 = propyl, R_2 = methyl.

Optically pure 2-alkyl-alkanoic acids are useful building blocks for the synthesis of biologically active compounds with branched structures. Kinetic resolution of racemic 2-substituted alkanoic acids bearing electron-withdrawing substituents at C-2 have been successfully achieved using lipase-catalyzed esterification (11, 101–104). However, the success of obtaining enantio-merically pure 2-methylalkanoic acids using lipase-catalyzed esterification has been, at best, modest. Engel demonstrated that lipase from *Candida rugosa* preferentially catalyzed the esterification of (S)-configurated 2-methylalkanoic

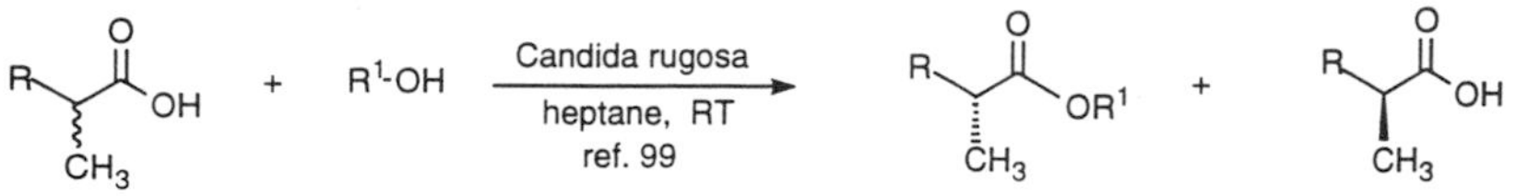

Entry	R	R^1-OH	time (h), % c	% ee	% ee	E
1	C$_2$H$_5$	methanol	4, 48	34.2	37.3	3
2	C$_2$H$_5$	ethanol	5, 50	34.4	37.4	3
3	C$_2$H$_5$	cyclohexanol	8, 54	45.4	45.4	3
4	C$_2$H$_5$	octanol	7, 53	48.9	43.4	4
5	C$_3$H$_7$	octanol	10, 46	79.5	93.3	70
6	C$_4$H$_9$	ethanol	24, 32	6.3	13.2	1.4
7	C$_4$H$_9$	octanol	22, 49	76.2	79.6	20
8	C$_4$H$_9$	octadecanol	18, 50	84.0	84.0	30

acids (105). The stereochemical preference of the esterification of 2-methyl-alkanoic acids was opposite to that observed for analogous acids with a C-2 halogen substituent (10). The enantioselectivity (E) of the reaction was markedly influenced by the structure of the substrate. The effect of the alcohol chain length on the discrimination of 2-methylhexanoic acid enantiomers (entries 6–8) was similar to results obtained for the *Candida rugosa* catalyzed esterifi-cation of 2-(4-chlorophenoxy)propanoic acid (106). On the other hand, the poor enantioselectivity observed with reactions of 2-methylbutanoic acid (entries 1–4) was not significantly improved by alterations of the alcohol structure. A methyl substituent at the C-2 position of acid substrates caused a sharp decrease of the esterification rate compared to the unbranched acid, e.g. octanol as the alcohol with hexanoic acid (100, relative reaction rate, rrr), 2-methylbutanoic acid (50, rrr), 2-methylpentanoic acid (30, rrr), and 2-methylhexanoic acid (14, rrr).

Since the enzymatic resolution of 2-phenoxypropanoic acids using lipases generally resulted in low enantioselectivities (107), a lipase-catalyzed irreversible transesterification procedure using vinyl esters was applied for the resolution of racemic 2-phenoxypropanoic acids (108). The enantioselectivity was found to be affected by both the alcohol as a nucleophile and the organic solvent used. The best condition was the use of *Aspergillus niger* lipase in cyclo-hexane solvent with methanol as the nucleophile. In ethers and halogenated methanes, the reaction rate and the enantioselectivity were moderate (% ee

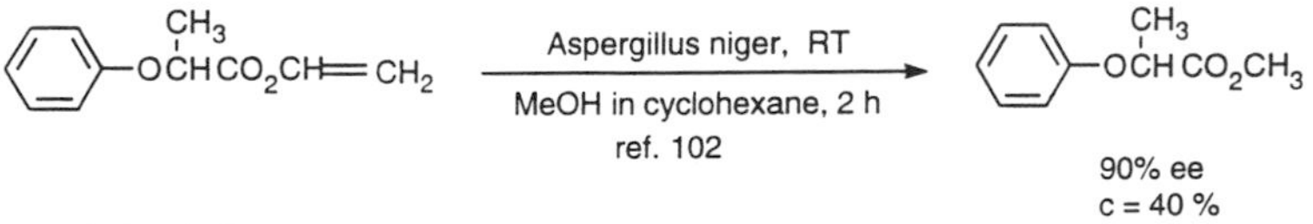

90% ee
c = 40 %

50–79), and in polar solvents such as dimethylformamide (DMF) and dimethyl sulfoxide (DMSO) the reaction became very slow and non-enantioselective (% ee 2.0 and 0.6).

The resolution of (R,S)-ibuprofen enantiomers by esterification in different organic solvents was studied using *Candida cylindracea* lipase (109). The enzyme preparation had high enantioselectivity for (S)-(+)-ibuprofen in the

R-(-)   S-(+)

ref. 103

esterification of racemic ibuprofen using primary alcohols. The synthesis of the esters was profoundly affected by the amount of water in the reaction mixture, and *Candida cylindracea* lipase was active only in very hydrophobic hydrocarbon solvents. The alcohol moiety had great influence on ester formation by *Candida cylindracea* lipase. Primary alcohols, especially the water-immiscible propanol, butanol and amyl alcohol, were good substrates for the esterification of ibuprofen. Aside from decreasing the rate of esterification, the water content of the reaction system also influenced the enantioselectivity of the *Candida cylindracea* lipase. The enantioselectivity decreased with increasing water content. With decrease in water content the conformation of the enzyme became more rigid. The substrate specificity was limited, but stereo/enantioselectivity was enhanced (110, 111). In the esterification of ibuprofen with amyl alcohol and *Candida cylindracea* lipase in hexane solvent, the (S)-(+)-enantiomer of ibuprofen disappeared completely from the reaction mixture (48 h, 30°C) before esterification of the other enantiomer started. The (S)-(+)-ester was formed without any detectable R-isomer when less than 50% of the ibuprofen was esterified. In practice, these results indicated that if the esterification reaction is monitored continuously, it can be stopped before the (R)-(−)-isomer started to react. The lipase from *Candida cylindracea* was found to preferentially esterify the S-isomer of Naproxen, the only member of the profen-based anti-inflammatory drugs which is sold as a single enantiomer. This was achieved by using trimethylsilyl methanol as the acyl acceptor; however, due to the low solubility of Naproxen in iso-octane, the esterification yield was low (111). By optimizing the reaction conditions (reaction temperature 53°C and an alcohol concentration between 20 mM and 40 mM in an 80% (v/v) iso-octane–20% (v/v) toluene mixture), a 72-times enhancement of yield of the desired S-ester was realized (75 h, c. 42.2%, % ee 98.8) (112).

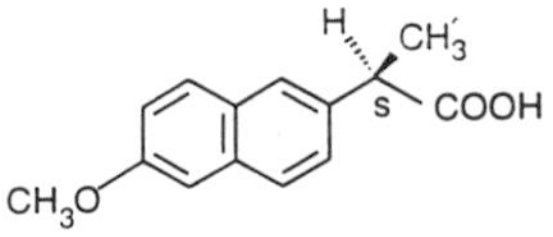

LIVERPOOL
JOHN MOORES UNIVERSITY
AVRIL ROBARTS LRC
TEL. 0151 231 4022

5.12 Lipase-mediated synthesis of nitrogen-containing compounds (amines, amides, amino acids, nitriles)

The ring opening of epoxides with 2-propylamine in the presence of PPL in organic solvents afforded exclusively (S)-propanol amines (113). There was

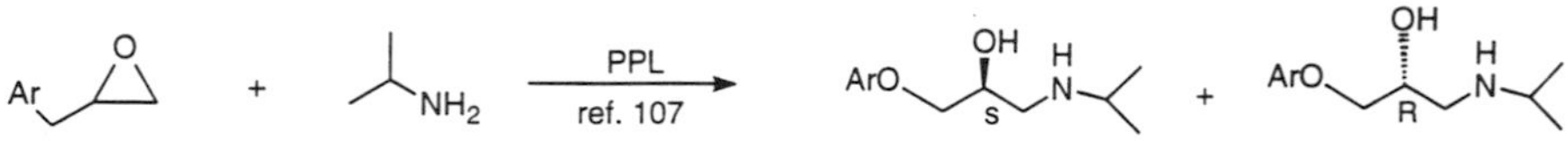

Ar	Solvent	% c	Enantiomeric ratio (%) (R)	(S)
C_6H_5	hexane	48	39	61
	toluene	42	0	>99
	ethyl acetate	36	40	60
$4\text{-}ClC_6H_4$	hexane	48	48	52
	toluene	45	7	93
	ethyl acetate	39	40	81
$2\text{-}CH_3OC_6H_4$	hexane	35	62	38
	toluene	39	35	65
	ethyl acetate	43	20	80
$4\text{-}NHCOCH_3C_6H_4$	toluene	47	0	>99
	ethyl acetate	40	20	80

clearly considerable influence of the solvents on the enantioselectivities. With the exception of the 4-methoxyphenyl compound in hexane, the ring opening to (S)-form appears to be preferred by PPL, particularly with toluene as the solvent. *Candida antarctica* lipase catalyzed the enantioselective acetylation of chiral primary amines, leading to an *ee* value of 90–98% (114).

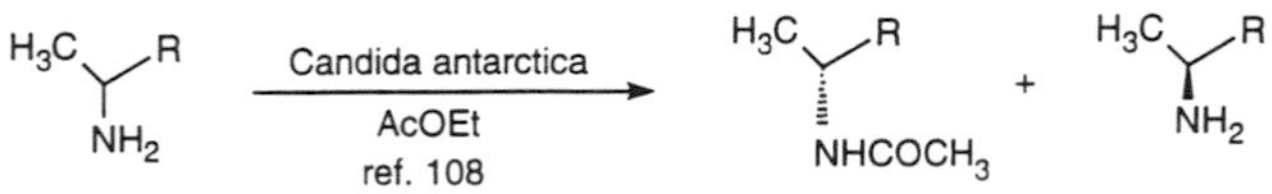

R	% c	% ee N-Acetyl
phenyl	25	97.6
phenyl	43	96.0
1-naphthyl	20	90.0
propyl	34	98.0
propyl	44	98.0

Enzymatic acylation of 2-amino-1-phenylethanol with PS-30 showed high enantioselectivity after the protection of the NH_2 group with CO_2Et or CO_2CH_2Ph was employed (115). However, direct acylation of 2-amino-1-phenylethanol or its N-alkyl substituted derivatives were slow, the reactions stopped before 40% conversion and non-enzymatic aminolysis was found to be significant. PS-30 exhibited higher enantioselectivity in catalyzing the deacylation than its acylation counterpart.

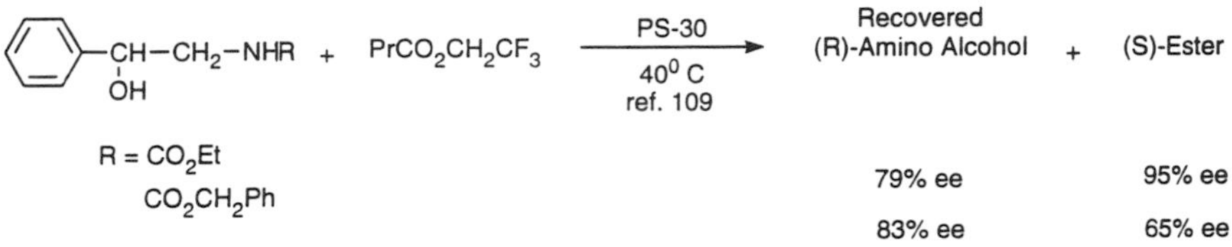

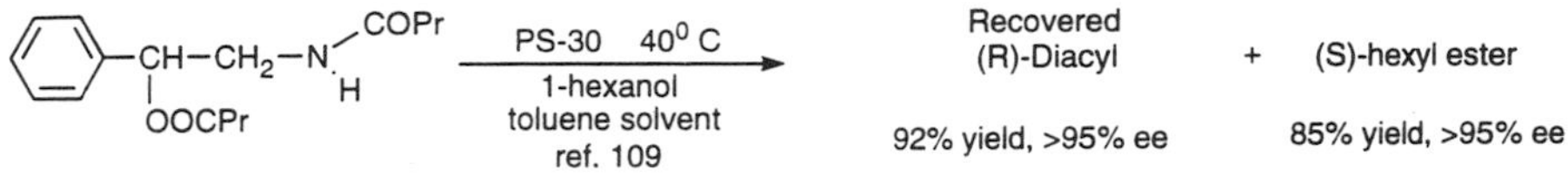

R = CO₂Et
CO₂CH₂Ph

Porcine pancreatic lipase (PPL) catalyzed the acylation of 2-aminoalkan-1-ols; the enantioselectivity depended on the starting amino alcohol (116). The catalytic activity of the enzyme was markedly improved when the benzyl carbamate derivatives were used as the substrates.

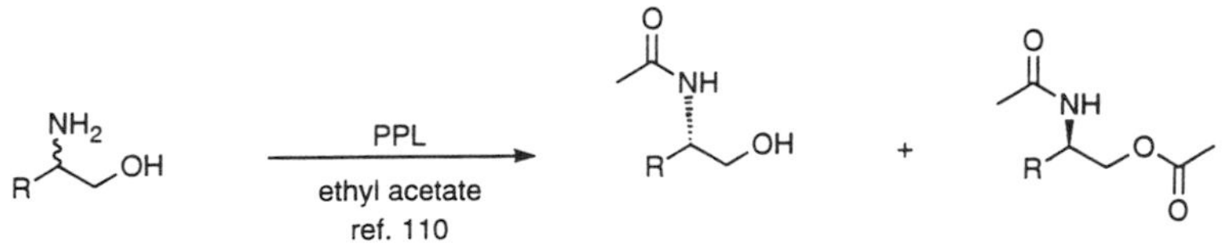

R	time, h	% c % ee	% c % ee
CH₃	24.5	52, 18	28, 67
C₂H₅	20	37, 95	37.6, 95
C₃H₇	24.5	37, 41	21, 0
C₄H₉	24	31, 32	34, 27

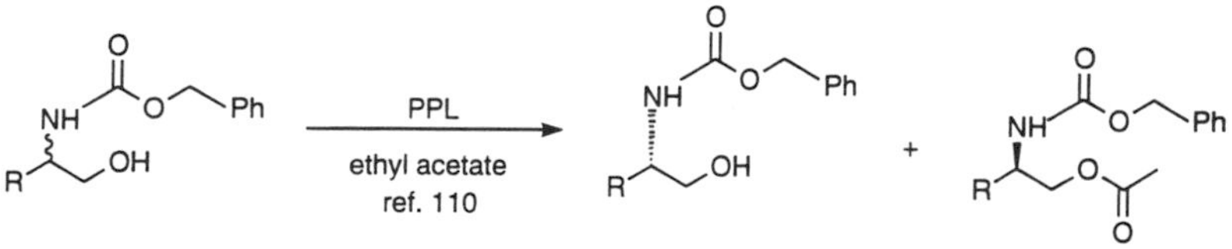

R	time, h	% c	% ee	% ee
CH₃	10	53	85	73
C₂H₅	16	54	83	78
C₃H₇	10.5	51	99	99
C₄H₉	9	51	95	95

α-Vinyl amino acids were reduced to produce 'neopentyl' alcohols that could be enriched in the L-antipode by lipase-mediated enantioselective acylation with vinyl acetate (117). Subsequent deacetylation, oxidation, and hydrolysis yielded enantiomerically enriched L-α-vinyl amino acids. Acyl transfer was quite slow for the side chains derived from phenylalanine, DOPA, and valine, and yet fast for the alanine side chain. Lipase AK appeared to be the most enantioselective catalyst for all the substrates tested. The best conditions found for the DOPA, phenylalanine and alanine side chains were to perform the lipase reaction using 80% wet benzene as the cosolvent.

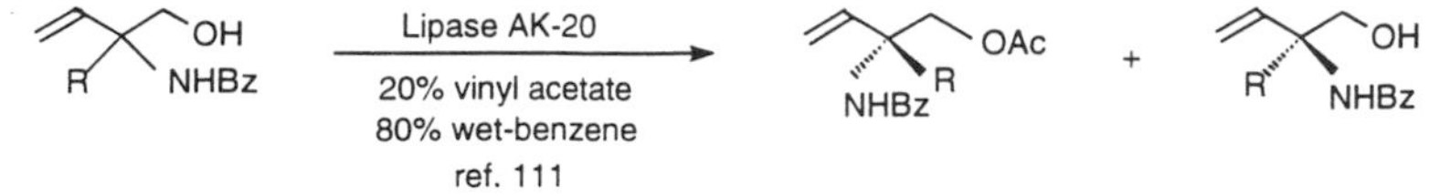

R	time	% Yield Acetate	% Yield Alcohol	% ee Acetate
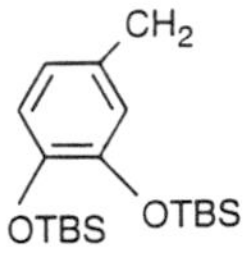 (3,4-bis-OTBS-benzyl) CH₂	1 week	32	64	67
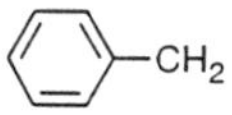 benzyl CH₂	2 days	42	55	86
CH₃	2.5 days	43	53	98

Lipases from *Candida cylindracea* (CCL) and *Candida antarctica* (CAL) catalyzed the aminolysis of activated and non-activated esters (118). The degree of enantioselectivity depended on the amine. Ethyl (±)-2-chloropropionate and the corresponding 2-bromo derivative were resolved by aminolysis of the ester to obtain the optically active (S)-amides. The bromo derivative required much longer reaction time than the chloro compound. The lipase reaction occurred much more rapidly using CAL and ethyl (±)-2-bromopropionate (60% conversion after 1.5 h in dioxan), but the reaction was not enantioselective. Aminolysis of ethyl (±)-2-methylbutyrate with CCL and butylamine as the nucleophile required long reaction times and heating at 50°C to reach 18% conversion. When the 2-methyl compound was subjected to CAL, a preference was found for the R-isomer of the ester, but the enantiomeric excesses were only moderate.

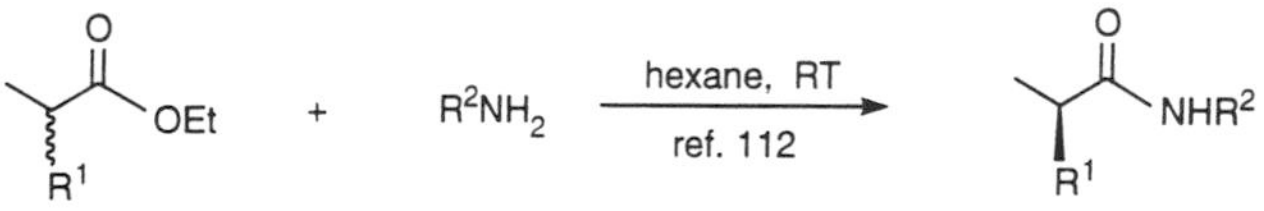

R¹	Lipase	R²	time, h	% c	% ee
Cl	CCL	decyl	5	35	92
Br	CCL	butyl	89	24	90
C₂H₅	CAL	benzyl	3 days	25	78
C₂H₅	CAL	butyl	3 days	20	40

Candida antarctica lipase (CAL) was a very efficient catalyst for the enantioselective aminolysis of different 3-hydroxy esters with aliphatic amines (119). The degree of enantioselectivity exhibited by the lipase depended on the substrate and nucleophile. In most cases the E values obtained were very satisfactory. Chemical reduction of the 3-hydroxyamides obtained by enzymatic aminolysis afforded a synthesis of the 1,3-aminoalcohols.

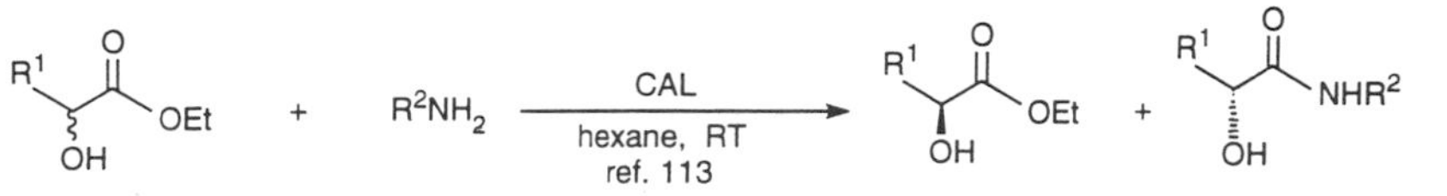

R^1	R^2	time, h	% c	% ee	% ee
CH_3	benzyl	21	45	82	>99
CH_3CH_2	*n*-octyl	1	29	40	>99
CH_3Cl	benzyl	2	50	>99	95

CAL efficiently catalyzed the aminolysis of different acrylic esters with aliphatic amines (120). Using racemic amines, the corresponding optically active acrylic amides were obtained in moderately high enantiomeric excesses.

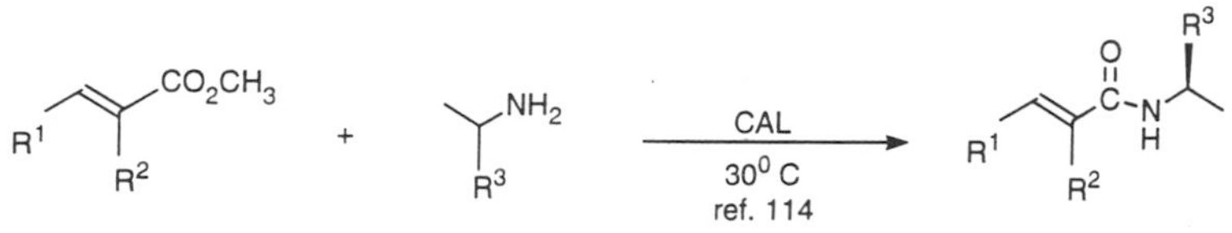

R^1	R^2	R^3	time, d	% Yield	% ee
H	H	C_2H_5	7	25	72
H	H	C_5H_{11}	7	20	95
CH_3	H	C_2H_5	7	30	72
CH_3	H	C_5H_{11}	7	40	74
H	CH_3	C_2H_5	11	27	95
H	CH_3	C_5H_{11}	11	22	95

Candida antarctica lipase efficiently catalyzed the aminolysis of non-activated esters with a β-furyl or β-phenyl group; use of racemic amines gave the corresponding optically active amides with high enantiomeric excess (121). Similar results were obtained with *Candida antarctica* and α,β-ethylenic esters of β-furyl and β-phenyl esters.

Pseudomonas cepacia lipase (PS-30) catalyzed the enantioselective methanolysis of a variety of 4-substituted 2-phenyloxazolin-5-one derivatives in a non-polar organic solvent to provide optically active *N*-benzoyl-L-α-amino acid methyl esters ($ee = 66-98\%$) (122). The enantioselectivity appeared to improve as the C-4 substituent increased in size. The reaction rate was somewhat

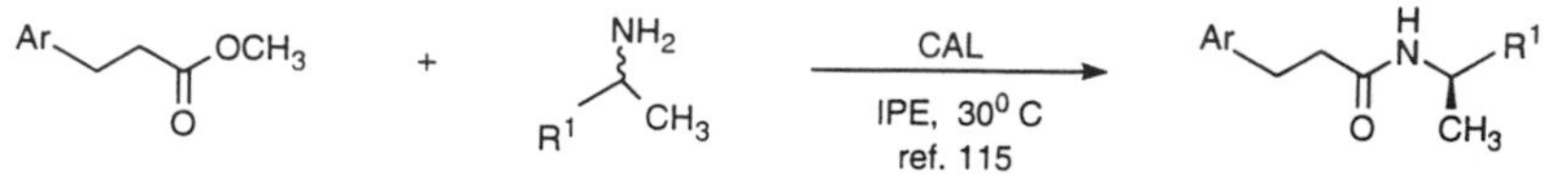

Ar	R¹	time, h	% Yield	% ee
2-Furyl	Et	7.5	81	63
2-Furyl	Pentyl	10	79	79
2-Furyl	Ph	8	83	>95
Ph	Et	9	77	68
Ph	Pentyl	13	79	>95
Ph	Ph	10	84	>95

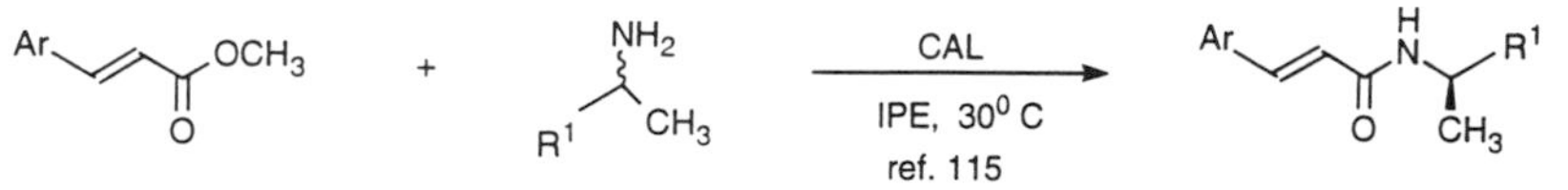

Ar = 2-Furyl, Phenyl

faster in the presence of 5 equivalents of water, but non-enzymatic hydrolysis also occurred under these conditions, resulting in a lowering of the product yield.

The *N*-benzoyl-L-α-amino acid methyl esters were subjected to a protease-catalyzed kinetic resolution to yield enantiomerically pure *N*-benzoyl-L-α-amino acids. A novel one-pot synthesis of optically active cyanohydrin acetates from aldehydes was accomplished by lipase-catalyzed kinetic resolution coupled with *in situ* formation and racemization of cyanohydrins in an organic

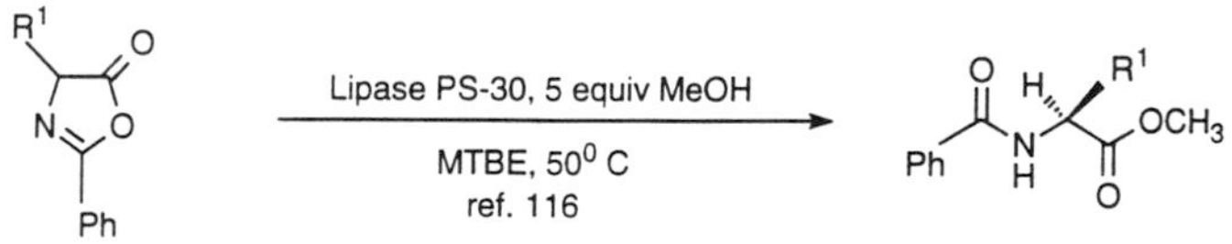

R¹	amt of H₂O (equiv)	time, h	% Yield	% ee	conf
Me₂CH	5	130	47	77	S
Me₂CHCH₂	5	72	85	90	S
	0	72	82	78	S
Me₂SCH₂CH₂	5	48	31	82	
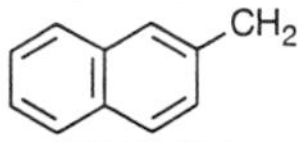 naphthyl-CH₂	0	114	56	71	S
	5	91	83	49	
	0	86	90	75	
p-CH₃PhCH₂	5	84	78	63	
	0	156	86	66	
Ph	5	84	46	75	
PhCH₂	5	35	80	78	S
	0	22	93	69	S
PhCH₂CH₂	5	42	61	93	
PhCH₂CH₂CH₂	5	72	76	95	
	0	72	91	84	

solvent (123). The racemic cyanohydrins were acetylated stereoselectively by *Pseudomonas cepacia* lipase (PS-30) with isopropenyl acetate as the acylating reagent. The authors observed that the *P. cepacia* lipase immobilized on Hyflo Super-Cel was much more active than the commercial grade of bulk enzyme (Amano) containing the lipase of the same origin. The (*S*)-isomer was preferentially acetylated by the lipase, while the unreacted (*R*)-isomer was continuously racemized through reversible transhydrocyanation catalyzed by the resin. This process enabled the one-stage conversion of various aldehydes into the corresponding (*S*)-cyanohydrin acetates to be accomplished with up to 94% *ee* in 63–100% conversion yields. In contrast to the reaction of aromatic

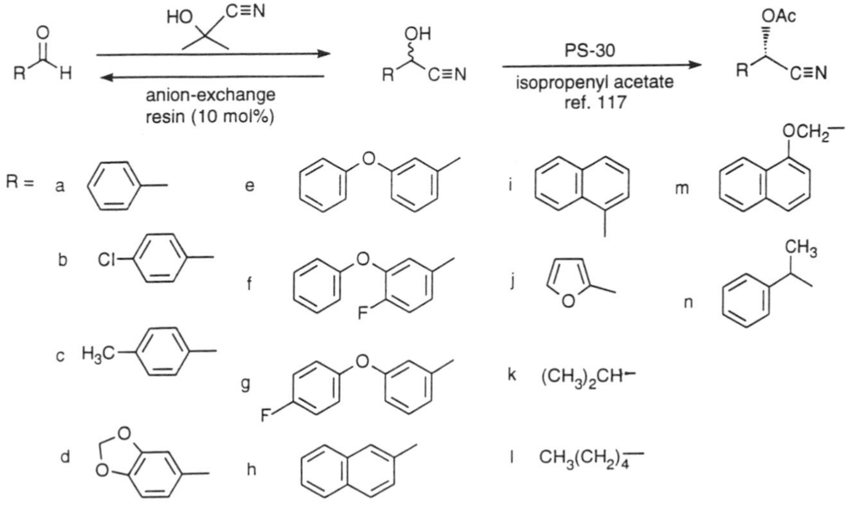

aldehydes, PS-30 did not discriminate the enantiomers of the cyanohydrins derived from the simple aliphatic aldehydes k and l, giving the corresponding acetates in moderate or low optical yields (51% *ee* and 15% *ee*, respectively). Introducing a large aromatic ring in the vicinity of the carbonyl group, however, caused a large increase in the product *ee* of compounds m and n (78% *ee* and 85% *ee*, respectively).

5.13 Conclusion

Although our fundamental understanding of enzyme catalysis in organic solvents is still limited, biocatalysis in organic solvents with low water content has now acquired an important foothold in preparative bio-organic synthesis. During the past ten years, we have witnessed a rapid growth in the use of hydrolytic enzymes such as lipases and proteases in organic solvents for the synthesis of enantiomerically pure compounds using acyl transfer reactions. This non-aqueous methodology not only removes some of the limitations of biocatalysis in aqueous systems but also provides the chemist with a new option for synthetic organic applications. While the lack of predictability of enzymatic properties in organic media and the low reaction rates have hindered their development for industrial applications, there is reason to believe that this situation will change as our basic understanding of enzyme catalysis in organic media improves. Greater insight into the relationship between enzyme structure and substrate binding will lead to improved predictability of the enantioselective preference of biocatalysts. This improved insight, together with recent advances in technologies for the screening of large libraries, may soon result in the creation of new, more efficient, enzymes possessing a broad range of useful properties for synthetic applications. Certainly, there are many chemists today utilizing biocatalytic methods in their investigations and their numbers are increasing. However, it is our feeling that biocatalysis in organic solvents is still an underutilized procedure for preparative bio-organic synthesis.

References

1. Chen, C.S., Fujimoto, Y., Girdaukas, G. et al. (1982) Quantitative analyses of biochemical kinetic resolutions of enantiomers. J. Amer. Chem. Soc., 104, 7294–7299.
2. Chen, C.S., Wu, S.H., Girdaukas, G. et al. (1987) Quantitative analyses of biochemical kinetic resolution of enantiomers. 2. Enzyme-catalyzed esterifications in water–organic solvent biphasic systems. J. Amer. Chem. Soc., 109, 2812–2817.
3. Degueil-Castaing, M., De Jeso, B., Drouillard, S. et al. (1987) Enzymic reactions in organic synthesis: ester interchange of vinyl esters. Tetrahedron Lett., 28, 953–956.
4. Wang, Y.F., Lalonde, J.J., Momongan, M. et al. (1988) Lipase catalyzed irreversible trans-esterifications using enol esters as acylating reagents: preparative enantio and regioselective syntheses of alcohols, glycerol derivatives, sugars and organometallics. J. Amer. Chem. Soc., 110, 7200–7205.
5. Sepulchre, M., Spassky, N. and Sigwalt, P. (1976) Stereoselective polymerization applied to heterocyclic compounds: a resolution method giving monomers of high optical purity. Israel J. Chem., 15, 33–38.
6. Kagan, H.B. and Fiaud, J.C. (1988) Kinetic resolution. Topics Stereochem., 18, 249–330.
7. Wallace, J.S. and Morrow, C.J. (1989) Biocatalytic synthesis of polymers. Synthesis of an optically active, epoxy-substituted polyester by lipase-catalyzed polymerization. J. Polym. Sci. A: Polym. Chem., 27, 2553–2567.
8. Puchot, C., Samuel, O., Dunach, E. et al. (1986) Nonlinear effects in asymmetric synthesis. Examples in asymmetric oxidations and aldolization reactions. J. Amer. Chem. Soc., 108, 2353–2357.
9. Charles, R. and Gil-Av, E. (1984) Self-amplification of optical activity by chromatography on an achiral adsorbent. J. Chromatogr., 298, 516–520.
10. Matori, M., Asahara, T. and Ota, Y. (1991) Positional specificity of microbial lipases. J. Ferment. Bioeng., 72, 397–402.
11. Grochulski, P., Li, Y., Schrag, J.D. et al. (1993) Insight into interfacial activation from an open structure of Candida rugosa lipase. J. Biol. Chem., 268, 12843–12847.
12. Murata, M., Terao, Y., Achiwa, K. et al. (1989) Efficient lipase-catalyzed synthesis of chiral glycerol derivatives. Chem. Pharm. Bull., 37, 2670–2672.
13. Scilimati, A., Ngooi, T.K. and Sih, C.J. (1988) Biocatalytic resolution of (±)-hydroxy-alkanoic esters. A strategy for enhancing the enantiomeric specificity of lipase-catalyzed ester hydrolysis. Tetrahedron Lett., 29, 4927–4930.

14. Bhalerao, U.T., Dasaradhi, L., Neelakantan, P. and Fadnavis, N.W. (1991) Lipase catalyzed regio- and enantio-selective hydrolysis: molecular recognition phenomenon and synthesis of (R)-dimorphecolic acid. *J. Chem. Soc., Chem. Commun.*, 1197–1198.

15. Deveer, A.M.Th.J., Dijkman, R., Leuveling-Tjeenk *et al.* (1991) A monolayer and bulk study on the kinetic behavior of *Pseudomonas glumae* lipase using synthetic pseudoglycerides. *Biochemistry*, **30**, 10034–10042.

16. Kamat, S.V., Beckman, E.J. and Russell, A.J. (1993) Control of enzyme enantio-selectivity with pressure changes in supercritical fluoroform. *J. Amer. Chem. Soc.*, **115**, 8845–8846.

17. Winkler, F.K., D'Arcy, A. and Hunziker, W. (1990) Structure of human pancreatic lipase. *Nature*, **343**, 771–774.

18. Brady, L., Brzozowski, A.M., Derewenda, Z.S. *et al* (1990) A serine protease triad forms the catalytic centre of a triacylglycerol lipase. *Nature*, **343**, 767–771.

19. Schrag, J.D., Li, Y., Wu, S. and Cygler, M. (1991) Ser–His–Glu triad forms the catalytic site of the lipase from *Geotrichum candidum*. *Nature*, **351**, 761–764.

20. Grochulski, P., Bouthillier, F., Kazlauskas, R.J. *et al.* (1994) Analogs of reaction intermediates identify a unique substrate binding site in *Candida rugosa* lipase. *Biochemistry*, **33**, 3494–3500.

21. Burgess, K. and Jennings, L.D. (1991) Enantioselective esterification of unsaturated alcohols mediated by a lipase prepared from *Pseudomonas* sp. *J. Amer. Chem. Soc.*, **113**, 6129–6139.

22. Kazlauskas, R.J., Weissfloch, A.N.E., Rappaport, A.T. and Cuccia, L.A. (1991) A rule to predict which enantiomer of a secondary alcohol reacts faster in reactions catalyzed by cholesterol esterase, lipase from *Pseudomonas cepacia* and lipase from *Candida rugosa*. *J. Org. Chem.*, **56**, 2656–2665.

23. Sih, C.J., Gu, R.-L., Crich, J.Z. and Brieva, R. (1994) Biocatalytic methods for enantioselective synthesis. In *Stereocontrolled Organic Synthesis* (ed. B.M. Trost), Blackwell Scientific Publications, Oxford, pp. 399–412.

24. Oberhauser, Th., Faber, K. and Griengl, H. (1989) A substrate model for the enzymatic resolution of esters of bicyclic alcohols by *Candida cylindracea* lipase. *Tetrahedron*, **45**, 1679–1682.

25. Ehrler, J. and Seebach, D. (1990) Enantioselective saponification of substituted achiral 3-acyloxy-propionates with lipases: synthesis of chiral derivatives of 'tris(hydroxymethyl)-methane'. *Liebigs Ann. Chem.*, 379–388.

26. Guo, Z.-W. and Sih, C.J. (1989) Enantioselective inhibition: A strategy for improving the enantioselectivity of biocatalytic systems. *J. Amer. Chem. Soc.*, **111**, 6836–6841.

27. Wu, S.-H., Guo, Z.-W. and Sih, C.J. (1990) Enhancing the enantioselectivity of *Candida* lipase catalyzed ester hydrolysis via noncovalent enzyme modification. *J. Amer. Chem. Soc.*, **112**, 1990–1995.

28. Parida, S. and Dordick, J.S. (1991) Substrate structure and solvent hydrophobicity control lipase catalysis and enantioselectivity in organic media. *J. Amer. Chem. Soc.*, **113**, 2253–2259.

29. Berger, B. and Faber, K. (1991) 'Immunization' of lipase against acetaldehyde emerging in acyl transfer reactions from vinyl acetate. *J. Chem. Soc., Chem. Commun.*, 1198–1200.

30. Kawase, M. and Tanaka, A. (1989) Effects of chemical modification of amino acid residues on the activities of lipase from *Candida cylindracea*. *Enzyme Microb. Technol.*, **11**, 44–48.

31. Borgtrom, B. and Brockman, H.L. (eds) (1984) *Lipases*, Elsevier, Amsterdam.

32. Longhi, S., Fusetti, F., Grandori, R. *et al.* (1992) Cloning and nucleotide sequences of two lipase genes from *Candida cylindracea*. *Biochim. Biophys. Acta*, **1131**, 227–232.

33. Shaw, J.F., Chang, C.H. and Wang, Y.J. (1989) Characterization of three distinct forms of lipolytic enzyme in a commercial *Candida* lipase preparation. *Biotechnol. Lett.*, **11**, 779–784.

34. Brahimi-Horn, M.-C., Guglielmino, M.L., Elling, L. and Sparrow, L.G. (1990) The esterase profile of a lipase from *Candida cylindracea*. *Biochim. Biophys. Acta*, **1042**, 51–54.

35. Grochulski, P., Li, Y., Scrag, J.D. and Cygler, M. (1994) Two conformational states of *Candida rugosa* lipase. *Protein Science*, **3**, 82–91.

36. Wang, Y.-F., Chen, C.-S., Girdaukas, G. and Sih, C.J. (1984) Bifunctional chiral synthons via biochemical methods. 3. Optical purity enhancement in enzymatic asymmetric catalysis. *J. Amer. Chem. Soc.*, **106**, 3695–3696.

37. Harris, K.J., Gu, Q.-M., Shih, Y.-E. *et al.* (1991) Enzymatic preparation of (3S,6R) and (3R,6S)-3-hydroxy-6-acetoxycyclohex-1-ene. *Tetrahedron Lett.*, **32**, 3941–3944.

38. Wang, Y.-F., Chen, C.-S., Girdaukas, G. and Sih, C.J. (1985) Extending the applicability of esterases of low enantioselectivity in asymmetric synthesis. In *Enzymes in Organic Synthesis* (ed. A.R. Battersby), Ciba Foundation Symposium 111, Pitman, London, pp. 128–145.

39. Wu, S.-H., Zhang, L.-Q., Chen, C.-S. *et al.* (1985) Bifunctional chiral synthons via biochemical methods. VII. Optically-active 2,2'-dihydroxy-1,1'-binaphthyl. *Tetrahedron Lett.*, **26**, 4323–4326.
40. Guo, Z.-.W., Wu, S.-H., Chen, C.-S. *et al.* (1990) Sequential biocatalytic kinetic resolutions. *J. Amer. Chem. Soc.*, **112**, 4942–4945.
41. Chen, C.-S. and Liu, Y.-C. (1991) Amplification of enantioselectivity in biocatalyzed kinetic resolution of racemic alcohols. *J. Org. Chem.*, **56**, 1966–1968.
42. Rakels, J.L.L., Wolff, A., Straathof, A.J.J. and Heijnen, J.J. (1994) Sequential kinetic resolution by two enantioselective enzymes. *Biocatalysis*, **9**, 31–47.
43. Faber, K. and Riva, S. (1992) Enzyme-catalyzed irreversible acyl transfer. *Synthesis*, 895–910.
44. Fang, J.M. and Wong, C.H. (1994) Enzymes in organic syntheses: alteration of reversible reactions to irreversible processes. *SynLetters*, **6**, 393–402.
45. Kirchner, G., Scollar, M.P. and Klibanov, A. (1985) Resolution of racemic mixtures via lipase catalysis in organic solvents. *J. Amer. Chem. Soc.*, **107**, 7072–7076.
46. Faber, K., Ottolina, G. and Riva, S. (1993) Selectivity-enhancement of hydrolase reactions. *Biocatalysis*, **8**, 91–132.
47. Secundo, F., Riva, S. and Carrea, G. (1992) Effects of medium and reaction conditions on the enantioselectivity of lipases in organic solvents and possible rationales. *Tetrahedron Asymmetry*, **3**, 267–280.
48. Hirose, Y., Kariya, K., Sasaki, I. *et al.* (1992) Drastic solvent effect on lipase-catalyzed enantio-selective hydrolysis of prochiral 1,4-dihydropyridines. *Tetrahedron Lett.*, **33**, 7157–7160.
49. Tawaki, S. and Klibanov, A. (1992) Inversion of enzyme enantioselectivity mediated by the solvent. *J. Amer. Chem. Soc.*, **114**, 1882–1884.
50. Russell, A.J. and Klibanov, A. (1988) Inhibitor-induced enzyme activation in organic solvents. *J. Biol. Chem.*, **263**, 11624–11626.
51. Ståhl, M., Jepsson-Wistrand, U., Månsson, M.O. and Mosbach, K. (1991) Induced stereo-selectivity and substrate selectivity of bio-imprinted α-chymotrypsin in anhydrous organic media. *J. Amer. Chem. Soc.*, **113**, 9366–9368.
52. Yennawar, H.P., Yennawar, N.H. and Farber, G.K. (1995) A structural explanation for enzyme memory in nonaqueous solvents. *J. Amer. Chem. Soc.*, **117**, 577–585.
53. Zaks, A. and Klibanov, A. (1986) Substrate specificity of enzymes in organic solvents vs. water is reversed. *J. Amer. Chem. Soc.*, **108**, 2767–2768.
54. Gololobov, M.Y., Voyushina, T.L., Stepanov, V.M. and Adlercreutz, P. (1992) Organic solvent changes the chymotrypsin specificity with respect to nucleophiles. *FEBS*, **307**, 309–312.
55. Laane, C., Boeren, S., Vos, K. *et al.* (1987) Rules for optimization of biocatalysis in organic solvents. *Biotechnol. Bioeng.*, **30**, 81–87.
56. Gorman, L.A.S. and Dordick, J.S. (1992) Organic solvents strip water off enzymes. *Biotechnol. Bioeng.*, **39**, 392–397.
57. Khmelnitsky, Y.L., Welch, S.H., Clark, D.S. and Dordick, J.S. (1994) Salts dramatically enhance activity of enzymes suspended in organic solvents. *J. Amer. Chem. Soc.*, **116**, 2647–2648.
58. Blackwood, A.D., Moore, B.D. and Halling, P.J. (1994) Are associated ions important for biocatalysis in organic media? *Biocatalysis*, **9**, 269–276.
59. Jongejan, J.A., Vantol, J.B.A. and Duine, J.A. (1994) Enzymes in organic media: new solutions for industrial enzymatic catalysis? *Chemistry Today*, **12**, 15–24.
60. Boland, W., Fröbl, C. and Lorenz, M. (1991) Esterolytic and lipolytic enzymes in organic synthesis. *Synthesis*, 1049–1072.
61. Santaniello, E., Ferraboschi, P., Grisenti, P. and Manzocchi, A. (1992) The biocatalytic approach to the preparation of enantiomerically-pure chiral building blocks. *Chem. Rev.*, **92**, 1071–1140.
62. Cesti, P., Zaks, A. and Klibanov, A.M. (1985) Preparative regioselective acylation of glycols by enzymatic transesterification in organic solvents. *Appl. Biochem. Biotechnol.*, **11**, 401–407.
63. Parma, V.S., Sinha, R., Bisht, K.S., Gupta, S., Prasad, A.K. and Taneja, P. (1993) Regioselective esterification of diols and triols with lipases in organic solvents. *Tetrahedron*, **49**, 4107–4116.
64. Jansen, A.J.M., Klunder, A.J.H. and Zwanenburd, B. (1991) PPL-catalyzed resolution of 1,2- and 1,3-diols in methyl propionate as solvent. An application of the tandem use of enzymes. *Tetrahedron*, **47**, 7409–7416.
65. Theil, F., Weidner, J., Ballschuh, S., Kunath, A. and Schick, H. (1994) Kinetic resolution

of acyclic 1,2-diols using sequential lipase-catalyzed transesterification in organic solvents. *J. Org. Chem.*, **59**, 388–393.

66. Herradon, B., Cueto, S., Morcuendo, A. and Valverde, S. (1993) Regio- and enantioselective esterifications of polyoxygenated compounds catalyzed by lipases. *Tetrahedron Asymmetry*, **4**, 845–864.

67. Hof, R.P. and Kellogg, R.M. (1994) Lipase catalyzed resolutions of some alpha,alpha-disubstituted 1,2-diols in organic solvents; near absolute regio and chiral recognition. *Tetrahedron Asymmetry*, **5**, 565-568.

68. Ferraboschi, P., Casati, S., Grisenti, P. and Santaniello, E. (1994) Enantioselective *Pseudomonas fluorescens* (*P. cepacia*) lipase-catalyzed irreversible transesterification of 2-methyl-1,2-diols in an organic solvent. *Tetrahedron Asymmetry*, **5**, 1921–1924.

69. Tombo, G.M.R., Schar, H.P., Fernandez, X., Busquets, I. and Ghisalba, O. (1986) Synthesis of both enantiomeric forms of 2-substituted 1,3-propanediol monoacetates starting from a common prochiral precursor, using enzymatic transformations in aqueous and in organic media. *Tetrahedron Lett.*, **27**, 5707–5710.

70. Tsuji, K., Terao, Y. and Achiwa, K. 91989) Lipase-catalyzed asymmetric synthesis of chiral 1,3-propanediols and its application to the preparation of optically pure building blocks for renin inhibitors. *Tetrahedron Lett.*, **30**, 6189–6192.

71. Guanti, G., Banfi, L. and Riva, R. (1994) Enzymatic asymmetrization of some prochiral and meso diols through monoacetylation with pig pancreatic lipase (PPL). *Tetrahedron Asymmetry*, **5**, 9–12.

72. Grisenti, P., Ferraboschi, P., Manzocchi, A. and Santaniello, E. (1992) Enantioselective trans-esterification of 2-methyl-1,3-propanediol derivatives catalyzed by *Pseudomonas fluorescens* lipase in an organic solvent. *Tetrahedron*, **48**, 3827–3834.

73. Eguchi, T. and Mochida, K. (1993) Lipase-catalyzed diacylation of 1,3-butanediol. *Biotechnol. Lett.*, **15**, 955–960.

74. Grisenti, P., Ferraboschi, P., Casati, S. and Santaniello, E. (1993) Studies on the enantio-selectivity of the transesterification of 2-methyl-1,4-butanediol and its derivatives catalyzed by *Pseudomonas fluorescens* lipase in organic solvents. *Tetrahedron Asymmetry*, **4**, 997–1006.

75. Ferraboschi, P., Grisenti, P., Manzocchi, A. and Santaniello, E. (1994) Regio- and enantio-selectivity of *Pseudomonas cepaica* lipase in the transesterification of 2-substituted 1,4-butane-diols. *Tetrahedron Asymmetry*, **5**, 691–698.

76. Kim, Mahn-Joo and Lee, I.S. (1993) Combined chemical and enzymatic synthesis of (*S,S*)-2,5-dimethylpyrrolidine. *SynLetters*, 767–768.

77. Nagai, H., Morimoto, T. and Achiwa, K. (1994) Facile enzymatic synthesis of optically active 2,5-hexanediol derivatives and its application to the preparation of optically pure cyclic sulfate for chiral ligands. *Synlett*, 289–290.

78. Allevi, P., Anastasia, M., Cajone, F., Ciuffreda, P. and Sanvito, A.M. (1993) Enzymatic resolution of (*R*)- and (*S*)-(*E*)-4-hydroxyalk-2-enals related to lipid peroxidation. *J. Org. Chem.*, **58**, 5000–5002.

79. Allevi, P., Anastasia, M., Cajone, F., Ciuffreda, P. and Sanvito, A.M. (1994) Enzymatic resolution of the ethyl acetals of (*R*)- and (*S*)-4-hydroxyalk-2-ynals. *Tetrahedron Asymmetry*, **5**, 13–16.

80. Boaz, N.W. and Zimmerman, R.L. (1994) Amine assisted enzymatic esterification of 1,2-diol monotosylates. *Tetrahedron Asymmetry*, **5**, 153–156.

81. Takahata, H., Uchida, Y., Ohkawa, Y. and Momose, T. (1992) Enzymatic resolution of *N,N*-dialkyl-3-hydroxy-4-pentenamides, unsuccessful in resolution by the Katsuki–Sharpless epoxidation. *Tetrahedron Asymmetry*, **4**, 1041–1042.

82. Takahata, H., Uchida, Y. and Momose, T. (1992) Transesterification-based enzymatic resolutions of racemic 3-hydroxy-4-pentenylurethanes in organic solvents. *Tetrahedron Lett.*, **33**, 3331–3332.

83. Burgess, K. and Henderson, I. (1990) Lipase catalyzed resolutions of Spac reaction products. *Tetrahedron Asymmetry*, **1**, 57–60.

84. Dominguez, E., Carretero, J.C., Fernandez-Mayoralas, A. and Conde, S. (1991) An efficient preparation of optically active (*E*)-γ-hydroxy-α,β-unsaturated phenyl sulfones using lipase-mediated acylations. *Tetrahedron Lett.*, **32**, 5159–5162.

85. Carretero, J.C. and Domingues, E. (1992) Lipase-catalyzed kinetic resolution of gamma-hydroxy phenyl sulfones. *J. Org. Chem.*, **57**, 3867–3873.

86. Sparks, M.A. and Panek, J.S. (1991) Lipase mediated resolution of chiral (*E*)-vinylsilanes: an improved procedure for the production of (*R*)- and (*S*)-(*E*)-1-trialkylsilyl-1-buten-3-ol derivatives. *Tetrahedron Lett.*, **32**, 4085–4088.
87. Takano, S., Setoh, M., Yamada, O. and Ogasawara, K. (1993) Synthesis of optically active 4-benzyloxymethyl- and 4-(4-methoxyphenoxy)methyl-buten-2-olides via lipase-mediated resolution. *Synthesis*, 1253–1256.
88. Kim, Mahn-Joo and Lee, I.S. (1993) Lipase-catalyzed transesterification as a route to each stereoisomer of 2,5-hexanediol and 3-hexyne-2,5-diol. Synthesis of (2*R*,5*R*)-2,5-dimethyltetrahydrofuran. *J. Org. Chem.*, **58**, 6483–6485.
89. Bosetti, A., Bianchi, D., Cesti, P., Golina, P. and Spezia, S. (1992) Enzymatic resolution of 1,2-diols: comparison between hydrolysis and transesterification reactions. *J. Chem. Soc., Perkin Trans I*, 2395–2398.
90. Deyo, D.T., Aebi, J.D. and Rich, D.H. (1988) Preparative-scale, facile synthesis of (2*R*,4*E*)-2-methyl-4-hexanal: A key intermediate of (2*R*,3*R*,4*R*,6*E*)-3-hydroxy-4-methyl-2-methylamino-6-octenoic acid (MeBmt). *Synthesis*, 608–610.
91. Ferraboschi, P., Brembilla, D., Grisenti, P. and Santaniello, E. (1991) Enzymatic synthesis of enantiomerically pure chiral synthons: Lipase-catalyzed resolution of (*R/S*,4*E*)-2-methyl-4-hexen-1-ol. *SynLetters*, 310–312.
92. Engel, K.H. (1991) Lipase-catalyzed enantioselective esterification of 2-methylalkanoic acids. *Tetrahedron Asymmetry*, **2**, 165–168.
93. Engel, K.H. (1992) Lipase-catalyzed enantioselective acidolysis of chiral 2-methylalkanoates. *J. Amer. Oil Chem. Soc.*, **69**, 146–150.
94. Barth, S. and Effenberger, F. (1993) Lipase-catalyzed resolution of racemic 2-alkyl substituted 1-alkanols. *Tetrahedron Asymmetry*, **4**, 823–833.
95. Ferraboschi, P., Grisenti, P., Manzocchi, A. and Santaniello, E. (1990) New chemoenzymatic synthesis of (*R*)- and (*S*)-4-(phenylsulfonyl)-2-methyl-1-butanol: A chiral C$_5$ isoprenoid synthon. *J. Org. Chem.*, **55**, 6214–6216.
96. Chong, J.M. and Mar, E.K. (1991) Preparation of enantiomerically enriched α-hydroxystannanes via enzymatic resolution. *Tetrahedron Lett.*, **32**, 5683–5686.
97. Bianchi, D., Moraschini, P., Bosetti, A. and Cesti, P. (1994) Enzymatic preparation of optically active 4-chloro-1-phenyl-1-butanol derivatives. *Tetrahedron Asymmetry*, **5**, 1917–1920.
98. Hanson, R.L., Banerjee, A., Comezöglu, F.T., Mirfakhrae, K.D., Patel, R.N. and Szarka, L.J. (1994) Resolution of α-(4-fluorophenyl)-4-(5-fluoro-2-pyrimidinyl)-1-piperazinebutanol (BMS 181100) and α-(3-chloropropyl)-4-fluorobenzenemethanol using lipase-catalyzed acetylation or hydrolysis. *Tetrahedron Asymmetry*, **5**, 1925–1934.
99. Naemura, K., Fukuda, R., Takahashi, N., Konishi, M., Hirose, Y. and Tobe, Y. (1993) Enzyme-catalyzed asymmetric acylation and hydrolysis of *cis*-2,5-disubstituted tetrahydrofuran derivatives: contribution to development of models for reactions catalyzed by porcine liver esterase and porcine pancreatic lipase. *Tetrahedron Asymmetry*, **4**, 911–918.
100. Guibe-Jampel, E. and Bassir, M. (1994) Lipase catalyzed resolution of chiral acids using their mixed carboxylic carbonic anhydrides. *Tetrahedron Lett.*, **35**, 421–422.
101. Kalaritis, P., Regenye, R.W., Partridge, J.J. and Coffen, D.L. (1990) Kinetic resolution of 2-substituted esters catalyzed by a lipase ex. *Pseudomonas fluorescens. J. Org. Chem.*, **55**, 812–815.
102. Gu, Q.-M., Chen, C.-S. and Sih, C.J. (1986) A facile enzymatic resolution process for the preparation of (+)-*S*-2-(6-methoxy-2-naphthyl)-propionic acid (Naproxen). *Tetrahedron Lett.*, **27**, 1763–1766.
103. Barton, M.J., Hamman, J.P., Fichter, K.C. and Calton, G.J. (1990) Enzymatic resolution of (*R,S*)-2-(4-hydroxyphenoxy)propionic acid. *Enzyme Microb. Technol.*, **12**, 577–583.
104. Gu, Q.-M., Reddy, D.R. and Sih, C.J. (1986) Bifunctional chiral synthons via biochemical methods VIII. Optically active 3-aroylthio-2-methylpropionic acids. *Tetrahedron Lett.*, **27**, 5203–5206.
105. Engel, K.-H. (1991) Lipase catalyzed enantioselective esterification of 2-methylalkanoic acids. *Tetrahedron Asymmetry*, **2**, 165–168.
106. Tanaka, A. and Sonomoto, K. (1990) Immobilized biocatalysts in organic solvents. *Chemtech*, **20**, 112–117.

107. Dernoncour, R. and Azerad, R. (1987) Enantioselective hydrolysis of 2-(chlorophenoxy)-propionic esters by esterases. *Tetrahedron Lett.*, **28**, 4661–4664,
108. Miyazawa, T., Kurita, S., Yamada, T. and Kuwata, S. (1992) Resolution of racemic carboxylic acids via the lipase-catalyzed irreversible transesterification using vinyl esters: Effects of alcohols as nucleophiles and organic solvents on enantioselectivity. *Biotechnol. Lett.*, **14**, 941–946.
109. Mustranta, A. (1992) Use of lipases in the resolution of racemic ibuprofen. *Appl. Microbiol. Biotechnol.*, **38**, 61–66.
110. Langrand, G., Baratti, J., Buono, G. and Triantaphylides, C. (1986) Lipase-catalyzed reactions and strategy for alcohol resolution. *Tetrahedron Lett.*, **27**, 29–32.
111. Zaks, A. and Klibanov, A.M. (1984) Enzymatic catalysis in organic media at 100°C. *Science*, **224**, 1249–1251.
112. Tsai, S.-W. and Wei, H.J. (1994) Effect of solvent on enantioselective esterification of naproxen by lipase with trimethylsilyl methanol. *Biotechnol. Bioeng.*, **43**, 64–68.
113. Kamai, A., Damayanthi, Y. and Rao, M.V. (1992) Stereoselective synthesis of (*S*)-propanol amines: lipase catalyzed opening of epoxides with 2-propylamine. *Tetrahedron Asymmetry*, **3**, 1361–1364.
114. Reetz, M.T. and Dreisbach, C. (1995) Highly efficient lipase-catalyzed kinetic resolution of chiral amines. *Chimia*, **48**, 570.
115. Kanerva, L.T., Rahiala, K. and Vanttinen, E. (1992) Lipase catalysis in the optical resolution of 2-amino-1-phenylethanol derivatives. *J. Chem. Soc., Perkin Trans I*, 1759–1762.
116. Fernandez, S., Brieva, R., Rebolledo, F. and Gotor, V. (1992) Lipase catalyzed enantioselective acylation of N-protected or unprotected 2-aminoalkan-1-ols. *J. Chem. Soc., Perkin Trans I*, 2885–2889.
117. Berkowitz, D.B., Pumphrey, J.A. and Shen, Q. (1994) Enantiomerically enriched α-vinyl amino acids via lipase-mediated 'reverse transesterification'. *Tetrahedron Lett.*, **35**, 8743–8746.
118. Quiros, M., Sanchez, V.M., Brieva, R., Rebolledo, F. and Gotor, V. (1993) Lipase catalyzed synthesis of optically active amides in organic media. *Tetrahedron Asymmetry*, **4**, 1105–1112.
119. Garcia, M.J., Rebolledo, F. and Gotor, V. (1993) Practical enzymatic route to optically active 3-hydroxyamides. Synthesis of 1,3-aminoalcohols. *Tetrahedron Asymmetry*, **4**, 2199–2210.
120. Puertas, S., Brieva, R., Rebolledo, F. and Gotor, V. (1993) Lipase catalyzed aminolysis of ethyl propiolate and acrylic esters. Synthesis of chiral acrylamides. *Tetrahedron*, **49**, 4007–4014.
121. Gotor, V., Memendez, E., Mouloungui, Z. and Gaset, A. (1993) Synthesis of optically active amides from beta-furyl and beta-phenyl esters by way of enzymatic aminolysis. *J. Chem. Soc., Perkin Trans I*, 2453–2456.
122. Crich, J.Z., Brieva, R., Marquart, P., Gu, R.L., Flemming, S. and Sih, C.J. (1993) Enzymatic asymmetric synthesis of alpha-amino acids. Enantioselective cleavage of 4-substituted oxazolin-5-ones and thiazolin-5-ones. *J. Org. Chem.*, **58**, 3252–3258.
123. Inagaki, M., Hiratake, J., Nishioka, T. and Oda, J. (1992) One-pot synthesis of optically active cyanohydrin acetates from aldehydes via lipase-catalyzed kinetic resolution coupled with *in situ* formation and racemization of cyanohydrins. *J. Org. Chem.*, **57**, 5643–5649.

6 Regioselectivity of hydrolases in organic media

S. RIVA

6.1 Introduction

According to IUPAC definition, 'a regioselective reaction is one in which one direction of bond making or breaking occurs preferentially over all the other possible directions' (1). The ability of hydrolases (lipases and proteases) to perform this kind of transformation has been extensively exploited in the modification of polyhydroxylated and polyfunctionalized compounds such as carbohydrates, alkaloids and steroids (2). Modification of one over several other groups with similar chemical reactivity has been used both as a protective step in a synthetic sequence (3) and as an efficient way to obtain specific compounds with potential applications as biodegradable surfactants, monomers suitable for radical polymerization, and pharmaceuticals. These findings further enlighten the versatility of lipases and proteases, enzymes able to interact with substrates that are structurally unrelated to their natural targets, triacylglycerides and proteins respectively.

This chapter will briefly cover the state of the art in the field of enzymatic acylation of polyhydroxylated compounds in organic solvents.

6.2 Enzymatic acylation of polyhydroxylated compounds

The pioneering work of Klibanov at MIT disclosed this area of synthetic application of hydrolytic enzymes. In 1985, he reported on the regioselective acylation of simple aliphatic glycols by the action of porcine pancreatic lipase (4). When suspended in solutions of various diols in ethyl carboxylates, this crude lipase powder produced primary monoesters of glycols on a multi-gram scale. Since then, this methodology has been successfully applied to other diols (5–7) and to other classes of polyhydroxylated compounds.

6.2.1 Carbohydrates

The intrinsic polarity of most of these molecules greatly narrows the choice of the solvent, quite often limited to pyridine, dimethylformamide (DMF) or dimethylacetamide. Consequently, a smaller number of enzymes that remain active in such a highly polar environment can be used, and some limitations arise

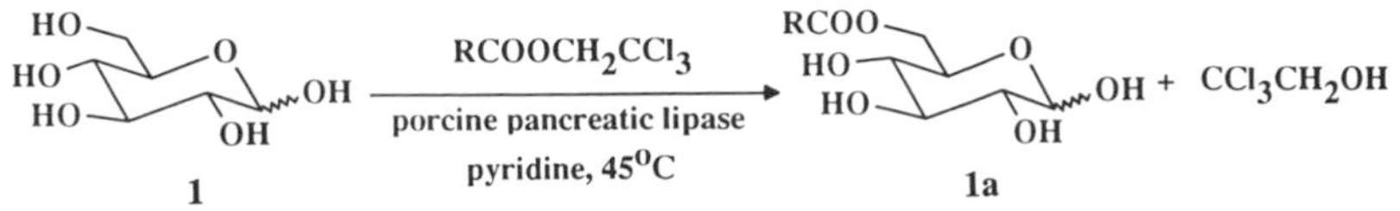

Figure 6.1 Acylation of glucose catalysed by porcine pancreatic lipase in pyridine (from Ref. 11).

also for the nature of the acylating agent (8–10). Anhydrides are usually too activated in dipolar aprotic solvents and give rise to aspecific chemical acylation. Sometimes this is also the case with vinyl esters, when used in large excess in solvents like pyridine and DMF. On the other hand, no spontaneous reaction takes place with activated 2-haloethyl esters (2).

In a first report applied to carbohydrates (11), porcine pancreatic lipase catalysed the regioselective esterifications of the primary hydroxyl groups of monosaccharides in pyridine. The reactions were conducted on a multigrams scale using 2,2,2-trichloroethyl esters of linear alyphatic acids as acylating agent. (Figure 6.1 exemplifies the results obtained with D-glucose, **1**.)

To broaden the spectrum of the enzymes that can be employed in the acylation of monosaccharides, Wong suggested the use of solvent mixtures in order to achieve both satisfactory lipase activities and good solubility of the substrates (12). For instance, a mixture of benzene and pyridine (2:1) allowed *Candida cylindracea* lipase to selectively acylate the 6-OH group of different sugars with vinyl acetate. Sugars' solubility problems can also be overcome by protecting the free anomeric hydroxyl group. For instance, Wong reported the regioselective acetylation of the primary hydroxy groups of furanosides by porcine pancreatic lipases in tetrahydrofuran (THF) (13).

Later on, Klibanov and coworkers suggested the use of another group of hydrolases, namely proteases (14). He found that subtilisin, a protease from *Bacillus licheniformis*, was both stable and active in numerous organic solvents, including pyridine and DMF. Consequently, a number of interesting primary monobutanoyl esters of mono- and even di- and oligosaccharides were enzymatically prepared on a preparative scale (Figure 6.2). In the case of sucrose (**2**), the unusual esterification of the 1′-OH (chemically less reactive than the other two primary hydroxyls) was observed.

Carrea *et al.* subsequently carried out a systematic investigation of the usefulness of this approach for the synthesis of a series of 1′-OH sucrose esters, bearing acyl groups of different sizes and types (15). These compounds could easily be converted, via α-glucosidase-promoted hydrolysis, into the corresponding 1-*O*-acyl fructose (Figure 6.3).

As a further example, subtilisin-catalysed esterification of lactosides (**3b**) was recently used for a chemo-enzymatic approach to 6′-deoxy-6′-fluoro- and 6-deoxy-6-fluoro-lactosides (16).

Regioselective discrimination of sugars' secondary hydroxyl groups is even more attractive and challenging from a synthetic point of view. The first report

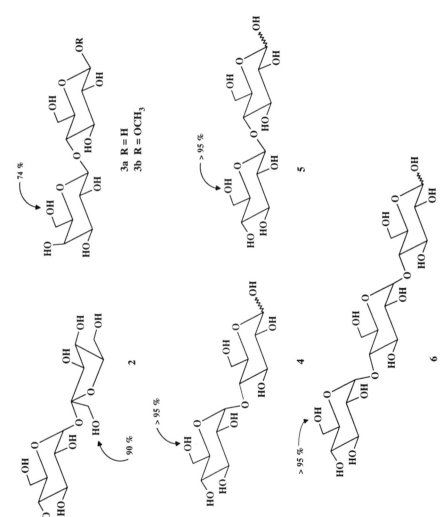

Figure 6.2 Acylation of di- and tri-saccharides catalysed by subtilisin in DMF (from Ref. 14).

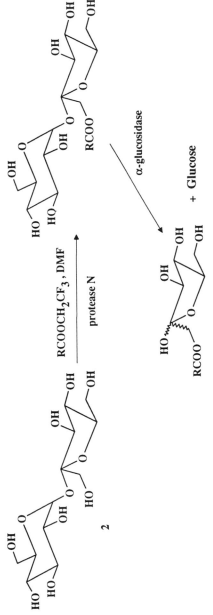

Figure 6.3 Enzymatic two-steps preparation of 1-*O*-acyl fructoses (from Ref. 15)

Table 6.1 Regioselective acylation of secondary hydroxyl groups in 6-O-butanoyl monosaccharides catalysed by different lipases in organic solvents (17)

6-O-Butanoyl sugar	Product composition: ratio of 3,6-di-O-butanoyl to the 2,6-isomer			
	Chromobacterium viscosum lipase	Porcine pancreatic lipase	*Candida cylindracea* lipase	*Aspergillus niger* lipase
Glucose	>50	0.1	0.4	>50
Galactose	1.5	1.5	0.5	4.9
Mannose	4.0	0.7	2.0	1.5

dealing with the enzymatic acylation of secondary hydroxyl groups was again published by Klibanov's group (17). For instance, using 6-O-butanoyl-glucose (**1a**, R = Pr) as substrate in THF, lipase from *Chromobacterium viscosum* showed an overwhelming preference for the 3-OH, while porcine pancreatic lipase directed its action to the 2-OH. The same results were obtained when the primary 6-OH was protected either as trityl ether or as *t*-butyl-diphenyl-silyl ether. The results obtained with four different lipases on **1a** and other 6-O-butanoyl-monosaccharides are reported in Table 6.1. The most striking point is the possibility of modulating the acylation position (2-OH or 3-OH) simply by changing lipases sources.

Analogously, Riva and coworkers reported an interesting example of complementary acylation by different lipases (18). Enzymatic esterification of 1,4-anhydro-5-O-hexadecyl-D-arabinitol (**7**) in benzene afforded alternatively the monoacylated products **7a** or **7b** (Figure 6.4).

Another Italian group described the enzymatic acylation of methyl α- and methyl β-L-arabinopyranoside and methyl α-D-xylopyranoside (**8**), yielding useful reference compounds for the structural elucidation of plant polysaccharides and saponins (19). The lipase was from *Candida cylindracea*, and ethyl acetate was both the solvent and the acylating agent. More recently, selective acylation of alkyl β-D-xylopyranosides by action of lipase PS was reported (20).

The acylation of sugars' benzylidene derivatives, useful intermediates in the synthesis of oligosaccharides, occurs with significant regioselectivity. For instance, acylation of 4,6-O-benzylidene-α-D-glucopyranoside (**9**) with vinyl acetate by action of *Pseudomonas cepacia* lipase gave quantitatively the 2-O-acetate **9a**. Results obtained with similar sugar derivatives are reported in Table

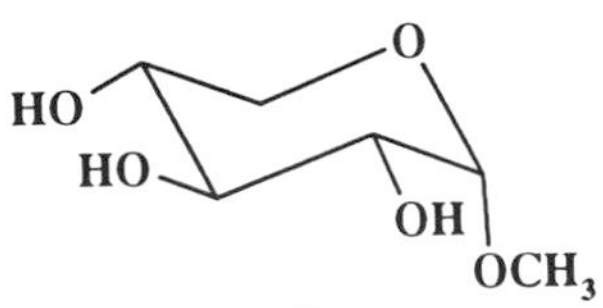

8

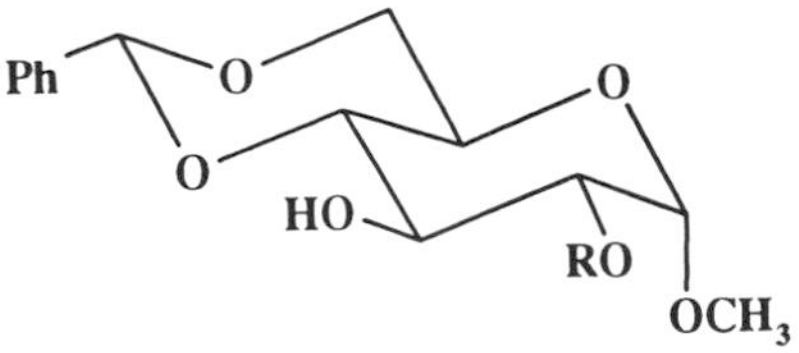

9   R = H
9a   R = Ac

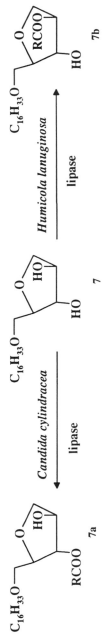

Figure 6.4 Complementary regioselective acylation of secondary hydroxyls (from Ref. 18).

Table 6.2 Lipase catalysed esterification of benzylidene glycopyranosides (21)

Substrate	% 2-O-Acetete	% 3-O-Acetate
Methyl 4,6-O-benzyliden-α-D-glucopyranoside (**9**)	100	0
Methyl 4,6-O-benzyliden-β-D-glucopyranoside	6	94
Allyl 4,6-O-benzyliden-α-D-galactopyranoside	0	0
Allyl 4,6-O-benzyliden-β-D-galactopyranoside	2	98
Methyl 4,6-O-benzyliden-α-D-mannopyranoside	97	3
Methyl 4,6-O-benzyliden-β-D-mannopyranoside	2	98

6.2. These data were obtained independently by Riva and coworkers (21) and by Roberts and coworkers (22, 23).

The results obtained by Ronchetti and coworkers were also quite interesting (24). They compared the acylation of methyl 6-deoxy-pyranosides belonging to opposite steric series by the action of three different lipases. Two sugars, the D-fucoside **10** and the L-rhamnoside **11**, were the best substrates with all the enzymes. This result was rationalized by hypothesizing that the orientation of the sugar into the acylated active site of the lipases was mainly determined by the three secondary hydroxyl groups. As indicated in Figure 6.5, **10** and **11** possess the same sequence, i.e. axial–equatorial–equatorial, the the acylation occurred at the same terminus of this hydroxyl triplet. Similar results were obtained by comparing the 'parent' sugars, methyl α-galactopyranosides and methyl α-mannopyranosides, previously protected at their primary hydroxyl groups (25), and with the corresponding glucopyranosides and 6-deoxygluco-pyranosides (26).

In closing this section on carbohydrates, attempts to increase sugar solubilities in more suitable non-polar solvents by the addition of organoboronic acids (27–29) should be mentioned. However, satisfactory results were obtained only with glucose.

6.2.2 Natural glycosides

Glycosides of various classes of natural products are widely distributed in nature, where they are often present esterified with aliphatic and aromatic acids (mainly

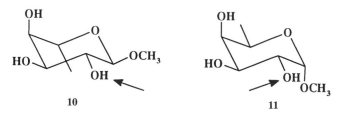

Figure 6.5 Ronchetti's rationale for the selective acylation of sugars' secondary hydroxyls (from Ref. 24).

acetic, malonic, *p*-coumaric and ferulic) at specific hydroxyl groups of their sugar moieties. Many of these compounds are bioactive molecules or possess other interesting properties. In nature, the formation of these esters is the last step in the biosynthetic pathway, and it is catalysed by different acyltransferases. These enzymes show relative flexibility towards the acyl group, but strict selectivity for the substrate to be esterified. Moreover, they require stoichiometric amounts of the corresponding acyl-coenzyme A, and therefore they are not very convenient for *in vitro* laboratory synthesis. On the other hand, direct selective chemical acylation of glycosides is still a distant target because of the present lack of suitable reagents and protocols.

In one of the above-mentioned papers (14), Klibanov and coworkers showed that subtilisin was also able to acylate natural glycosides regioselectively. The reaction was performed on four model substrates: salicin (**12**), riboflavin (**13**), and the nucleosides adenosine (**14**) and uridine (**15**). Not only was subtilisin reactive with **12–15**, but it also showed absolute selectivity for their sugar moieties even in the presence of reactive functional groups on the aglycones (as in **12** and **14**, Figure 6.6).

The same protocol was later applied to other natural glycosides.

6.2.2.1 Flavonoid glycosides. Flavonol glycosides and their esters are an important group of natural compounds widely distributed in the plant kingdom. In a first report (30), Danieli and coworkers applied the usual methodology (subtilisin suspended in a pyridine solution of the substrate and of the activated

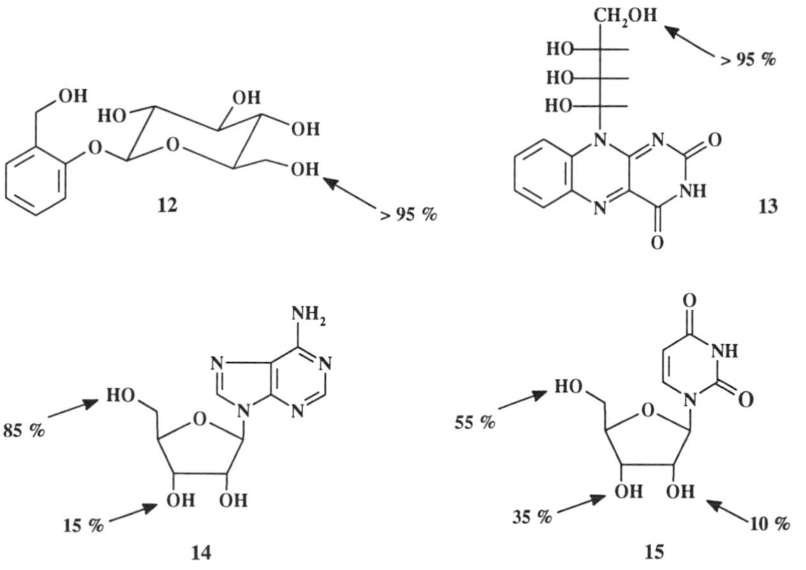

Figure 6.6 Acylation of natural glycosides catalysed by subtilisin in DMF (from Ref. 14).

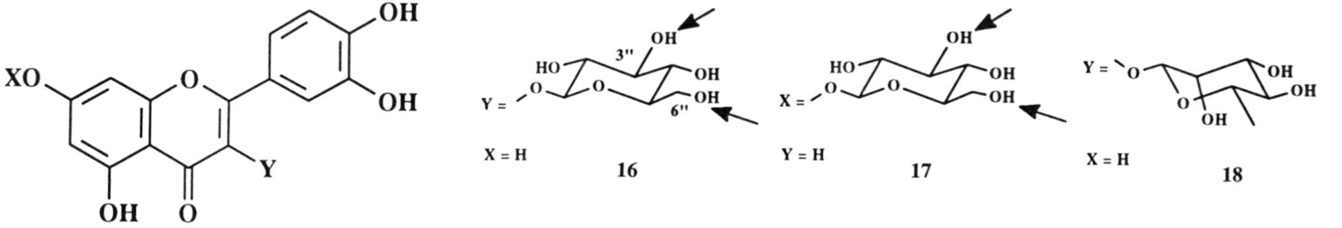

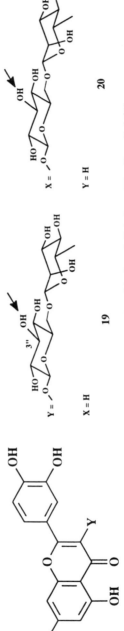

Figure 6.7 Acylation of flavonol monosaccharides monoglycosides by subtilisin in pyridine (from Ref. 30).

Figure 6.8 Acylation of flavonol disaccharides monoglycosides by subtilisin in pyridine (from Ref. 31).

ester trifluoroethyl butanoate) to three flavonol monoglycosides: isoquecritrin **16**, luteolin 7-glucoside **17**, and quercitrin **18**. The two glucosides **16** and **17** were acylated to afford the corresponding 6″-O-butanoates, accompanied by smaller amounts of the 3″-O-mono- and 3″,6″-O-di-butanoates. In contrast, the rhamnoside **18** was completely unreactive (Figure 6.7).

More complex flavonol disaccharide monoglycosides were then considered (31), and subtilisin showed a remarkable behaviour towards the regioselectivity of acylation. The results were particularly impressive with rutin **19** and hesperidin **20**, giving single monoesters at the 3″-OH of their glucose moieties. It should be emphasized that only one hydroxyl group was acylated, in preference to six secondary and four phenolic hydroxyl groups (Figure 6.8).

Later on, a chemo-enzymatic approach to some phenylpropenoate esters (cinnamate, coumarate, feruloate) of **16** was used to overcome the inability to introduce these acyl moieties directly, by transesterification with the corresponding activated esters (32). The target products **16c–16e** were obtained through initial subtilisin-catalysed introduction of a methyl malonate residue at the 6″-OH, enzymatic chemoselective hydrolysis of the mixed diester intermediate **16a** by the action of biophine esterase, and Knoevenagel-type condensation of the malonic monoester **16b** with the appropriate aromatic aldehydes.

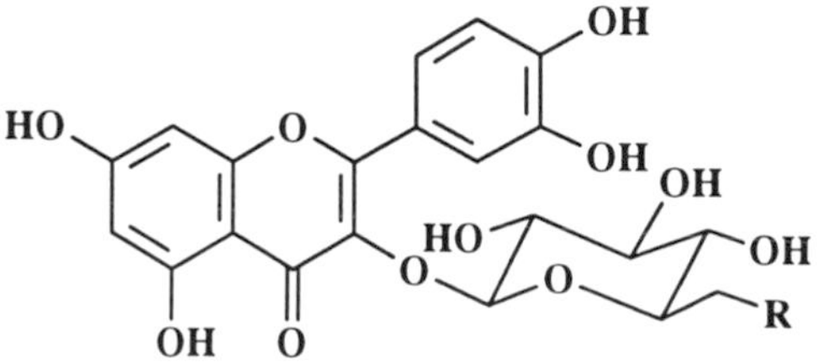

16a R = OCOCH$_2$COOMe
16b R = OCOCH$_2$COOH

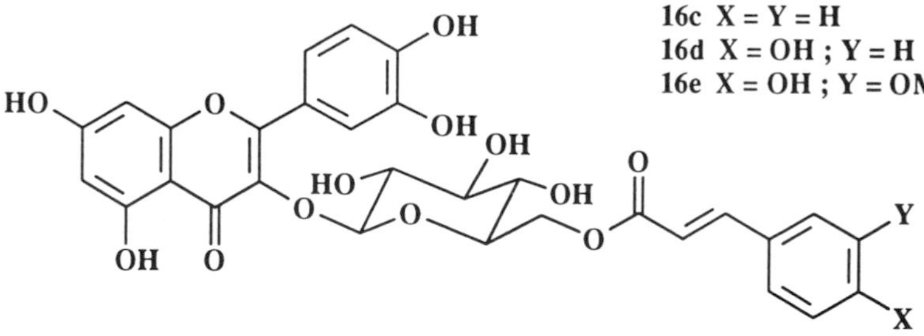

16c X = Y = H
16d X = OH ; Y = H
16e X = OH ; Y = OMe

6.2.2.2 Terpene and alkaloid glycosides. Ginsenosides are dammarane-type triterpene oligoglycosides which are isolated from *Panax ginseng* C.A. Meyer, a plant widely used in the traditional Chinese medicine. Some of these ginsenosides are present as monoesters of malonic acid, the acylation site occurring invariably at one of the primary hydroxyl groups of their sugar moieties. With these substrates, e.g. ginsenoside Rg_1 **21**, subtilisin gave unsatisfactory results, both in terms of reaction rate and selectivity. More suitable conditions were investigated, and the best results were obtained with the lipase from *Candida antarctica* in *t*-amyl alcohol (33). Using vinyl acetate as acyl donor, a complete conversion to only two products in a 22:1 ratio took place, the main compound being identified as $6'$-O-acetylginsenoside Rg_1 **21a**. Subsequently, a new chemoenzymatic approach allowed the preparation of the corresponding $6'$-O-malonate **21b**.

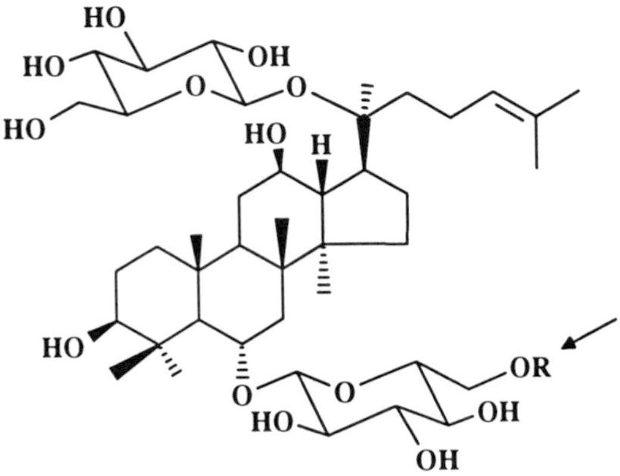

21   R = H
21a R = COCH$_3$
21b R = COCH$_2$COOH

Excellent results were obtained with the more complex ginsenosides Rb_1 **22** (containing four glucoses) and Rg_3 **23**, which gave with good selectivity the $6''''$-O-acetate **22a** and $6''$-O-acetate **23a** respectively (34).

Using the same lipase, the sweet diterpene glucoside stevioside **24** gave an acetylation at the outer glucose to form the $6''$-O-acetyl derivative **24a**. Quite surprisingly, the steviol bioside **25** underwent acetylation at the primary hydroxyl group of the inner glucose to afford **25a** (35).

The natural alkaloid colchicoside **26** and its sulphur synthetic analogue **26a** have been acylated by subtilisin in pyridine. High degrees of conversion with excellent selectivity for the primary hydroxyl group of the glucose moiety have been obtained with isopropenyl acetate and trifluoroethyl butanoate (36).

6.2.2.3 Nucleosides. Following the first examples by Klibanov (14), this area was carefully investigated by Gotor and Moris. They used oxime esters as irreversible acyl transfer agents (37). With lipase PS, the regioselective acylation of $2'$-deoxynucleosides on the secondary alcohol group, instead of the more chemical reactive primary hydroxyl group, was obtained (Figure 6.9) (38). On

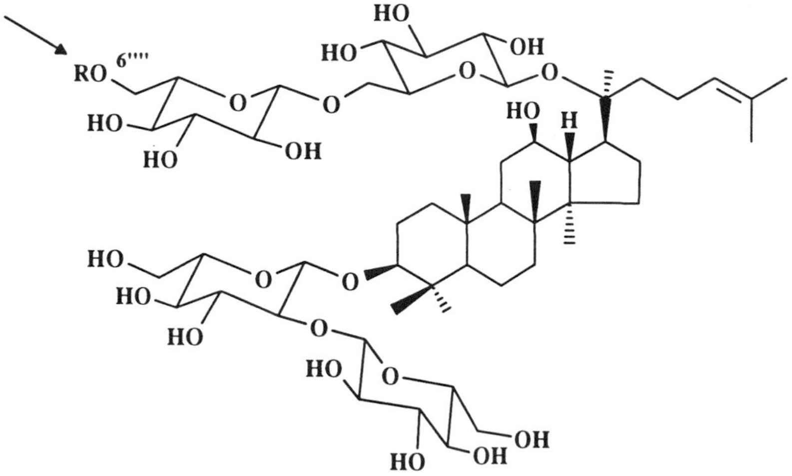

22   R = H
22a R = Ac

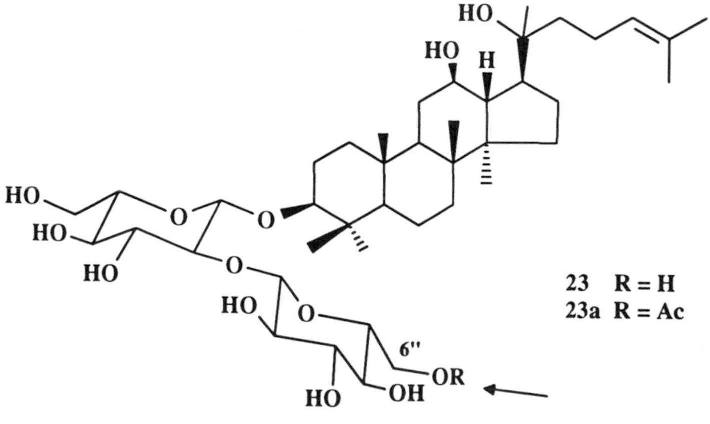

23   R = H
23a R = Ac

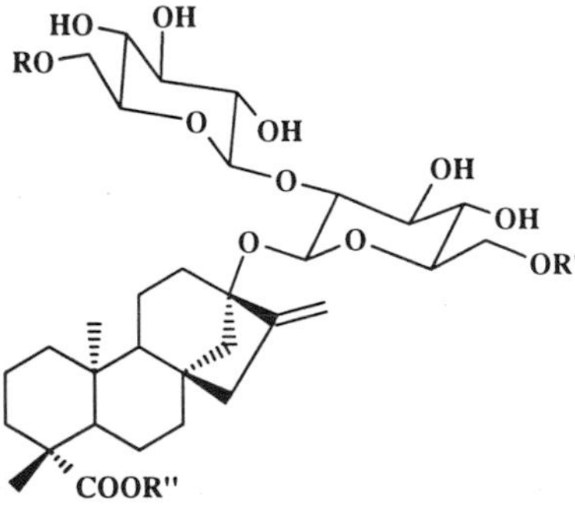

24   R = R' = H ; R'' = β-glc
24a R = Ac ; R' = H ; R'' = β-glc
25   R = R' = R'' = H
25a R = R'' = H ; R' = Ac

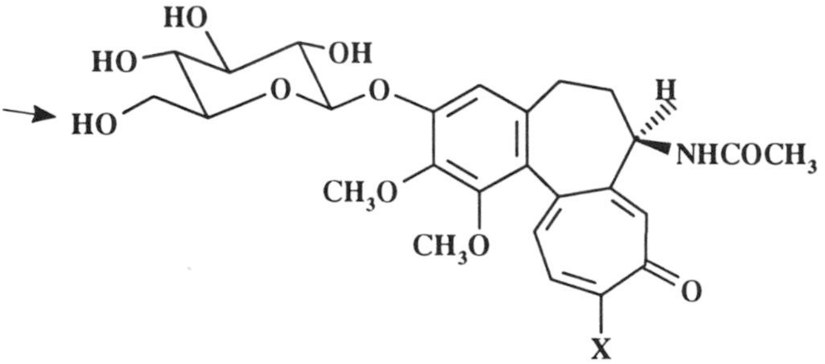

26   X = OCH₃

26a X = SCH₃

the other hand, lipase from *Candida antarctica* gave the expected 5'-*O*-acyl derivatives of numerous nucleosides (39, 40). Excellent results were also obtained by Wong and coworkers by exploiting mutants of subtilisin obtained by site-directed mutagenesis (41, 42).

6.2.2.4 Acylation of aglycones. In all the examples reported so far, enzymatic esterification occurred at the sugar moiety of natural glycosides. However, there are two examples in the literature in which the acylation occurred on the aglycone.

In the first one, the cytokinins zeatin riboside (**27**) and 1″-methylzeatin riboside (**28**) were acylated by subtilisin in pyridine at the 4″-OH, furnishing the corresponding 4″-*O*-acetates **27a** and **28a** (43).

More recently, lipase-catalysed acetylation of 2-*O*-(β-D-glucopyranosyl)-glycerol (**29**) furnished alternatively the (2*S*)-1-*O*-acetyl- (**29a**) or the (2*R*)-1-*O*-acetyl derivative (**29b**, the natural compound lilioside A), depending on the enzyme used (lipases from *Pseudomonas cepacia* and from *Candida antarctica*, respectively) (44).

6.2.3 Other polyols

Obviously, hydrolases' ability to discriminate among different hydroxyl groups is not limited to sugar structures. The following paragraphs will exemplify regioselective transformations of other classes of polyols.

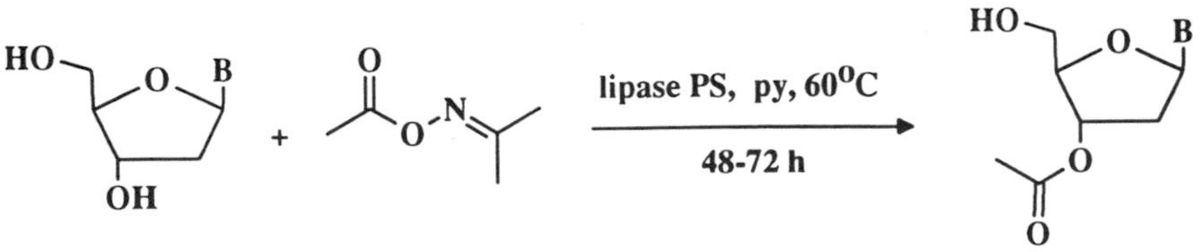

Figure 6.9 Acylation of 2'-deoxy-nucleosides (from Ref. 38).

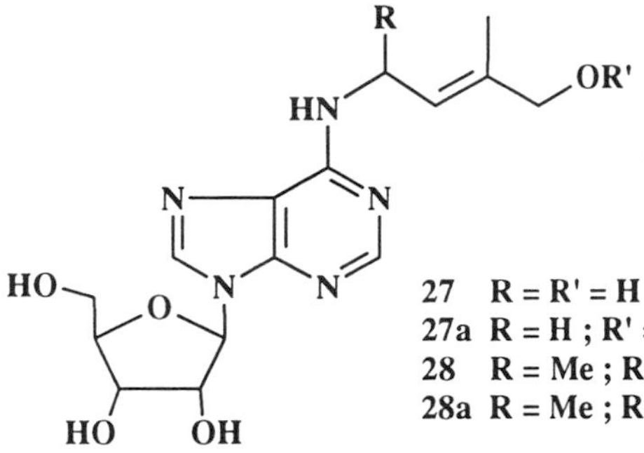

27   R = R' = H
27a R = H ; R' = Ac
28   R = Me ; R' = H
28a R = Me ; R' = Ac

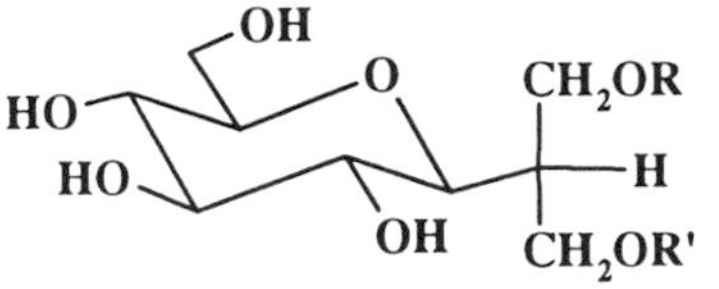

29   R = R' = H
29a R = Ac ; R' = H
29b R = H ; R' = Ac

6.2.3.1 Cyclitols. Quinic acid (**30**) is an important secondary metabolite widely distributed in higher plants and is receiving growing attention as useful chiral synthetic precursor for natural compounds. Utilization of **30** takes advantage of the selective manipulation of the functional groups on the cyclohexane ring. However, while 5-OH can be easily protected by lactonization to quinolactone, only two chemical processes are known to discriminate between 3-OH and 4-OH in a quite inefficient way (either low yield or multi-step procedure). Compound **30** and its methyl and benzyl esters (**30a**, **30b**) were acylated in organic solvents by several lipases and by subtilisin. The most satisfactory results were obtained with methyl (or benzyl) quinate and lipase from *Chromobacterium viscosum*, which showed an overwhelming preference towards the acylation of 4-OH (45). Under optimized conditions, the synthetically useful 4-*O*-acetylquinate **31b** was obtained in about 90% yield (Figure 6.10). Other examples of acylation of cyclic polyols are shown in Figure 6.10 (46–48).

Myo-Inositol (**32**) is another very interesting cyclitol. Various inositol phosphates and related inositol phospholipids display a wide variety of biological activities, e.g. as second messengers in cell regulation processes. Chiral derivatives of **32** are among the most important central intermediates for the synthesis of biologically active inositol polyphosphates, compounds that otherwise are only accessible with difficulty from scarce natural sources, and different groups have investigated chemo-enzymatic approaches to some of these intermediates (49–51). Recently, Schneider and Ardersch suggested a very

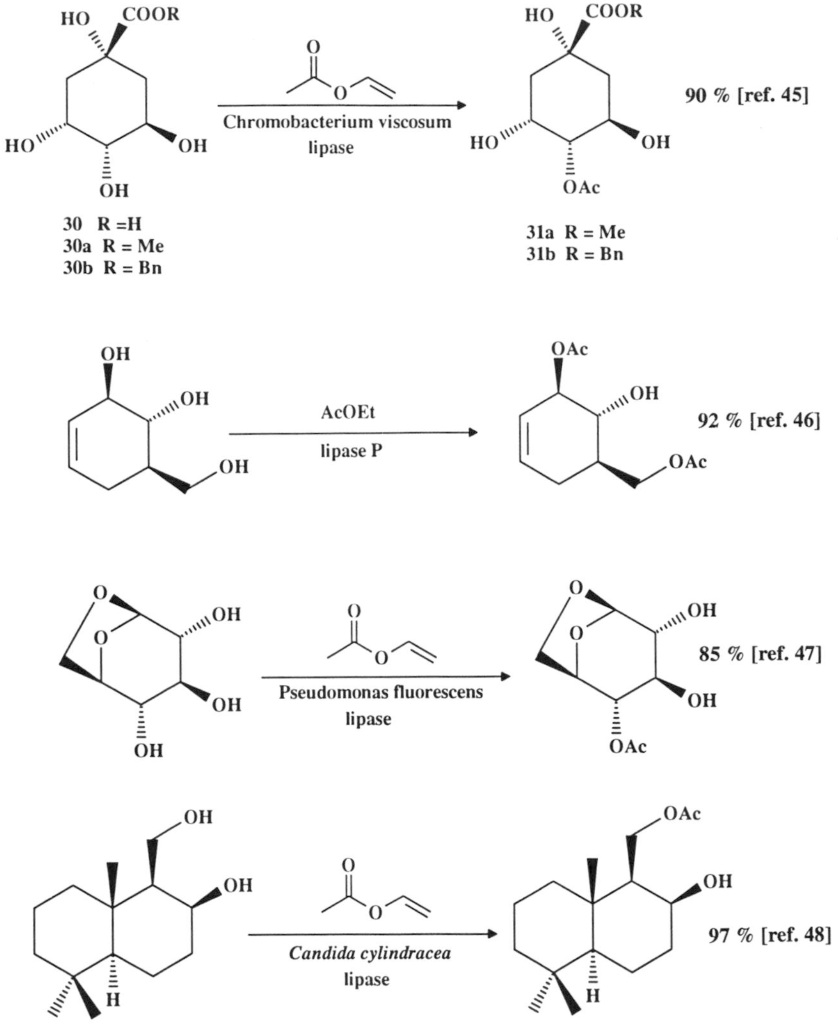

Figure 6.10 Acylation of cyclic polyols.

elegant and efficient solution to this problem (52). As shown in Figure 6.11, the key compound is the meso-derivative **33**, easily obtained from **32** on a multi-gram scale through simple standard procedures. Lipase from *Pseudomonas* acylated **33** regio- and enantioselectively, giving the enantiomerically pure (−)-**34**. This compound can be manipulated to obtain all the different inositol phosphates in optically pure forms, e.g. D-*myo*-inositol 1,4,5-triphosphate (**35**).

6.2.3.2 The 'meso trick'.   Enzymatic acylation of the inositol derivatives **33** is a brilliant example of the so-called 'meso trick', that is, the enantioselective

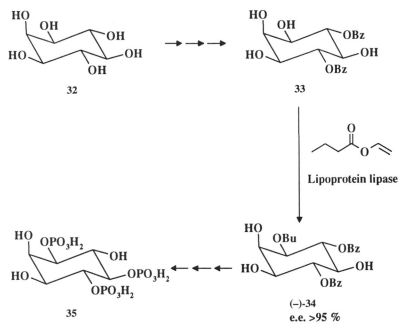

Figure 6.11 Preparation of D-*myo*-inositol 1,4,5-triphosphate via Schneider's intermediate (from Ref. 52).

transformation of a meso substrate by the differentiation of its enantiotropic group to give, theoretically, only one enantiomer in quantitative chemical and optical yield. This aspect has been discussed in detail in the previous chapter. Here it is just worth remembering that several examples deal with the selective esterification of cyclic and acyclic polyols derivatives, such as for instance **36** (53), **37** (54), and **38** (55).

Several examples of the resolution of racemic polyols have also been reported. Just to give one example, Figure 6.12 shows the resolution of the mucolytic drug (±)-*trans*-sobrerol (**39**) by action of lipase PS. In *t*-amyl alcohol the acetylation of the secondary hydroxyl group spontaneously stopped at 50% conversion, leaving the two compounds (+)-**39** and (−)-**39a** in practically 100% optical purity (56).

6.2.3.3 Steroids.   In a first report by Riva and Klibanov (57), *Chromobacterium viscosum* lipase and subtilisin have been used to acylate the model dihydroxy-steroid 5α-androstane-3β,17β-diol (**40**) in dry acetone. The two hydroxyl groups in **40** were esterified with opposite regioselectivities: while the lipase trans-formed exclusively the 3-OH, subtilisin showed a marked preference for the 17-OH. The reactivities of numerous hydroxysteroids (**40–52**) with *Chromobacterium viscosum* lipase and subtilisin were examined and the results are reported in Table 6.3. It was evident that the lipase was very sensitive to the immediate environment of the 3-OH group, accepting only the steroids with an A/B ring fusion in the *trans* configuration and the 3-OH in the equatorial (β) position, and was quite insensitive to variations of the side chain. Conversely, changes in the A or B ring did not dramatically affect the acylation of steroids by subtilisin, while very low reaction rates were observed with compounds **47–50**, confirming the preference of this protease for C-17 or side-chain hydroxyls. Reactions were then scaled up, and the monoacylated steroids were isolated in multi-gram amounts.

Their strict regioselectivity prevented these two hydrolases from acylating hydroxysteroids with different stereochemical characteristics, such as, for instance, deoxycholic acid methyl ester (**53**), which has a *cis* A/B ring fusion and no hydroxyl groups on the D ring or on the side chain. However, simply

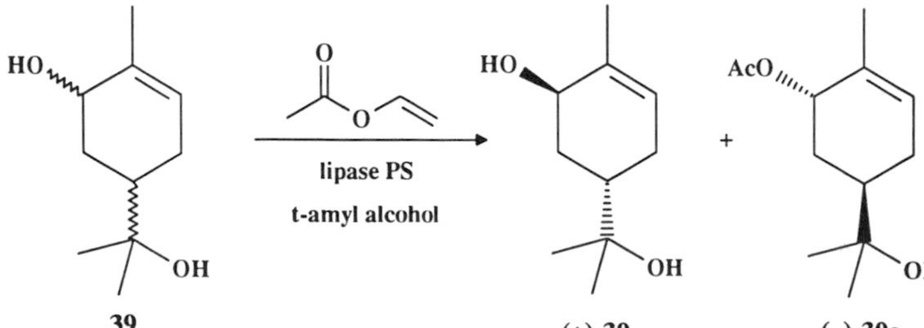

Figure 6.12 Resolution of the mucolytic drug (±)-*trans*-sobrerol (from Ref. 56).

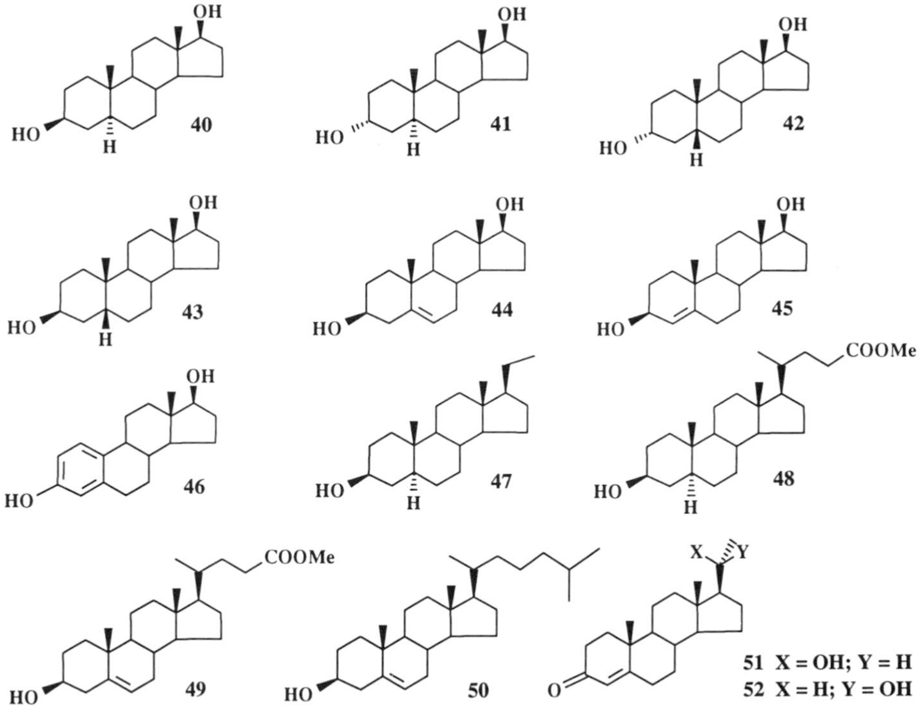

by changing the solvent and moving from acetone to the more hydrophobic benzene, a different lipase (from *Candida cylindracea*) was able to acylate **53** and several other bile acid derivatives regioselectively (58). The substrate specificity of *Candida cylindracea* lipase was investigated similarly, and the

Table 6.3 Initial rates of acetylation of various hydroxysteroids catalysed by *Chromobacterium viscosum* lipase and subtilisin in acetone (57)

Steroid		Initial rate (µmol/h)	
		Ch.v. lipase	Subtilisin
3β,17β-Dihydroxy-5α-androstane	(**40**)	3.30	0.63
3α,17β-Dihydroxy-5α-androstane	(**41**)	0	0.53
3β,17β-Dihydroxy-5β-androstane	(**42**)	0	0.41
3α,17β-Dihydroxy-5β-androstane	(**43**)	0	0.63
3β,17β-Dihydroxy-5-androstene	(**44**)	1.72	0.67
3β,17β-Dihydroxy-4-androstene	(**45**)	0	0.32
17β-Estradiol	(**46**)	0	0.63
3β-Hydroxy-5α-pregnane	(**47**)	6.26	0.06
3β-Hydroxy-5α-cholanic acid methyl ester	(**48**)	8.32	0
3β-Hydroxy-5-cholenic acid methyl ester	(**49**)	2.75	0
3β-Hydroxy-5-cholestene (cholesterol)	(**50**)	3.00	0
3β,20β-Dihydroxy-5α-pregnane	(**51**)	4.54	0.10
3β,20α-Dihydroxy-5α-pregnane	(**52**)	4.34	0.55

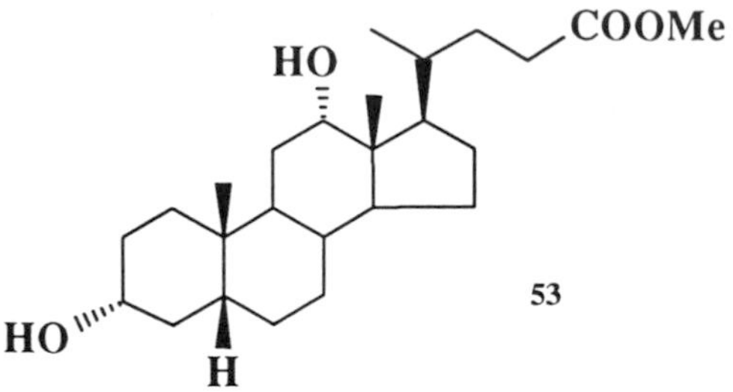

results showed that the stereochemical requirements of this enzyme were less strict than those of *Chromobacterium viscosum* lipase. Different A/B ring junctions were accepted by the enzyme, and different 3-*O*-acyl derivatives were isolated. Again, no acylation occurred on the other limb of the steroid skeleton or on the side chain.

Finally, very recently the selectivity of a third lipase, from *Candida antarctica*, was studied (59). As shown in Table 6.4, hydroxyl groups at the C-3 position were always esterified, though at different rates: the best substrate (3α,5β) was acylated about 170 times faster than the worst one (3β,5β). On the other hand, this enzyme was inactive with the alcoholic moieties on the internal B and C rings; as none of the hydrolases investigated so far has been found to be able to interact with these hydroxyl groups, it is likely that this is an exclusive prerogative of specific enzymes belonging to other classes of activity, e.g. the hydroxysteroid dehydrogenases. Finally, the data relative to the D ring and the side chain were quite intriguing. While C-17 hydroxyl groups were not substrates, the enzyme showed an interesting stereoselectivity for the C-20 hydroxyl groups (entries 10 and 11 of Table 6.4). The primary C-21 hydroxyl

Table 6.4 Initial rates of acetylation of various hydroxysteroids catalysed by *Candida antarctica* lipase (59)

Steroid	Initial rate	Relative rate
3β-Hydroxy-5α-androstane-17-one	1.12	17
3α-Hydroxy-5α-androstane-17-one	0.13	2
3α-Hydroxy-5β-androstane-17-one	12.06	172
3β-Hydroxy-5β-androstane-17-one	0.07	1
3β-Hydroxy-4-androstene-17-one	0.41	6
3β-Hydroxy-5-androstene-17-one	1.11	16
6β-Hydroxy-4-androstene-3,17-dione	0	—
17β-Hydroxy-5β-androstane-3-one	0	—
17β-Hydroxy-5α-androstane-3-one	0	—
20α-Hydroxy-4-pregnene-3-one	0	—
20β-Hydroxy-4-pregnene-3-one	0.44	6
21-Hydroxy-4-pregnene-3,20-dione	37.06	529
7α-Hydroxy-3-oxo-5β-cholan-24-oic acid methyl ester	0	—
7β-Hydroxy-3-oxo-5β-cholan-24-oic acid methyl ester	0	—
12α-Hydroxy-3-oxo-5β-cholan-24-oic acid methyl ester	0	—

group was also easily acetylated, being the best substrates of the series. Therefore, contrary to the behaviour of the other three hydrolases, *Candida antarctica* lipase does not have an exclusive preference for a portion of the steroid skeleton, but it accommodates these molecules in its active site in different ways, even if with distinctive preferences towards the positional substitutions.

Two more papers dealing with the enzymatic modifications of steroids in organic solvents will be discussed later (60).

6.2.3.4 Alkaloids. The α-glucosidase-I inhibitors castanospermine (**54**) and 1-deoxynojirimicin (**55**) affect the processing of glycoproteins. Recent data suggested that these alkaloids have potential anti-HIV activity. As it has been reported that esters of **54** and **55** are more active than the parent compounds in inhibiting HIV replication, Margolin and coworkers exploited the regio-selectivity of hydrolases to get numerous specific ester derivatives of these alkaloids. As it is shown in Figure 6.13, subtilisin acylated castanospermine (a molecule possessing four secondary hydroxyl groups with similar chemical reactivity) quite efficiently in pyridine, giving the corresponding 1-*O*-acyl derivatives in good yield (61). It was possible to regulate the hydrophobicity of the acylating group and, besides butanoate, other 1-*O*-aliphatic, aromatic, and aminoacidic esters were produced. In a subsequent step, the more soluble 1-*O*-butanoyl castanospermine (**54a**) was dissolved in THF and acylated with several lipases. The best results were obtained with the lipase from *Chromobacterium viscosum*, which gave the 1,7-dibutanoyl derivative **54b** in high yield. Finally, regioselective hydrolysis of **54b** by action of subtilisin allowed the isolation of the pure 7-*O*-butanoate **54c** in pure form.

Unlike castanospermine, 1-deoxynojirimycin has primary and secondary hydroxyl groups and also a more reactive amino function. As shown in Figure

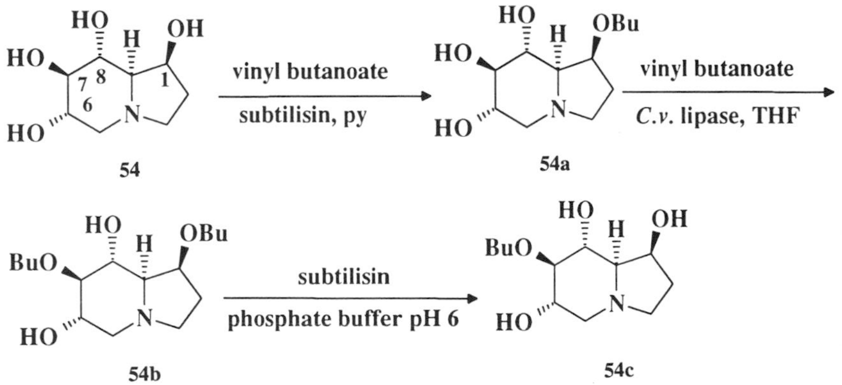

Figure 6.13 Synthesis of various esters of the alkaloid castanospermine by different hydrolases (from Ref. 61).

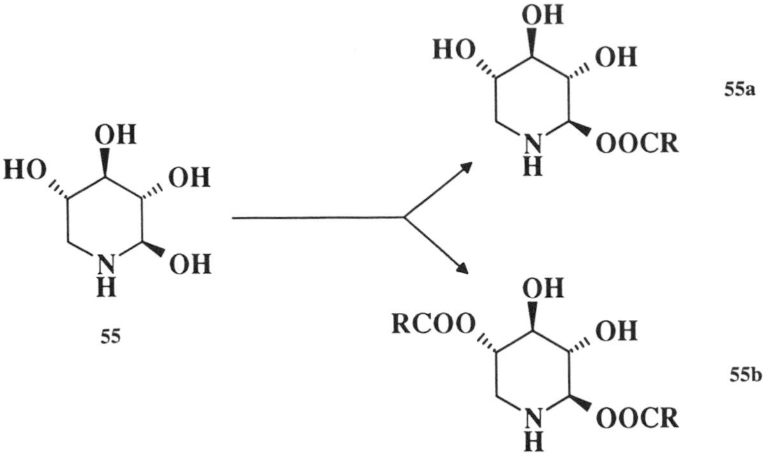

Figure 6.14 Synthesis of various esters of the alkaloid 1-deoxynojirimicin by subtilisin in pyridine (from Ref. 62).

6.14, acylation of **55** by subtilisin in pyridine in the presence of a small excess of trichloroethyl butanoate led predominantly to the 6-*O*-monobutanoate **55a**. A large excess of the acylating agent resulted mostly in the 2,6-*O*-dibutanoyl derivative **55b**. Neither the more reactive amino group nor the 3-OH or the 4-OH was acylated (62).

6.2.4 *Esterification with different acyl moieties*

In the large majority of the examples discussed so far, the acylating agents were simple aliphatic acids, almost always acetate and butanoate. However, hydrolases can catalyse transesterifications even with different acyl moieties, some of which are reported in Figure 6.15.

Compound **7** was enzymatically acylated with the esters **56–59** (63). Specifically, it has been pointed out that ω-halo esters are suitable for subsequent chemical manipulation of their alyphatic chain (64).

4,6-*O*-Benzyliden-α-D-glucopyranoside (**9**), as well as its β analogue, have been acylated with different synthetically useful esters, such as benzoate (**60**), chloroacetate (**61**), pivaloate (**62**), and levulinate (**63**) (65). Alkoxycarbonylation (introduction of Cbz) of 2-deoxysugars has been reported by Gotor, using his usual oxime derivatives, specifically (**64**) (66). Selective acylation of sugars with amino acids has been initially reported by Klibanov and coworkers (14), and more recently by Monsan and coworkers (67). Finally, regioselective preparation of hemisuccinates of polyhydroxylated steroids has been obtained using succinic anhydride and the above described hydrolases (68). All of them showed the same preference for 3-OH groups. This finding was quite surprising for subtilisin, which showed a selectivity opposed to that previously found with trifluoroethyl

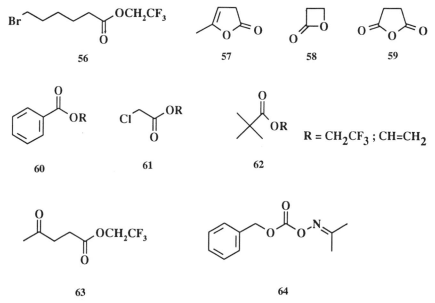

Figure 6.15 Different activated esters that have been used as acyl donors.

butanoate (57). A similar influence of the nature of the acylating agent on regio-selectivity has been observed in the acylation of butyl α-D-glucopyranoside with anhydrides (69).

As a general remark, it can be said that hydrolases show a high versatility towards the acylating agent. However, generally speaking, it is not true that it is possible to acylate any substrate with any kind of ester. As has been pointed out by Riva and coworkers studying the acylation of sugars (15) and of flavonoid glycosides (32), it is likely that a reciprocal steric hindrance occurs between large acyl groups and large nucleophiles, the latter being excluded by the catalytic site and thus prevented from attacking the acyl-enzyme intermediate.

6.2.5 Miscellaneous

Besides the above-mentioned use of organoboronic acids (27–29), another approach has been suggested for the acylation of hydrophylic polyols (glycols and sugars) in hydrophobic solvents. According to this methodology, the substrates have been preadsorbed on silica gel (1 or 2% weight of polyol). Then the solid mixtures have been reacted with fatty acid vinyl esters in an appropriate organic solvent (e.g. di-isopropyl ether or methyl *t*-butyl ether) and in the presence of either *Mucor miehei* (70) or porcine pancreatic (71) lipase.

Conversely, besides side-directed mutagenesis, stability of hydrolases in polar organic solvents such as DMF has been increased by reductive alkylation with acetaldehyde, propionaldehyde, octaldehyde, and benzaldehyde. These alkylated enzymes were more stable in DMF compared to the native unmodified enzyme at temperatures between 26 and 60°C (72).

Finally, while the use of solvent engineering to control enzyme enantio-selectivity has been described in great detail, there are only a few studies on controlling regioselectivity. In a first report (73), Rubio and coworkers studied the alcoholysis of octylhydroquinone, butyrylated at both phenolic groups (**65**), with different lipases in different organic solvents. Using *Pseudomonas cepacia* lipase, they found a correlation between the regioselectivity of alcoholysis and the hydrophobicity of the reaction medium, the 4-*O*-butanoate being preferentially hydrolized in hydrophobic solvents.

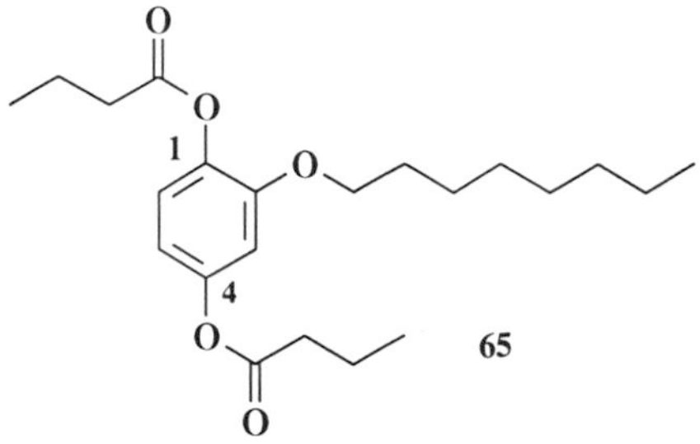

More recently, the influence of solvent on the regioselective acylation of nucleosides (74) and sugars (20) has been reported. In the latter paper, the acylation of octyl β-D-xylopyranoside (**66**) was studied in different solvents. As a general trend, it was observed that esterification at 2-OH was preferred in hydrophobic solvents, while regioselectivity dropped on moving to more hydrophilic media.

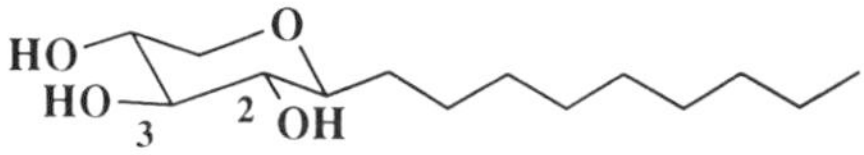

6.3 Enzymatic hydrolysis of peracylated polyhydroxylated compounds

Selective enzymatic hydrolyses are usually performed in water or biphasic systems, and therefore they are beyond the scope of this chapter. However, there are some examples in which these reactions have been performed in organic solvent either at controlled water activity (in this case water is still the nucleo-phile) or containing a suitable alcohol to accomplish a transesterification reaction. As a general remark it has to be pointed out that in these cases the peracylated compounds give a different tetrahedric intermediate and therefore, contrary to what happens in the resolution of racemic mixtures, enzyme selectivity might be completely different in hydrolysis versus transesterification. In the latter case, reaction might even not take place at all.

Regioselective deacetylation of sucrose octa-acetate (**67**) has been obtained with different hydrolases (75). Their positional specificity is indicated in Figure 6.16.

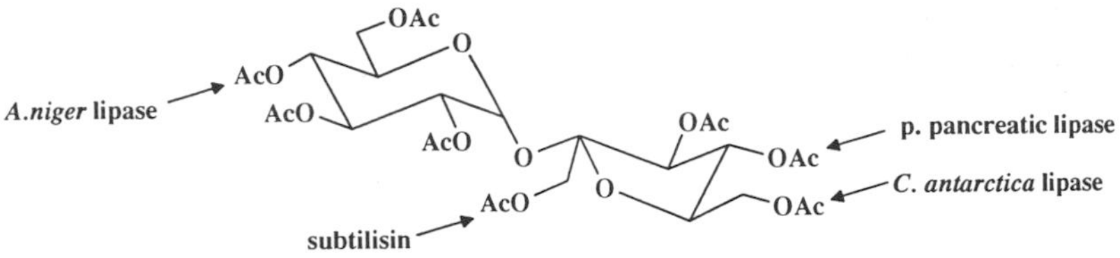

Figure 6.16 Regioselective deacetylation of sucrose octaacetate by different hydrolases (from Ref. 75).

Regioselective deprotections of polyacylated steroids by *Candida cylindracea* lipase-catalysed transesterification with octanol in either di-isopropyl ether or acetonitrile have also been reported (76).

Finally, alcoholysis of flavones acetates has been investigated by Nicolosi (77, 78). Several flavone esters, such as luteolin tetra-acetate (**68**), were subjected to transesterification with *n*-butanol by the action of *Pseudomonas cepacea* lipase in THF. The products, e.g. **68a**, are useful intermediates in the synthesis of flavonoid derivatives rarely occurring in nature and/or possessing biological activity. Similarly, alcoholyses of peracetylated benzopyranones, acetophenones, and chalcones have been reported (79, 80).

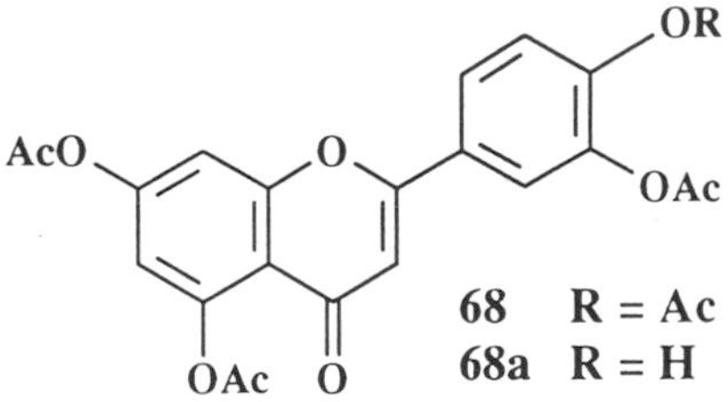

68   R = Ac
68a R = H

6.4 Scaled-up procedures

Enzymatic esterification of polyhydroxylated compounds can be a useful methodology for the preparation of compounds in industrial-scale amounts. Specifically, in recent years the production of sugar-based polymers and surfactants has been investigated. As the former compounds are described in detail in a following chapter, this paragraph will be limited to biosurfactants (surfactants synthesized with biocatalysts) and food emulsifiers.

Fatty acid esters of carbohydrates are biodegradable non-ionic surfactants which are, in principle, obtainable from cheap raw materials. They are used in the food, detergent and cosmetic industries as they compare well with other compounds in emulsification, detergency, and related properties. 'Biosurfactants' can also be considered 'natural' compounds when they are used as food additives.

Enzymatic esterification of sucrose and of various monosaccharides with fatty acids in aqueous media has been tried in the early 1980s, but the selectivity and the yields obtained were low (81). More recently, Chopineau *et al.* described the production of biosurfactants from sugar alcohols and vegetable oils by lipase catalysis in organic media (82). Specifically, porcine pancreatic and *Chromobacterium viscosum* lipases were found able to catalyse regioselective transesterification reactions between a number of sugar alcohols and various plant and animal oils in dry pyridine. The surfactant properties of the monoesters so obtained were compared with the corresponding chemically prepared compounds, and it was found that, thanks to their structural homogeneity, they showed superior surface-active properties in term of reduction of interfacial tension between xylenes and water, reduction of surface tension of DMF, and stabilization of water-in-xylene emulsions. The bottlenecks of this procedure, as evidenced by the authors themselves, were the slow conversion rates and the need to use organic solvents. Other examples of biosurfactant syntheses in organic media have been reported in the following years (83, 84).

To avoid the use of toxic organic solvents such as pyridine, Bjorkling and coworkers developed an enzymatic solvent-free process for the acylation of sugar glycosides (85–87). The reactions were carried out by simply mixing the starting sugar with a fatty acid in the presence of immobilized *Candida antarctica* lipase, the water generated in the acylation reaction being removed *in vacuo* as 70°C. As shown in Table 6.5, alkyl glycopyranosides with longer aliphatic chains were acylated faster, probably due to their increased solubility in the molten fatty acids.

This procedure was scaled up. In a pilot reactor, the reaction was performed on a 20 kg scale using a mixture of α- and β-ethyl-D-glucopyrosides, prepared by glycosylation in ethanol by ion-exchange resin catalysis, and different fatty acids. The reaction yields of 6-*O*-monoester were 85–90% within 24 h, and the enzyme was recycled several times with no noticeable loss of activity.

To overcome the low solubility of sugars in fatty acids, Vulfson and coworkers suggested the use of isopropylidene sugars according to the scheme of Figure 6.17 (88, 89). According to the authors, this process is a slightly more expensive option than the previous one, but it is more versatile and offers a route to monosaccharide fatty acid esters that can be readily extended to disaccharides.

Table 6.5 Acylation of various alkyl D-glucopyranosides with dodecanoic acid by action of *Candida antarctica* lipase (85)

Sugar	Time for 50% conversion	% Conversion after 24 h
Glucose	>1 week	5
Methyl-α-D-glucopyranoside	22 h	53.3
Ethyl-D-glucopyranoside	2.5 h	92.5
Isopropyl-D-glucopyranoside	2.1 h	93.2
n-Propyl-D-glucopyranoside	1.4 h	95.6
n-Butyl-D-glucopyranoside	1.0 h	94.4

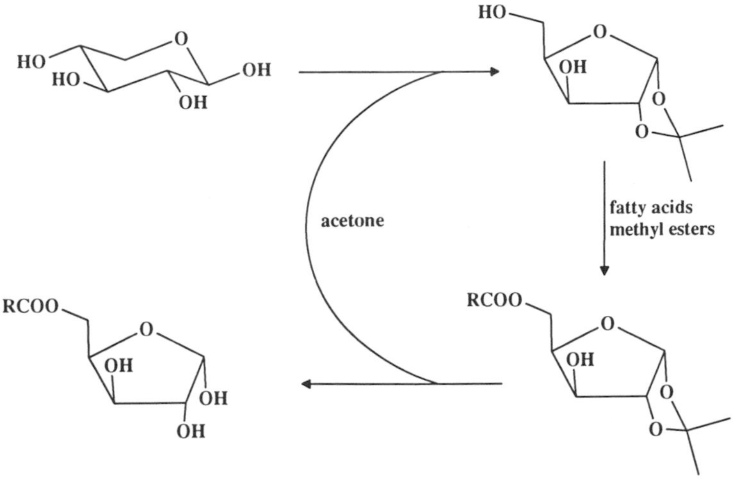

Figure 6.17 Large-scale preparation of long-chain sugar esters (from Refs 88, 89).

As a general consideration, it has to be pointed out that production cost should always be considered in conjunction with actual product performance. Obviously, additional costs would be justified if the emulsifying properties of these biosurfactants were found to be superior in specific applications compared with currently used surfactants.

6.5 Closing remarks

To the best of the author's knowledge, this chapter has described the 'state of the art' in the enzymatic regioselective acylations of polyhydroxylated compounds, at the end of 1994. The first report by Klibanov's group was published only ten years ago (4), and in such a short time this methodology has shown an unexpected versatility and efficiency. Further developments will probably strengthen the synthetic assistance to carbohydrate and natural-products chemists, particularly in the selective discrimination among secondary hydroxyl groups. The solvent-free processes developed by Bjorkling (85–87) and Vulfson (88, 89) seem to be promising for obtaining structurally-definite long-chain sugar esters with surfactant properties. Similarly, the chemoenzymatic synthesis of sugar-containing polymers (90) is receiving increasing attention. It is likely that new enzymes, either natural or genetically engineered, will be screened and utilized. Moreover, up to now the effect of the nature of the organic solvent on the regioselective outcome of the enzymatic reaction (20, 73, 74) has only been addressed superficially. The so called 'medium engineering' might offer new opportunities for synthetic applications.

So far, what is completely missing is a rationalization of the results obtained. In the last few years great attention has been paid to investigations aimed at explaining hydrolases' enantioselectivity and rationales based on X-ray analysis (91) or computer-assisted molecular modelling (92) have been suggested. Conversely, similar studies have not been performed to explain regioselectivity. It is likely that this will be an even more difficult task to accomplish. In fact, enantioselectivity can usually be explained in terms of bulkiness of the substituents at the chiral centre and of their interactions with the amino acids of the active site, interactions which allow the preferential accommodation of one of the two enantiomers in a productive way. It is likely that regioselectivity is mainly determined by electrostatic interactions of the different hydroxyl groups of the nucleophile (which is bulky in any case) with the amino acids of the active site. Therefore, a correct analysis of these interactions requires a detailed knowledge of the enzyme structure and, more specifically, of the active site solvated with the organic solvent. In any case, this effort of rationalization will have to be accomplished in order to overcome the present 'black-box' status of hydrolases concerning their regioselective performances.

References

1. Muller, P. (1994) Glossary of terms used in physical organic chemistry (IUPAC recommendations 1994). *Pure Appl. Chem.*, **66**, 1077–1184.
2. Faber, K. and Riva, S. (1992) Enzyme-catalyzed irreversible acyl transfer. *Synthesis*, 895–910.
3. Waldmann, H. and Sebastian, D. (1994) Enzymatic protecting group techniques. *Chem. Rev.*, **94**, 911–937.
4. Cesti, P., Zaks, A. and Klibanov, A.M. (1985) Preparative regioselective acylation of glycols by enzymatic transesterification in organic solvents. *Appl. Biochem. Biotechnol.*, **11**, 401–407.
5. Ottolina, G., Carrea, G. and Riva, S. (1990) Synthesis of ester derivatives of chloramphenicol by lipase-catalyzed esterification in organic solvents. *J. Org. Chem.*, **55**, 2366–2369.
6. Ramaswamy, S., Morgan, B. and Oehlschlager, A.C. (1990) Porcine pancreatic lipase mediated selective acylation of primary alcohols in organic solvents. *Tetrahedron Lett.*, **31**, 3405–3408.
7. Parmar, V.S., Sinha, R., Bisht, K.S. *et al.* (1993) Regioselective esterification of diols and triols with lipases in organic solvents. *Tetrahedron*, **49**, 4107–4116.
8. Riva, S. and Secundo, F. (1990) Selective enzymatic acylations and deacylation of carbohydrates and related compounds. *Chimica Oggi*, (6), 9–16.
9. Drueckhammer, D.G., Hennen, W.J., Pederson, R.L. *et al.* (1991) Enzyme catalysis in synthetic carbohydrate chemistry. *Synthesis*, 499–525.
10. Riva, S. (1994) Enzymatic synthesis of carbohydrate esters. In *Carbohydrate Polyesters as Fat Substitutes* (Eds C.C. Akoh and B.C. Swanson), Marcel Dekker, New York, pp. 37–64.
11. Therisod, M. and Klibanov, A.M. (1986) Facile enzymatic preparation of monoacylated sugars in pyridine. *J. Amer. Chem. Soc.*, **108**, 5638–5640.
12. Wang, Y.F., Lalonde, J.J., Momongan, M. *et al.* (1988) Lipase-catalyzed irreversible transesterifications using enol esters as acylating agent: preparative enantio- and regioselective syntheses of alcohols, glycerol derivatives, sugars, and organometallics. *J. Amer. Chem. Soc.*, **110**, 7200–7205.
13. Jennen, W.J., Sweers, H.M., Wang, Y.F. *et al.* (1988) Enzymes in carbohydrate synthesis: lipase-catalyzed selective acylation and deacylation of furanose and pyranose derivatives. *J. Org. Chem.*, **53**, 4939–4945.
14. Riva, S., Chopineau, J., Kieboom, A.P.G. *et al.* (1988) Protease-catalyzed regioselective

esterification of sugars and related compounds in anhydrous dimethylformamide. *J. Amer. Chem. Soc.*, **110**, 584–589.

15. Carrea, G., Riva, S., Secundo, F. *et al.* (1989) Enzymatic synthesis of various 1′-*O*-sucrose and 1-*O*-fructose esters. *J. Chem. Soc., Perkin Trans. I*, 1057–1061.

16. Cai, S., Hakomori, S. and Toyokuni, T. (1992) Application of protease-catalyzed regioselective esterification of 6′-deoxy-6′-fluoro- and 6-deoxy-6-fluorolactosides. *J. Org. Chem.*, **57**, 3431–3437.

17. Therisod, M. and Klibanov, A.M. (1987) Regioselective acylation of secondary hydroxyl groups in sugars catalyzed by lipases in organic solvents. *J. Amer. Chem. Soc.*, **109**, 3977–3981.

18. Nicotra, F., Riva, S., Secundo, F. *et al.* (1989) An interesting example of complementary regioselective acylation of secondary hydroxyl groups by different lipases. *Tetrahedron Lett.*, **30**, 1703–1704.

19. Carpani, G., Orsini, F., Sisti, M. *et al.* (1989) Lipase-catalyzed transesterification of methyl α- and β-pentapyranosides. *Gazz. Chim. Ital.*, **119**, 463–465.

20. Lopez, R., Sanchez, F. and Fernandez-Mayoralas, A. (1994) Regioselective acetylations of alkyl-β-D-xylopyranosides by use of lipase PS in organic solvents and application to the chemoenzymatic synthesis of oligosaccharides. *J. Org. Chem.*, **59**, 7027–7032.

21. Panza, L., Luisetti, M., Crociati, E. *et al.* (1993) Selective acylation of 4,6-*O*-benzylidene glycopyranosides by enzymatic catalysis. *J. Carbohydr. Chem.*, **12**, 125–130.

22. Chin, M.J., Iacazio, G., Spackman, D.G. *et al.* (1992) Regioselective enzymatic acylation of methyl-α- and β-glucopyranoside. *J. Chem. Soc., Perkin Trans. I*, 661–662.

23. Iacazio, G. and Roberts, S.M. (1993) Investigation of the regioselectivity of some esterifications involving methyl 4,6-*O*-benzylidene D-pyranosides and *Pseudomonas fluorescens* lipase. *J. Chem. Soc., Perkin Trans. I*, 1099–1101.

24. Ciuffreda, P., Colombo, D., Ronchetti, F. *et al.* (1990) Regioselective acylation of 6-deoxy-L- and -D-hexosides through lipase-catalyzed transesterification. *J. Org. Chem.*, **55**, 4187–4190.

25. Colombo, D., Ronchetti, F. and Toma, L. (1991) Enzymic acylation of sugars. Rationale of the regioselective butyrrylation of secondary hydroxy groups of D- and L-galacto- and mannopyranosides. *Tetrahedron*, **47**, 103–110.

26. Colombo, D., Ronchett, F., Scala, A. *et al.* (1992) Enzymic acylation of methyl D- and L-glucopyranosides and 6-deoxy-glucopyranosides. *J. Carbohydr. Chem.*, **11**, 89–94.

27. Schlotterbeck, A., Lang, S., Wray, V. *et al.* 1993) Lipase-catalyzed monoacylation of fructose. *Biotechnol. Lett.*, **15**, 61–64.

28. Oguntimein, G.B., Erdmann, H. and Schmid, R.D. (1993) Lipase catalysed synthesis of sugar ester in organic solvents. *Biotechnol. Lett.*, **15**, 175–180.

29. Ikeda, I. and Klibanov, A.M. (1993) Lipase-catalyzed acylation of sugars solubilized in hydrophobic solvents by complexation. *Biotechnol. Bioeng.*, **42**, 788–791.

30. Danieli, B., DeBellis, P., Carrea, G. *et al.* (1989) Enzyme-mediated acylation of flavonoid monoglycosides. *Heterocycles*, **29**, 2061–2064.

31. Danieli, B., DeBellis, P., Carrea, G. *et al.* (1990) Enzyme-mediated acylation of flavonoid disaccharide monoglycosides. *Helv. Chim. Acta*, **73**, 1837–1844.

32. Danieli, B., Bertario, A., Carrea, G. *et al.* (1993) Chemo-enzymatic synthesis of 6″-*O*-(3-arylprop-2-enoyl) derivatives of the flavonol glucoside isoquercitrin. *Helv. Chim. Acta*, **76**, 2981–2991.

33. Danieli, B., Luisetti, M., Riva, S. *et al.* (1995) Regioselective enzyme-mediated acylation of polyhydroxy natural compounds. A remarkable highly efficient preparation of 6′-*O*-acetyl- and 6′-*O*-carboxyacetyl ginsenoside Rg₁. *J. Org. Chem.*, **60**, 3637–3642.

34. Danieli, B., Riva, S., Bertinotti, A. *et al.* (1993) Remarkable regioselective enzyme-mediated acylation of ginsenoside Rg₁ and ginsenoside Rb₁. *Proc. 6th Intl. Ginseng Symposium*, pp. 195–206.

35. Danieli, B. and Riva, S. (1994) Enzyme-mediated regioselective acylation of polyhydroxylated natural products. *Pure Appl. Chem.*, **66**, 2215–2218.

36. Danieli, B., DeBellis, P., Carrea, G. *et al.* (1991) Regioselective enzyme-mediated acylation of colchicoside and thiocolchicoside. *Gazz. Chim. Ital.*, **121**, 123–125.

37. Gotor, V. and Pulido, R. (1991) An improved procedure for regioselective acylation of carbohydrates: novel enzymatic acylation of α-D-glucopyranose and methyl α-D-glucopyranoside. *J. Chem. Soc., Perkin Trans. I*, 491–492.

38. Gotor, V. and Moris, F. (1991) Regioselective acylation of 2′-deoxynucleosides through an enzymatic reaction with oxime esters. *Synthesis*, 626–628.

39. Gotor, V. and Moris, F. (1993) Enzymatic acylation and alkoxycarbonylation of α-, xylo-, anhydro-, and arabino-nucleosides. *Tetrahedron*, **44**, 10089–10098.
40. Moris, F. and Gotor, V. (1993) A useful and versatile procedure for the acylation of nucleosides through an enzymatic reaction. *J. Org. Chem.*, **58**, 653–660.
41. Wong, C.H., Chen, S.T., Hennen, W.J. *et al.* (1990) Enzymes in organic synthesis: use of subtilisin and a highly stable mutant derived from multiple site-specific mutations. *J. Amer. Chem. Soc.*, **112**, 945–953.
42. Zong, Z., Liu, J.L.C., Dinterman, L.M. *et al.* (1991) Engineering subtilisin for reaction in dimethylformamide. *J. Amer. Chem. Soc.*, **113**, 683–684.
43. Evidente, A., Fujii, T., Iacobellis, N.S. *et al.* (1991) Structure–activity relationships of zeatin cytokinins produced by plant pathogenic *Pseudomonades*. *Phytochemistry*, **30**, 3505–3510.
44. Colombo, D., Ronchetti, F., Scala, A. *et al.* (1994) Regio- and diastereoselective lipase catalyzed preparation of acetylated 2-*O*-glucosylglycerols. *Tetrahedron Asymmetry*, **5**, 1377–1384.
45. Danieli, B., DeBellis, P., Barzaghi, L. *et al.* (1992) Studies on the enzymatic acylation of quinic acid, shikimic acid and their derivatives in organic solvents. *Helv. Chim. Acta*, **75**, 1297–1304.
46. Holla, W.E. (1989) Enzymatic synthesis of selectively protected glycals. *Angew. Chem. Int. Edn Engl.*, **28**, 220–221.
47. Chon, C., Heisler, A., Junot, N. *et al.* (1993) Regioselective acylation of 1,6-anhydro-β-D-glucopiranose catalysed by lipase. *Tetrahedron Asymmetry*, **4**, 2441–2444.
48. Nair, M.S., Anikumur, A.T. (1994) Lipase catalyzed regioselective acylation: a facile method for the synthesis of commercially important Ambrox™ intermediate. *Biotechnol. Lett.*, **16**, 161–162.
49. Baudin, G., Glanzer, B.I., Swaminathan, K.S. *et al.* (1988) A synthesis of 1D-and 1L-*myo*-inositol 1,3,4,5-tetraphosphate. *Helv. Chim. Acta*, **71**, 1367–1378.
50. Liu, Y.C. and Chen, C.S. (1989) An efficient synthesis of optically active D-*myo*-inositol 1,4,5-triphosphate. *Tetrahedron Lett.*, **30**, 1617–1620.
51. Ling, L. and Ozaki, S. (1994) A chemoenzymatic synthesis of D-*myo*-inositol 1,4,5-triphosphate. *Carbohydr. Res.*, **256**, 49–58.
52. Andersch, P. and Schneider, M.P. (1993) Enzyme assisted synthesis of enantiomerically pure *myo*-inositol derivatives. Chiral building blocks for inositol polyphosphates. *Tetrahedron Asymmetry*, **4**, 2135–2138.
53. Burgess, K. and Henderson, I. (1991) Biocatalytic desymmetrizations of pentitol derivatives. *Tetrahedron Lett.*, **32**, 5701–5704.
54. Bonini, C., Racioppi, R., Viggiani, L. *et al.* (1993) Enzyme-catalyzed desymmetrization of *meso*-skipped polyols to useful chiral building blocks. *Tetrahedron Asymmetry*, **4**, 793–805.
55. Breuilles, P., Schmittberger, T. and Uguen, D. (1993) The double-meso trick. *Tetrahedron Lett.*, **34**, 4205–4208.
56. Bovara, R., Carrea, G., Ferrara, L. *et al.* (1991) Resolution of (±)-*trans*-sobrerol by lipase PS-catalyzed transesterification and effects of organic solvents on enantioselectivity. *Tetrahedron Asymmetry*, **2**, 931–938.
57. Riva, S. and Klibanov, A.M. (1988) Enzymochemical regioselective oxidation of steroids without oxidoreductases. *J. Amer. Chem. Soc.*, **110**, 3291–3295.
58. Riva, S., Bovara, R., Ottolina, G. *et al.* (1989) Regioselective acylation of bile acids derivatives with *Candida cylindracea* lipase in anhydrous benzene. *J. Org. Chem.*, **54**, 3161–3164.
59. Bertinotti, A., Carrea, G., Ottolina, G. *et al.* (1994) Regioselective esterification of poly-hydroxylated steroids by *Candida antarctica* lipase B. *Tetrahedron*, **50**, 13165–13172.
60. Riva, S. (1991) Enzymatic modification of steroids. In *Applied Biocatalysis*, vol. 1 (Eds H.W. Blanch and D.S. Clark), Marcel Dekker, New York, pp. 179–220.
61. Margolin, A.L., Delinch, D.L. and Whalon, M.R. (1990) Enzyme-catalyzed regioselective acylation of castanospermine. *J. Amer. Chem. Soc.*, **112**, 2849–2854.
62. Delinck, D.L. and Margolin, A.L. (1990) Enzyme-catalyzed acylation of castanospermine and 1-deoxynojirimycin. *Tetrahedron Lett.*, **31**, 3093–3096.
63. Nicotra, F., Riva, S., Secundo, F. *et al.* 91990) ω-Functionalized esters by enzymatic acylation. *Synthetic Commun.*, **20**, 679–685.
64. Nicotra, F., Panza, L., Russo, G. *et al.* (1992) Chemoenzymatic approach to carbohydrate-derived analogues of platelet-activating factor. *J. Org. Chem.*, **57**, 2154–2158.
65. Panza, L., Brasca, S., Riva, S. *et al.* (1993) Selective lipase-catalyzed acylation of 4,6-*O*-benzylidene-D-glucopyranosides to synthetically useful esters. *Tetrahedron Asymmetry*, **4**, 931–932.
66. Pulido, R. and Gotor, V. (1994) Towards the selective acylation of secondary hydroxyl groups

of carbohydrates using oxime esters in an enzyme-catalyzed process. *Carbohydr. Res.*, **252**, 55–68.

67. Fabre, J., Paul, F., Monsan, P. *et al.* (1994) Enzymatic synthesis of amino acid ester of butyl α-D-glucopyranoside. *Tetrahedron Lett.*, **35**, 3535–3536.

68. Ottolina, G., Carrea, G. and Riva, S. (1991) Regioselective enzymatic preparation of hemi-succinates of polyhydroxylated steroids. *Biocatalysis*, **5**, 131–136.

69. Fabre, J., Betbeder, D., Paul, F. *et al.* (1993) Versatile enzymatic diacid ester synthesis of butyl α-D-glucopyranoside. *Tetrahedron*, **47**, 10877–10882.

70. Berger, M., Laumen, K. and Schneider, M.P. (1992) Lipase catalyzed esterification of hydrophilic diols in organic solvents. *Biotechnol. Lett.*, **14**, 553–558.

71. Sharma, A. and Chattopadhyay, S. (1993) Lipase catalyzed acetylation of carbohydrates. *Biotechnol. Lett.*, **15**, 1145–1146.

72. Salleh, A.K., Teoh, A.B., Yunus, W. *et al.* (1991) Sugar esterification catalysed by alkylated trypsin in dimethylformamide. *Biotechnol. Lett.*, **13**, 25–30.

73. Rubio, E., Fernandez-Mayorales, A. and Klibanov, A.M. (1991) Effect of the solvent on enzyme regioselectivity. *J. Amer. Chem. Soc.*, **113**, 695–696.

74. Singh, H.K., Cote, G.L. and Hadfield, T.M. (1994) Manipulation of enzyme regioselectivity by solvent engineering: enzymatic synthesis of 5′-*O*-acylribonucleosides. *Tetrahedron Lett.*, **35**, 1353–1356.

75. Palmer, D.C. and Terradas, F. (1994) Regioselective enzymatic deacetylation of sucrose octa-acetate in organic solvents. *Tetrahedron Lett.*, **35**,1673–1676.

76. Njar, V.C.O. and Caspi, E. (1987) Enzymatic transesterification of steroid esters in organic solvents. *Tetrahedron Lett.*, **28**, 6549–6552.

77. Natoli, M., Nicolosi, G. and Piattelli, M. (1990) Enzyme-catalyzed alcoholysis of flavone acetates in organic solvent. *Tetrahedron Lett.*, **31**, 7371–7374.

78. Natoli, M., Nicolosi, G. and Piattelli, M. (1992) Regioselective alcoholysis of flavonoid acetates with lipase in organic solvent. *J. Org. Chem.*, **57**, 5776–5778.

79. Parmar, V.S., Prasad, A.K., Sharma, N.K. *et al.* (1992) Potential applications of enzyme-mediated transesterifications in the synthesis of bioactive compounds. *Pure Appl. Chem.*, **64**, 1135–1139.

80. Parmar, V.S., Prasad, A.K., Sharma, N.K. *et al.* (1993) Lipase-catalysed selective deacetylation of peracetylated benzopyranones. *J. Chem. Soc., Chem. Commun.*, 27–29.

81. Seino, H., Uchibori, T., Nishitani, T. *et al.* (1984) Enzymatic synthesis of carbohydrate esters of fatty acid (1). Esterification of sucrose, glucose, fructose and sorbitol. *J. Amer. Oil Chem. Soc.*, **61**, 1761–1765.

82. Chopineau, J., McCafferty, F.D., Therisod, M. *et al.* (1988) Production of biosurfactants from sugars alcohols in vegetable oils catalyzed by lipases in nonaqueous medium. *Biotechnol. Bioeng.*, **31**, 208–214.

83. Mutua, L.N. and Akoh, C.C. (1993) Synthesis of alkyl glycoside fatty acid esters in non-aqueous media by *Candida* sp. lipase. *J. Amer. Oil Chem. Soc.*, **70**, 43–46.

84. Fabre, J., Betbeder, D., Paul, F. *et al.* (1993) Regiospecific enzymatic acylation of butyl α-D-glucopyranoside. *Carbohydr. Res.*, **243**, 407–411.

85. Bjorkling, F., Godtfredsen, S.E. and Kirk, O. (1989) A highly selective enzyme-catalysed esterification of simple glucosides. *J. Chem. Soc., Chem. Commun.*, 934–935.

86. Adelhorst, K., Bjorkling, F., Godtfredsen, S.E. *et al.* (1990) Enzyme catalysed preparation of 6-*O*-acylglucopyranosides. *Synthesis*, 112–115.

87. Kirk, O., Bjorkling, F., Godtfredsen, S.E. *et al.* (1992) Fatty acid specificity in lipase-catalyzed synthesis of glucoside esters. *Biocatalysis*, **6**, 127–134.

88. Fregapane, G., Sarney, D.B. and Vulfson, E.N. (1991) Enzymic solvent-free synthesis of sugar acetal fatty acid esters. *Enzyme Microb. Technol.*, **13**, 796–800.

89. Fregapane, G., Sarney, D.B., Greenberg, S.G. *et al.* (1994) Enzymatic synthesis of mono-saccharides fatty acid esters and their comparison with conventional products. *J. Amer. Oil Chem. Soc.*, **71**, 87–91.

90. Dordick, J.S., Linhardt, R.J. and Rethwisch, D.G. (1994) Chemical and biochemical catalysis to make swellable polyers. *Chemtech*, **24**, 33–39.

91. Cygler, M., Grochulski, P., Kazlauskas, R.J. *et al.* (1994) A structural basis for the chiral preferences of lipases. *J. Amer. Chem. Soc.*, **116**, 3180–3186.

92. Fitzpatrick, P.A., Ringe, D. and Klibanov, A.M. (1992) Computer-assisted modeling of subtilisin enantioselectivity in organic solvents. *Biotechnol. Bioeng.*, **40**, 735–742.

7 Hydrolase-catalysed asymmetric and other transformations of synthetic interest

L.T. KANERVA

As a definition, enzyme catalysis in an organic solvent means enzymatic reactions in a medium which contains only few per cent (1–2%) of water or in an anhydrous organic solvent (1). Dry enzyme powder is insoluble in such a system, and consequently the biotransformation proceeds by heterogeneous catalysis. Polyethylene glycol (PEG)-modified enzymes are able to dissolve in organic solvents, making homogeneous catalysis feasible as well (2, 3).

Hydrolases are the most useful enzymes in the biotransformations of non-natural substrates. This is mainly due to the wide substrate specificity, the high stability and the novel reactions which they catalyse. Moreover, the commercial availability of many hydrolases is good and there is no need for expensive cofactors. Their most important application certainly concerns the resolution of racemic compounds. Another type of application is the remarkable regioselectivity which hydrolases exert in the acylation and deacylation of polyfunctional compounds such as carbohydrates and steroids. Moreover, chemoselectivity — selectivity towards one kind of a functional group in a bifunctional molecule — is also exploited in organic chemistry. It is possible that two or all three of these selectivities work simultaneously. Biotransformations by hydrolases concerning peptide synthesis, modifications of sugars as well as dissymmetrization of *meso* or prochiral compounds are dealt with elsewhere in this book.

7.1 Hydrolases

According to the nomenclature committee of the International Union of Biochemistry, hydrolases (or hydrolytic enzymes) belong to the third main class of the enzymes (Table 7.1), the code number (EC 3.*a.b.c*) further characterizing an individual enzyme (4). In the code number, *a* refers to the type of the reaction, *b* specifies the nature of the substrate and *c* is for the individual enzyme number. Ester and peptide hydrolases are the most commonly used enzymes in organic solvents. Further, biotransformations catalysed by lipases are of special value, the most important lipases being listed in Table 7.2. Hydrolases belonging to other groups (EC 3.2, 3.3 and 3.5–3.11) have had only limited use in organic

Table 7.1 Classification of hydrolases

Hydrolase	Common representatives used in organic solvents
(EC 3.1) acting on ester bonds	*Carboxylic ester hydrolases*: Carboxylesterase (EC 3.1.1.1) Lipase (EC 3.1.1.3) Lipoprotein lipase (EC 3.1.1.34)
(EC 3.2) acting on *O*-, *N*- and *S*-glycosyl compounds	
(EC 3.3) acting on ether bonds	
(EC 3.4) acting on peptide bonds	*Serine proteases*: Chymotrypsin (EC 3.4.21.1) Subtilisin (EC 3.4.21.14) *Cysteine proteases*: Papain (EC 3.4.22.2) *Aspartyl proteases*: Pepsin (EC 3.4.23.1) *Metalloproteases*: Thermolysin (EC 3.4.24.4)
(EC 3.5) acting on C–N bonds (other than peptide bonds)	Penicillinase (EC 3.5.2.6)
(EC 3.6) acting on acid anhydrides	
(EC 3.7) acting on C–C bonds	
(EC 3.8) acting on halide bonds	
(EC 3.9) acting on P–N bonds	
(EC 3.10) acting on S–N bonds	
(EC 3.11) acting on C–P bonds	

media. The opening of *cis*-1-benzoyl-3-acetoxy-4-phenylazetidin-2-one in the presence of penicillinase in *tert*-butyl methyl ether was reported, but the product was racemic (5). Peptide hydrolases often have esterolytic action as well.

The reactions catalysed by serine hydrolases (carboxylesterases, lipases, penicillin acylase and serine and cysteine proteases) proceed through a tetra-hedral acyl-enzyme intermediate (RCO-E) where the acyl group (RCO) of the substrate (RCONu$_1$) is covalently bound to the serine hydroxyl (to the cysteine HS in the case of cysteine proteases) at the active site of the enzyme (E) (Scheme 7.1) (6). In the hydrolytic process water acts as a natural nucleophile (Nu$_2$H) which reacts with the intermediate, forming the product and liberating the enzyme. Working in organic media, novel reactions become possible because water is replaced by other nucleophilic reagents such as alcohols or amines. In the cases of aspartyl groups and metallo-proteases, the nucleophile reacts with the substrate without the existence of the acyl-enzyme intermediate on the reaction path. Generally speaking, hydrolase-catalysed hydrolysis in aqueous solutions and novel reactions such as esterification (RCO$_2$H + R$_1$OH), alcoholysis (RCO$_2$R′ + R$_1$OH), aminolysis (RCO$_2$R′ + R$_1$NH$_2$) and so on in organic solvents are simply acylations of a nucleophile Nu$_2$H (water, R$_1$OH, R$_1$NH$_2$, etc.) or deacylations of a substrate RCONu$_1$. In addition to these reactions, acidolysis (RCO$_2$R′ + R*CO$_2$H) and interesterification

Table 7.2 Main lipases used in biotransformations in organic solvents

Lipase source	Abbreviation	Supplier
Pseudomonas aeruginosa	LPL	Amano
	TE3285	Toyo Jozo
Pseudomonas cepacia	Lipase PS	Amano
Pseudomonas fluorescens	Lipase P	Amano
	SAM-2	Fluka
Pseudomonas sp.	Lipase AK	Amano
Candida antarctica B[a]	SP 525	Novo
	Novozym 435	Novo
Candida antarctica A[a]	SP 526	Novo
Candida cylindracea (Candida rugosa)[b]	CCL	Sigma
	Lipase AY-30	Amano
	Lipase OF	Toyo Jozo
	Lipase MY	Meito Sangyo
Mucor miehei	Lipase MAP-10	Amano
	Lipozyme	Novo
Chromobacterium viscosum	CV	Biocatalysis[c]
Aspergillus niger	Lipase A or K-30	Amano
Porcine pancreatic	PPL	Sigma

[a] Abbreviation CAL does not distinguish between *Candida antarctica* lipases A and B. [b] The name now replacing the old name, *Candida cylindracea*. [c] CV lipase mostly used was delivered by Finnsugar Biochemical; this is no longer commercially available.

($RCO_2RR' + R*CO_2R''$) are possible. Especially lipase-catalysed alcoholysis, but also aminolysis and acidolysis reactions, are typical transesterification reactions considered in this chapter.

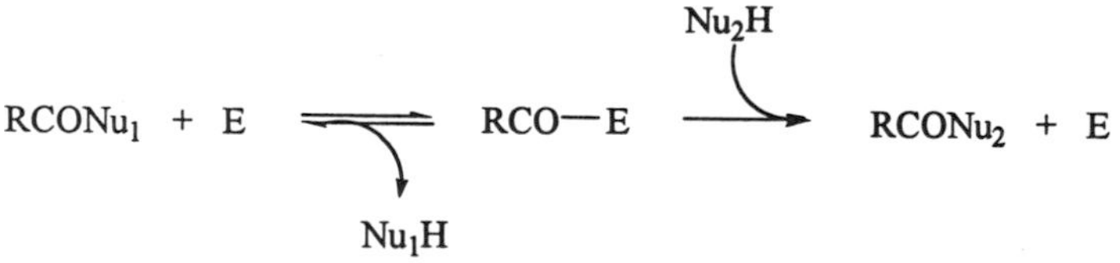

$$RCONu_1 + E \rightleftharpoons RCO{-}E \xrightarrow{\ Nu_2H\ } RCONu_2 + E$$

$$Nu_1H$$

Scheme 7.1

7.2 Lipases versus esterases: mechanistic models

Lipases and carboxylesterases both work on ester bonds, triglycerides being the natural substrates of lipases. In nature, the involvement of a lipid–water interface in the catalytic process of a lipase is the fundamental feature which distinguishes lipases from esterases. Carboxylesterases, such as pig liver esterase, have been widely used for biotransformations of *meso* and prochiral diesters.

Serine hydroxyl, imidazole of histidine and an acidic amino acid residue (the so-called catalytic triad) at the active site of serine hydrolases are mainly responsible for the catalytic process itself (6). On the other hand, to achieve high selectivity the substrate must be firmly bound at the active site of the enzyme prior to catalysis in such a way that the reactive group is correctly oriented to the chemical operator of the enzyme. (At least three of the groups at a chiral or prochiral centre must interact with the enzyme: the multi-point attachment theory.) For the resolution of carboxylic acids or esters, the serine hydroxyl of the triad is a chemical operator, and enantiodiscrimination takes place in the formation of the acyl-enzyme intermediate and liberation of the first product (Nu_1H) (step 1, Scheme 7.1). For the resolution of racemic Nu_2H, enantio-discrimination takes place when this nucleophile reacts with the acyl-enzyme intermediate (step 2, Scheme 7.1), forming the new reaction product ($RCONu_2$) and releasing the enzyme. In this case, the acylated serine hydroxyl is the operator at the active site.

In an enzymatic kinetic resolution, it is of importance for a synthetic chemist to predict which enantiomer reacts faster. Although the X-ray structures of many hydrolases are already known, the mechanism of enantiorecognition is still mainly probed by substrate mapping. This has been a basis for the active-site models and empirical rules. The most valid model for the alcohol binding site for secondary alcohols predicts that there is a large and a small binding pocket at the active site of a hydrolase capable of binding the groups R_{large} and R_{small} of the alcohol (Scheme 7.2a) (7). When the substituents R_{large} and R_{small} differ significantly in size, efficient resolution can be expected. In the case of lipases and esterases, an empirical rule predicts that drawing the alcohol with the hydroxyl group forward, out of the plane, a favoured enantiomer bears a large substituent on the right. In other words, the enzymes should show preferred (R)-selectivity when the priority of the group R_{large} is higher than that of the group R_{small}. The X-ray crystal structures of covalent complexes of *Candida cylindracea* lipase with transition state analogues have further confirmed that the model is a good low-resolution description of the alcohol binding site of the enzyme (8). Although this model was originally proposed for cholesterol esterase and *Pseudomonas cepacia* and *Candida cylindracea* lipases, the resolution results of this article clearly favour its more general validity.

In the case of *Candida cylindracea* lipase-catalysed esterification of 2-hydroxy acids ($RCH(OH)CO_2H$), there is a switch of enantioselectivity with increasing acyl chain length. The model predicts the existence of large (for R_{long}) and small (for R_{short}) acyl binding pockets, a short (but not a long) chain acyl group R fitting to the smaller pocket, and the switch taking place when the carbon chain becomes long enough (Scheme 7.2b) (9).

There is not one simple active-site model in the case of porcine pancreatic lipase. This may be at least partly due to the presence of other enzymes such as cholesterol esterase in the crude enzyme preparation usually used for synthetic purposes. A four-site tetrahedral model is among the best models to explain enantioselectivity for porcine pancreatic lipase-catalysed resolution of primary alcohols through ester hydrolysis in water (Scheme 7.2c) (10). This model contains the pockets for a small, possibly polar (R_{small}), and large, hydrophobic

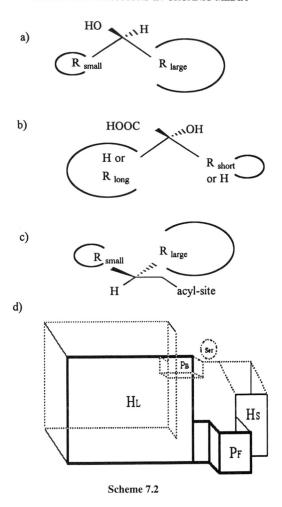

Scheme 7.2

(R_{large}) groups, acyl-site representing the catalytic region. The most useful model in the case of pig liver esterase is the box model of Jones, where the specificity is interpreted in terms of substrate interactions with two polar binding sites (P_F and P_B) and two hydrophobic binding pockets (large H_L and small H_S) (Scheme 7.2d) (11, 12).

7.3 Principles of enzymatic kinetic resolutions

Enzymes as chiral catalysts often selectively react more rapidly with one of the enantiomers. Thermodynamically, the enantiomers — the (R)- and (S)-isomers — of a racemic substrate are two reagents A and B which compete for the same

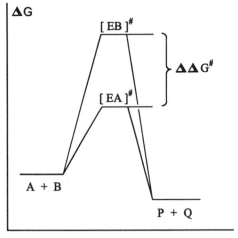

Figure 7.1 Free-energy profile in the case of the enzymatic reaction for two competitive substrates, A being more reactive than B.

enzyme (Figure 7.1). Thus, the Gibbs free energy difference ($\Delta\Delta G^{\#}$) between the transition states for the faster and lower reacting enantiomer is responsible for the effectiveness of the kinetic resolution. When A is a fast reacting enantiomer and $\Delta\Delta G^{\#}$ is high enough (> 19 kJ mol^{-1}), the reaction stops at 50% conversion. At this point, the new product P and the unreactive starting material B are enantiopure. Usually, however, the reaction proceeds enantioselectively and the resolution products are enriched with respect to the enantiomers which represent P and B. The enantiopurities of these products depend on conversion (c): the purity of the less reactive enantiomer B is best at conversions over and that of the product P at conversions less than 50%.

Enantiomeric excess (ee) of the enriched enantiomer is generally used as a measure of the optical purity of the compound. This quantity is expressed for the enantiomers P and B through equations (7.1) and (7.2), respectively, where A, B, P and Q represent the amounts of substrates and products.

$$ee_P = \frac{[P] - [Q]}{[P] + [Q]} \tag{7.1}$$

$$ee_B = \frac{[B] - [A]}{[A] + [B]} \tag{7.2}$$

7.3.1 Quantitative analysis of irreversible kinetic resolution

Enantiomeric ratio (E), which is independent of conversion, was introduced by Chen *et al.* (13, 14) (equation 7.3) to depict enantioselectivity.

$$E = \frac{(k_{cat}/K_M)_A}{(k_{cat}/K_M)_B} = \frac{(V/K_M)_A}{(V/K_M)_B} = \frac{(v_0)_A[B]}{(v_0)_B[A]} = e^{-\Delta\Delta G^{\#}/RT} \qquad (7.3)$$

Assuming the enzymatic reaction is irreversible, there is no product inhibition and Michaelis–Menten kinetics is valid, the constant E is the ratio of the specificity constants (V/K_M or k_{cat}/K_M) between the two competing enantiomers A and B. The constant E is also roughly the initial rate ratio $(v_0)_A/(v_0)_B$ of the two enantiomers, because the racemic mixture contains equal quantities of A and B. In equation (7.3), V denotes maximal velocity, k_{cat} first-order rate constant and K_M the Michaelis constant for the given enantiomer in a simple three-step kinetic mechanism. The ratio of the specificity constants can be used to describe chemo- and regioselectivity as well, the competing substrates A and B in that case being 'the two functional groups' of the molecule.

In order to be able to terminate a reaction at a conversion where the chemical and optical yields of either P or B are as high as possible, the integrated equation (7.4) was introduced (13, 14). The corresponding relation between E, c and ee_P is shown in equation (7.5). In equations (7.4) and (7.5), conversion $c = ee_B/(ee_B + ee_P) = 1 - (A + B)/(A_0 + B_0)$.

$$E = \frac{\ln[(1-c)(1-ee_B)]}{\ln[(1-c)(1+ee_B)]} \qquad (7.4)$$

$$E = \frac{\ln[(1-c)(1+ee_P)]}{\ln[(1-c)(1-ee_P)]} \qquad (7.5)$$

Based on these equations at fixed E, the theoretical plots between ee_B, ee_P and c are obtained. According to such plots, it is easy to see that in a kinetic enzymatic resolution the less reactive enantiomer B is always obtained with higher enantiopurity than the product P. Example 7.1 (below) describes this in practice. Moreover, the E value of 10 is necessary for the preparation of the less reactive enantiomer with reasonable chemical and optical yields (Table 7.3). Thus, when $E = 10$ the theoretical yield of B is 30% at 98% ee.

The validity of equations (7.4) and (7.5) to obtain the theoretical plots of ee versus c is restricted by the accuracy in determining the E value (15, 16). When the pure enantiomers of a chiral compound are available, the E value with high

Table 7.3 Conversion corresponding to $ee_B = 0.98/0.95$ at different E values according to equation (7.4)

E	5	10	25	50	100	200	500
Conversion (%)	84/79	70/66	58/56	54/52	52/51	51/50	50/50

accuracy is obtained by the laborious determination of the specificity constants through equation (7.3) (13, 14). Another method for the determination of E is based on a linear correlation between $1/(v_{rac} - v_B)$ and $1/x$ or $1/v_A - v_{rac})$ and $1/(1 - x)$ (15). In this method, the initial rate v_B or v_A must be determined separately, and x is the molar fraction of the enantiomer which represents A. In the third method, the non-linear correlation between the measured ee_B and ee_P gives the E value (17). However, equation (7.4) or (7.5) is most often applied for the determination of E. For the sake of accuracy, it is firmly recommended that the values of ee_B (equation 7.1) and ee_P (equation 7.2) should be first obtained simultaneously, e.g. from the same chromatogram, and the conversion should then be calculated using these ee values (13, 14). The E values >100 are rather inaccurate and even small errors in ee may cause a significant variation in E.

When the substrate of an irreversible enzymatic reaction visits the active site of the enzyme twice, the second step may improve the enantioselectivity introduced by the first. The quantitative expressions (equations 7.6 and 7.7) which govern *sequential kinetic resolutions* of axially disymmetric compounds have been developed using a biocatalytic enantioselective acylation–acylation sequence of racemic pentan-2,4-diol as a model reaction (Scheme 7.3) (18). The complicated regio- and enantioselective alcoholysis–alcoholysis sequence of 2-methylpentanedioic anhydride with butan-2-ol to enantiopure 1-(2-methyl-propyl) 5-hydrogen (R)-2-methylpentanedioate was based on the rapid esterification of the (S)-monoester enantiomer to the corresponding diester (19). This kind of sequential resolution often works also in the desymmetrization of *meso*-compounds. Another type of sequential kinetic resolution is shown in Scheme 7.4. The hydrolysis–acylation reaction of some bicyclo[3.2.0]hept-2-ene esters in

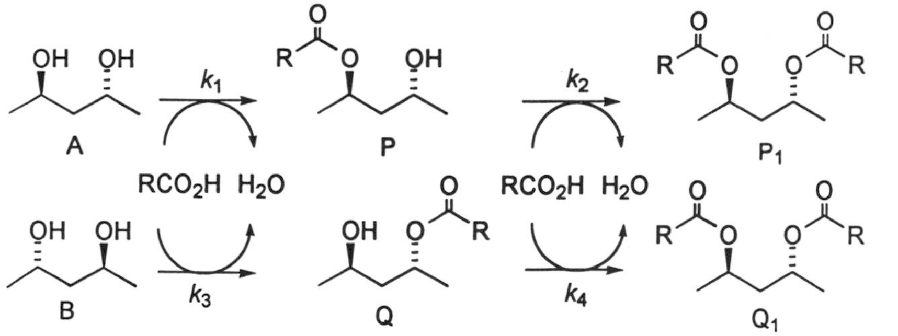

$$(7.6) \quad P = \frac{A_o}{1 - E_2} = [(A/A_o)^{E_2} - A/A_o]$$

$$(7.7) \quad Q = \frac{B_o}{1 - E_3} = [(B/B_o)^{E_3} - B/B_o]$$

$$E_1 = k_1/k_2 \; ; \; E_2 = k_2/k_1 \; ; \; E_3 = k_4/k_3$$

Scheme 7.3

LIVERPOOL
JOHN MOORES UNIVERSITY
AVRIL ROBARTS LRC
TEL. 0151 231 4022

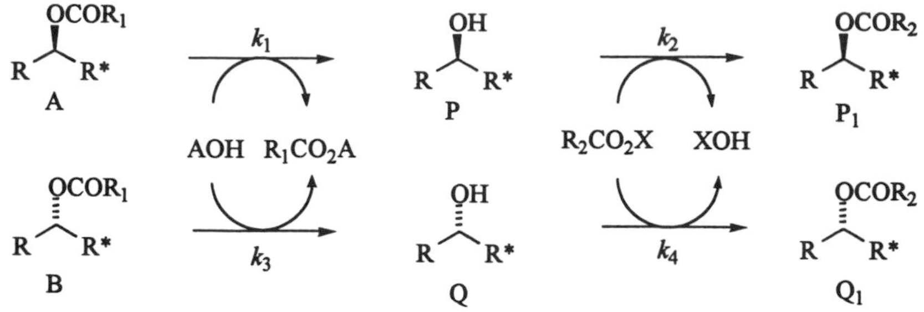

$$OCOR_1 \qquad\qquad OH \qquad\qquad OCOR_2$$

$$\begin{array}{ccc} R \quad R^* & \xrightarrow{k_1} & R \quad R^* \\ A & & P \end{array} \xrightarrow{k_2} \begin{array}{c} R \quad R^* \\ P_1 \end{array}$$

AOH R_1CO_2A $\qquad\qquad$ R_2CO_2X XOH

$$OCOR_1 \qquad\qquad OH \qquad\qquad OCOR_2$$

$$\begin{array}{ccc} R \quad R^* & \xrightarrow{k_3} & R \quad R^* \\ B & & Q \end{array} \xrightarrow{k_4} \begin{array}{c} R \quad R^* \\ Q_1 \end{array}$$

(7.8) $\quad |1-c[(1+ee)/(1+ee_o)]| = |1-c[(1-ee)/(1-ee_o)]|^{E}$

$c = 1-[(P+Q)/(P_o+Q_o)]$; $ee_o = [(P_o -Q_o)/(P_o +Q_o)]$; $ee = [(P_1 -Q_1)/(P_1 +Q_1)]$

Scheme 7.4

water-saturated hexane $(Y = H, X = H \text{ or } R_3)$ (20–22), and the butanolysis $(Y = Bu)$ of racemic carboxylic acid esters in hexane followed by the addition of vinyl acetate $(X = CH=CH_2)$ at 50% conversion (33), both represent the cases where two sequential kinetic resolutions *in situ* led to the product P_1 with high optical purity. In practice, the last-mentioned case corresponds to the normal enzymatic resolution where the enantiomerically enriched product (with ee_0 and the quantities P_0 and Q_0) is first separated from the reaction mixture and there-after subjected to another enzymatic resolution as such or after regeneration of enantiomerically enriched starting material (see Example 7.1) (13,14,23,24). The quantitative expression for the correlation between ee and c at the known E is obtained through equation 7.8.

Example 7.1.   Quantitative analysis of kinetic resolution. Practical production of (*R*)- and (*S*)-solketal (Scheme 7.5) (24). Racemic solketal (**1**) (99 g, 0.75 mol) and butyric anhydride (77 g, 0.49 mol) in 1.0 dm^3 of di-isopropyl ether were added on 12.5 g of the enzyme preparate (8.6% lipase AK from *Pseudomonas* sp. in the presence of sucrose on Celite) in a round-bottomed flask at 0–1°C. The reaction proceeded under vigorous stirring.

Determination of E.   A sample was withdrawn from the reaction mixture and filtered. The unreacted solketal in the sample was derivatized with acetic anhydride prior to the simultaneous ee determination of acetate (**2**) and butyrate (**3**) by GLC equipped with a cyclodextrin column, which perfectly separated the enantiomers of (**2**) and (**3**). The ee values 0.23 and 0.90 obtained, respectively, corresponded to $c = 0.20$ $[c = ee_B/(ee_B + ee_P)]$. The value of $E = 24$ was calculated by using equation (7.4).

Optimal conversion.   The theoretical plots of ee versus c at $E = 24$ were obtained according to equations (7.4) and (7.5), and are shown in Figure 7.2.

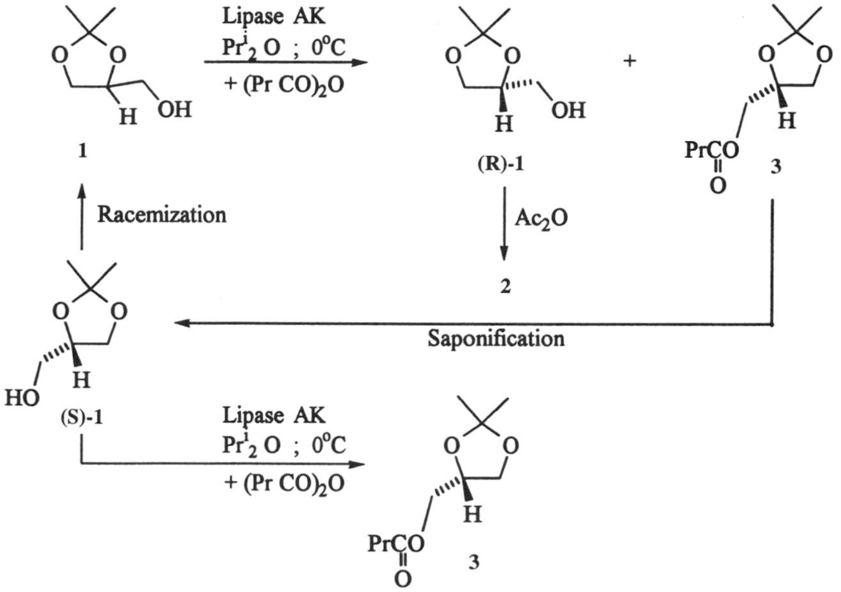

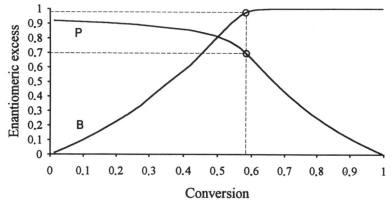

Scheme 7.5

Consequently, the reaction terminated at *c.* 60% conversion should give the less reactive alcohol (*R*)-(**1**) and the (*R*)-butyrate (**3**) with ee ≥ 0.98 and < 0.70, respectively.

In practice, the above reaction was terminated at 62% conversion after 25 h by filtering off the enzyme. The solvent was evaporated. Unreacted solketal (**1**) and solketal butyrate (**3**) were separated by extracting with hexane–water. The hexane phase was evaporated, leaving butyrate (*R*)-(**3**) (95.6 g, 63%, ee 0.59). The free solketal was extracted from the water phase into ethyl acetate and after

Figure 7.2 Correlations between enantiomeric excess and conversion at $E = 24$ for the less reactive enantiomer B and product P.

drying with Na_2SO_4 the solvent was evaporated. (*R*)-Solketal (**1**) (37.8 g, 38%, ee > 0.99) was obtained.

Double (sequential) resolution of enriched enantiomer (R)-(3).   The saponi-fication of butyrate (**3**) (289 g, ee 0.64) resulted in (*S*)-solketal (**1**) (165 g, 87%, ee 0.64). The theoretical plot of ee versus *c* for the (*R*)-butyrate produced was obtained using equation (7.8) at $E = 24$ and $ee_0 = 0.64$ (Figure 7.3). Thus, the optically enriched solketal should produce optically purified butyrate (**3**) (ee > 0.95) at conversions < 70%.

In practice, the mixture of (*S*)-solketal (**1**) (100 g, 0.76 mol, ee 0.64) and butyric anhydride (93 g, 0.59 mol) in di-isopropyl ether was subjected to a second resolution in the presence of the lipase preparate (17.8 g). The reaction was stopped at 71% conversion after 8.5 h. After work-up, (*R*)-butyrate (**3**) (111 g, 0.55 mol, ee 0.94) was separated.

7.3.2 Quantitative analysis of reversible kinetic resolution

In hydrolase-catalysed transformations, the gradual accumulation of products $RCONu_2$ and Nu_1H (Scheme 7.1) will easily give rise to reverse catalysis which becomes important at $c \geq 40\%$. In kinetic resolutions, the reverse catalysis should be reduced because it will lead to decreasing enantiopurities (Example 7.2, below). Namely if A is the fast-reacting enantiomer in the forward direction, the P produced is the fast-reacting enantiomer in the reverse reaction. As a result, the optical purity of the less reactive enantiomer decreases with increasing conversion. The optical purity of the product fraction decreases at the same time due to the accumulation of the product Q. In such a case, the equilibrium constant *K* must be incorporated in the expression of *E* (equations 7.9 and 7.10) (25) in order to get a correct correlation between ee and *c*.

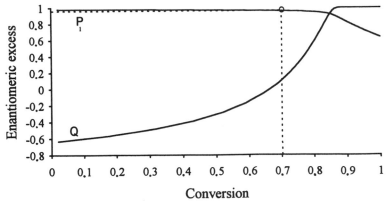

Figure 7.3 Correlations between enantiomeric excess and conversion at $E = 24$ and $ee_0 = 0.64$ for the less reactive Q and product P_1.

$$E = \frac{\ln\{1 - (1 + K)[c + ee_B(1 - c)]\}}{\ln\{1 - (1 + K)[c - ee_B(1 - c)]\}} \tag{7.9}$$

$$E = \frac{\ln[1 - (1 + K)c(1 + ee_p)]}{\ln[1 - (1 + K)c(1 + ee_p)]} \tag{7.10}$$

Example 7.2. Effect of an equilibrium on enantioselectivity. Butyrylation of racemic (**1**) with 2,2,2-trifluoroethyl butyrate (**24**). The mixture of solketal (**1**) (0.1 M), 2,2,2-trifluoroethyl butyrate (0.2 M) and lipase AK from *Pseudomonas* sp. on Celite in di-isopropyl ether was shaken in an orbit shaker at 25°C. The value of $E = 15$ was determined at low conversion as in Example 7.1. The plots of ee versus c according to equations (7.4) and (7.5) are shown in Figure 7.4 (dotted lines). However, the experimental points (the circles) do not follow the theoretical plots, but show lower ee values than was expected for an irreversible reaction. In fact, the reaction stopped at an equilibrium at 96–98% conversion. The solid lines which go through the experimental points were obtained according to equations (7.9) and (7.10) with $K = 0.04$.

7.3.3 Methods of irreversible hydrolase-catalysed resolutions

For hydrolase-catalysed ester hydrolysis *in water*, the thermodynamic equilibrium is shifted to the hydrolysis products due to the high excess of the solvent water. For hydrolase-catalysed conversions *in water-free environments*, this situation can be more or less imitated by using an achiral reagent as a solvent. High concentrations of alkan-1-ols seem, however, to decrease enzymatic activity and enantioselectivity. The use of an achiral acyl donor (RCONu$_1$; Scheme 7.1) as a solvent is more successful. This procedure was first introduced for the hog liver carboxyl esterase-catalysed acylations of many racemic

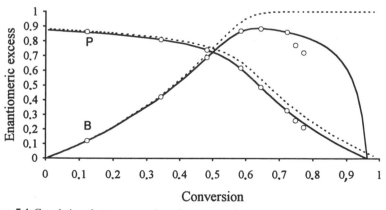

Figure 7.4 Correlations between enantiomeric excess and conversion at $E = 15$. Irreversible case: dotted lines. Reversible case ($K = 0.04$): solid lines. Experimental poinsts (○).

alkan-1-ols in methyl propionate, as well as for the *Candida cylindracea* lipase-catalysed acylations of various secondary alcohols and 1,2-butanediol in tributyrin in a biphasic system, where the aqueous phase was a solution of the enzyme confined to the pores of Sepharose or Chromosorb (26). After that, lipase- and esterase-catalysed acylations of racemic primary and secondary alcohols have been performed in various alkyl carboxylates (27–39). In fact, the demand for irreversibility in these reactions is often poorly fulfilled. Namely, an alcohol product (Nu_1H = MeOH, EtOH, etc.) which is released from the useful alkyl carboxylates easily reacts with a product P ($RCONu_2$) in the reverse enzymatic alcoholysis. This can be deduced from the results of Table 7.4. Vinyl esters as solvent and an acyl donor are more advantageous because the vinyl alcohol formed then leaves the reaction path as acetaldehyde (39).

A classical method to shift the reaction equilibrium in favour of the forward reaction is *the removal of one of the reaction products* as it is formed. The addition of molecular sieves to the reaction mixture achieves the adsorption of small nucleophiles (Nu_1H = H_2O, MeOH, EtOH, etc.) liberated in hydrolase-catalysed hydrolyses, transesterifications and esterifications in organic solvents (29, 38). The favourable effect of molecular sieves on conversion and on the optical purity of the less reactive enantiomer is evident for the resolutions of 1-phenylethanol in methyl propionate (Table 7.4) (29). Another simple and

Table 7.4 Acylation in alkyl carboxylates by porcine pancreatic lipase (PPL), *Mucor* esterase (ME) and *Candida antarctica* lipase (CAL)

Solvent	Enzyme	Time (h)	Conversion (%)	ee_B	ee_P	E	Ref.
Octan-2-ol							
$EtCO_2Me^a$	PPL	68	49	82(S)	86(R)	35	29
$EtCO_2Me^a$	ME	68	44	69(S)	85(R)	25	29
$CH_3(CH_2)_6CO_2Et^b$	CAL	7	45	80(S)	97(R)	43	35, 36
$CH_3(CH_2)_6COSEt$	CAL	0.9	52	>98(S)	>97(R)	67	37
(±)-Phenylethanol							
$AcOMe^a$	PPL	68	19	23(S)	>98(R)	—	29
$EtCO_2Me^a$	PPL	68	33	48(S)	>98(R)	—	29
$EtCO_2Me^c$	PPL	68	45	79(S)	97(R)	—	29
$EtCO_2Me^a$	ME	68	44	76(S)	>98(R)	—	29
$EtCO_2Me^c$	ME	68	45	82(S)	98(R)	—	29
$CH_3(CH_2)_6CO_2Et^b$	CAL	12	42	71(S)	97(R)	>100	35, 36
$CH_3(CH_2)_6COSEt$	CAL	2.5	51	98(S)	97(R)	>200	37
(±)-1-Cyclohexylethanol							
$EtCO_2Me^a$	PPL	68	43	71(S)	94(R)	70	29
$EtCO_2Me^a$	ME	68	34	46(S)	89(R)	25	29
$CH_3(CH_2)_6CO_2Et^b$	CAL	19	42	80(S)	96(R)	70	35, 36
$CH_3(CH_2)_6COSEt$	CAL	4.4	52	>98(S)	95(R)	>130	37

[a] At 40°C. [b] EtOH distilled (25°C/15 mmHg) when it is formed. [c] Molecular sieves (4 Å) added in the reaction mixture; at 40°C.

useful method to achieve high equilibrium conversion is the distillation of volatile alcohols produced by transesterification. Thus, the *Candida antarctica* lipase-catalysed acylations of secondary alcohols (Table 7.4) were successfully conducted with ethyl octanoate as an acyl donor and a solvent by evaporating the ethanol produced, under reduced pressure (35, 36). When *S*-ethyl thio-octanoate replaced the *O*-analogue, the ethanethiol produced was evaporated at atmospheric pressure (37). Moreover, in classical esterification by lipase catalysis a distinct improvement of ester formation was observed by removal of water under azeotropic distillation (40).

Commonly, enzymatic reactions are performed by using low concentrations of substrates in an inert organic solvent. Irreversible reactions are then ensured by paying attention to *the structural features of the reagents*. The use of alkyl-activated esters as acyl donors is advantageous because a week nucleophile (Nu_1H) which is liberated (step 1; Scheme 7.1) is not supposed to react with a new product ($RCONu_2$) in the backward reaction. An additional advantage of activated esters over normal alkyl carboxylates is enhanced reaction rates (41–50). Thus, for the PPL-catalysed acylation of octan-2-ol with butyrates $PrCO_2CH_2R_1$ (R_1 is Et, CH_2Cl, $CHCl_2$, CCl_3 and CF_3) in diethyl ether, the initial rates ($v_0/\mu mol\ min^{-1}$) increase from the value of 0.07 to 13 at 25°C across the series, resulting in a relatively good Taft correlation (46). The unexpectedly high rate difference of 2,2,2-trifluoroethyl over 2,2,2-trichloroethyl butyrate (v_0 13 and 3.9 $\mu mol\ min^{-1}\ g^{-1}$, respectively) with almost the same σ values (2.60 and 2.65, respectively), testifies to the importance of steric over polar effects of the alkyl group.

Esters of phenol and naphthols (44, 45) and especially esters of differently halogenated ethanols (41, 46), enols (47–49) and oximes (50) usually serve as activated achiral acyl donors which react with a chiral nucleophile, such as octan-2-ol (**4**) (Scheme 7.6) (41). An activated acyl donor can also be the ester of a chiral carboxylic acid, such as amino acid ester (**6**) which reacts with an achiral nucleophile (Scheme 7.7) (42). Some words of precaution as to the irreversibility are, however, appropriate. Thus, the reactions of trihaloethanol esters can be quasi-irreversible, leading in some cases to a depletion of the enantiopurity of the enantiomers produced at higher conversions (Figure 7.4)

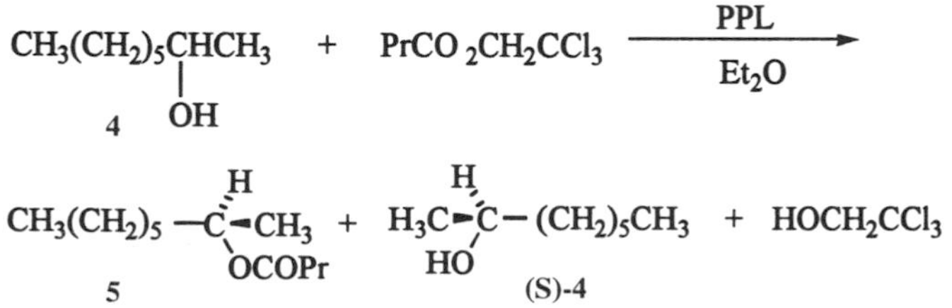

Conversion 47% (130h); $ee_P = 0.95$; $ee_B = 0.90$

Scheme 7.6

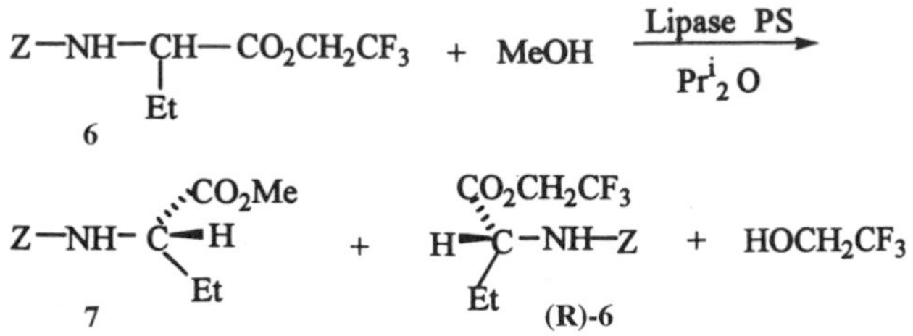

$$Z-NH-\underset{\underset{6}{\overset{|}{Et}}}{CH}-CO_2CH_2CF_3 \quad + \quad MeOH \quad \xrightarrow[Pr^i_2O]{Lipase\ PS}$$

$$Z-NH-\underset{\underset{7}{}}{\overset{\overset{,,,CO_2Me}{}}{C}}\overset{H}{\underset{Et}{\diagdown}} \quad + \quad \underset{\underset{Et}{\overset{CO_2CH_2CF_3}{|}}}{H-C-NH-Z} \quad + \quad HOCH_2CF_3$$

(R)-6

Conversion 38% (2,5h); ee$_p$ = 0.92; E = 42
Z = benzoyloxycarbonyl

Scheme 7.7

(24). The possibility of quasi-irreversible acyl transfers also exists in the case of oxime esters (51). Moreover, co-substrate inhibition has been reported.

Real irreversible transesterification with activated esters is achieved when enol esters are used as acyl donors because the liberating alcohol (Nu$_1$H = vinyl or isopropenyl alcohol) then decomposes immediately to acetaldehyde or acetone, respectively. As a drawback, however, the Schiff base formation (Scheme 7.8) between these co-products and free amino groups (particularly terminal amino residues of lysine) of the enzyme may lead to a partial deactivation and loss of selectivity of a hydrolase. This can be avoided by binding the free ε-amino residues covalently on a solid support prior to an enzymatic reaction or by trapping the acetaldehyde formed in molecular sieves (52, 53).

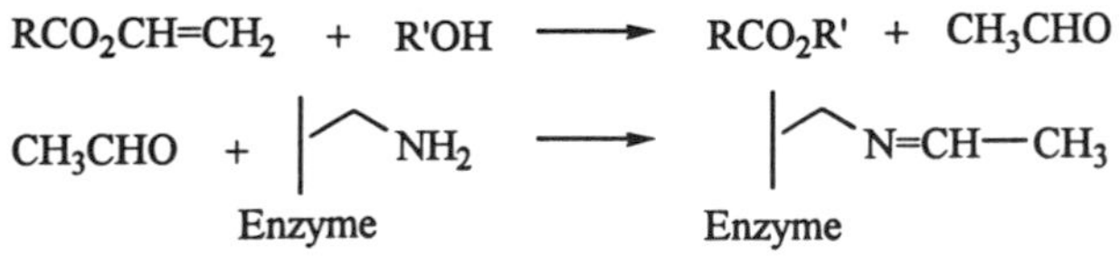

$$RCO_2CH=CH_2 \quad + \quad R'OH \quad \longrightarrow \quad RCO_2R' \quad + \quad CH_3CHO$$

$$CH_3CHO \quad + \quad \Big|{\sim}NH_2 \quad \longrightarrow \quad \Big|{\sim}N{=}CH-CH_3$$

Enzyme   Enzyme

Scheme 7.8

Effective acyl donors are needed especially when sterically hindered secondary alcohols are resolved (Table 7.5) (55–58). Acid anhydrides are the most effective irreversible acyl donors whenever the enzyme tolerates the presence of the liberated acid and non-enzymatic acylation is not disturbing (54). Removal of carboxylic acids by adding a base or by adsorbing the enzyme onto diatomaceous earth can significantly improve enantioselectivity. This is clearly shown for the *Candida cylindracea* lipase-catalysed esterification of compound **12** (Scheme 7.9) (59).

A different way of ensuring irreversible acyl transfers is based on steric arguments (60–63). Thus, when the ester of a bulky secondary alcohol, such as propionate (**14**) (Scheme 7.10) is subjected to hydrolase-catalysed alcoholysis, the secondary alcohol released (**15**) cannot be acylated in the reverse enzymatic reaction with the unactivated ester produced (EtCO$_2$(CH$_2$)$_5$CH$_3$) (62). In order to enhance reactivity, O-formyl esters of sterically hindered alcohols were recommended as substrates (61).

Table 7.5 Resolution of methyl mandelate (reaction A) and *threo*-2-hydroxy-3-(4-methoxyphenyl)-3-(2-X-phenylthio)propionates (reaction B) by *Pseudomonas cepacia* lipase in organic solvents at 22°C

Reaction A

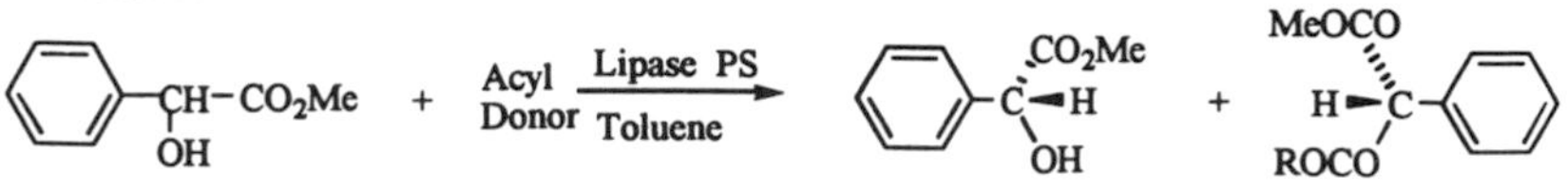

Reaction B

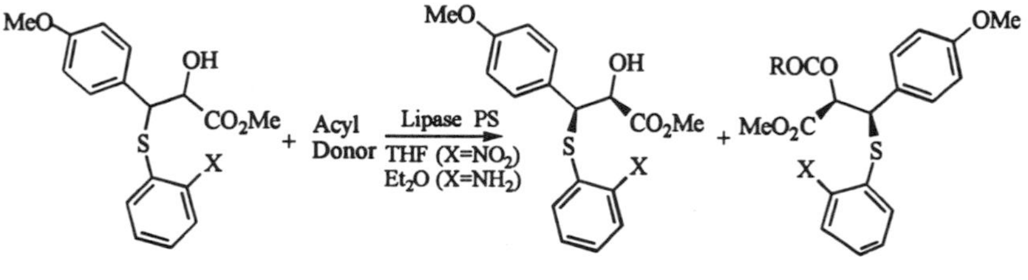

Acyl donor	Time (h)	Conversion (%)	ee$_B$	ee$_P$	Ref.
A					
(PrCO)$_2$O	30	54	0.98(R)	0.87(S)	55
PrCO$_2$CH=CH$_2$	44	47	0.68(R)	0.83(S)	55
PrCO$_2$CH$_2$CF$_3$	72	8	—	—	55
AcOCH=CH$_2$[a]	12	57	>0.99(R)	—	56
B					
Ac$_2$O[b]	48	50	≫95(2S,3S)	≫95(2R,3R)	57
(PrCO)$_2$O[b]	96	50	≫95(2S,3S)	≫95(2R,3R)	57
AcOCH=CH$_2$[b]	168	48	—	—	57
AcOCH=CH$_2$[c]	72	51	—	—	58
AcON=CMe$_2$[c]	48	52	≫95(2S,3S)	≫95(2R,3R)	58
AcOCMe=CH$_2$[c]	48	50	—	—	58

[a] Butyl mandelate was resolved using AcOCH=CH$_2$ as a solvent and acyl donor. [b] X = NO$_2$. [c] X = NH$_2$.

Previously some novel, more exotic acyl donors such as cyclohexyl and 2,2'-biphenyl dipalmitates, 3-acyloxypyridines and 4-acetyl-1,2,4-triazole were reported (64,65). Moreover, alkoxycarbonylation of butan-2-ol, octan-2-ol and 1-phenylethanol in the presence of *Candida antarctica* lipase in anhydrous organic solvents was shown to result in the formation of the corresponding enantiomerically enriched (R)-carbonates (66). As a curiosity, lipases do not cleave amide bonds, although their formation is catalysed. Based on this, the *Candida antarctica* lipase-catalysed aminolysis of ethyl 3-hydroxybutyrate (16) with aniline was successfully performed (Scheme 7.11) (67).

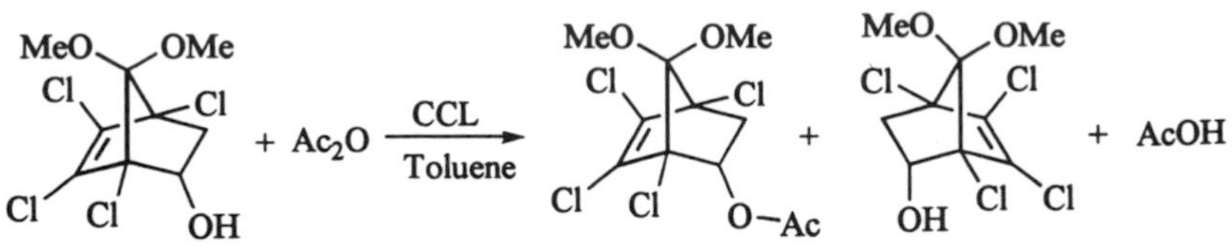

12   **13**   **(1R,2S,4S)-12**

Conversion 54%; $ee_B = 0.87$; $ee_P = 0.74$; $E = 19$
2,6-Lutidine added: Conversion 47%; $ee_B = 0.86$; $ee_P = 0.97$; $E = 180$
KHCO$_3$ added: Conversion 45%; $ee_B = 0.80$; $ee_P = 0.98$; $E = 240$
Lipase on Celite: Conversion 45%; $ee_B = 0.80$; $ee_P = 0.99$; $E = 490$

Scheme 7.9

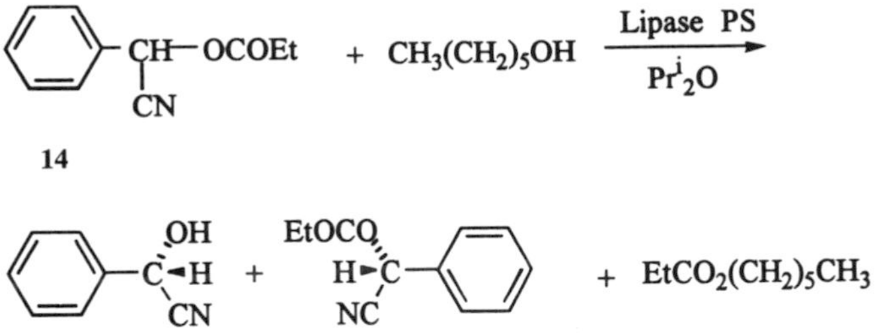

14

15   **(R)-14**

Conversion 54% (48h); $ee_B \gg 0.95$

Scheme 7.10

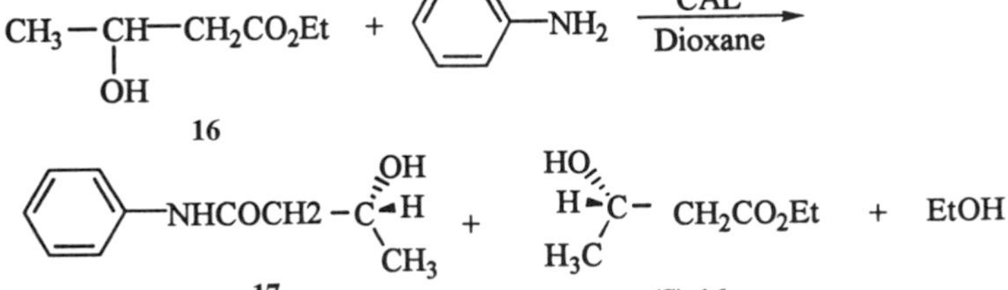

16

17   **(S)-16**

Conversion 45% (21h); $ee_p > 0.99$

Scheme 7.11

7.3.4 Possibilities affecting enantioselectivity

There are certain possibilities of affecting enzymatic enantioselectivity. It is clear that any factor which prevents an equilibrium enhances enantiopurities. Thus, as was discussed in the previous section, the removal of nucleophilic products (Nu_1H) with molecular sieves or by distillation, as well as the elimination of hazardous byproducts, such as carboxylic acids, has a positive effect on enantioselectivity. Stabilization of the enzyme through immobilization may also, although not necessarily, result in enantioselectivity enhancement (24, 68).

Temperature affects enantioselectivity according to equation (7.11) (69).

$$E_1^{T_1} = E_2^{T_2} \qquad (7.11)$$

Enantioselectivity will increase when the temperature is decreased, this effect being more pronounced the higher the enantioselectivity is (Figure 7.5). Experimental data for lipase-catalysed transesterification reactions seem to follow equation (7.11) relatively well (24, 69, 70). In this connection, possible inaccuracy of the fixed E value (e.g. E_1 at the temperature T_1) should be remembered.

The water level associated with the enzyme is often critical to its activity and enantioselectivity in organic solvents. The optimal water level varies from one enzyme to another (1, 71). It also depends on the reaction medium. For the acylation of sulcatol the dehydration of porcine pancreatic lipase to constant weight under vacuum was reported to cause an almost threefold increase in enantioselectivity (72). Evidently, the enzyme had adsorbed water during storage. Dehydrated *Candida cylindracea* lipase in the same reaction lost part of its activity without an effect on enantioselectivity. This is in accordance with the observation that the enzyme very easily loses its essential water (71). An elegant way to control and maintain constant water activity is to perform the reaction in the presence of the binary mixture of anhydrous and the corresponding hydrated

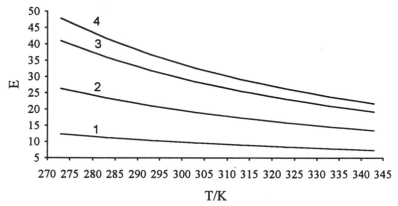

Figure 7.5 Temperature dependence of E. 1: $E_1 = 10$; 2: $E_1 = 20$; 3: $E_1 = 30$; 4: $E_1 = 35$; $T_1 = 298$ K.

inorganic salts (73, 74). In transesterifications, enantioselectivities of *Pseudomonas* lipases were rather insensitive to the addition of solid salt hydrates (75). In *Candida cylindracea* lipase-catalysed esterifications, on the other hand, the enhancement of enantioselectivity at constant water activity caused by salt hydrates was evident (74). It was further shown that part of the water necessary to the enzyme can be replaced by water-mimicking organic compounds (76, 77).

For enzyme-catalysed reactions in organic solvents the possibility of manipulating reactivity and enantioselectivity as well as other selectivities by *the proper choice of the solvent* has been among the most fundamental discoveries (1, 78). Catalytic activity of a hydrolase usually increases with the increase in the solvent hydrophobicity as measured by log P (P is the partition coefficient of the solvent between water and octan-1-ol (79)). Solvent effects on enantio-selectivity often take an opposite course. Thus, for the subtilisin Carlsberg-catalysed propanolysis of N-acetylalanine chloroethyl ester, enantioselectivity increased 35-fold almost linearly when going from hydrophobic cyclohexane (log $P = 3.2$) to hydrophilic dimethylformamide (log $P = -1.0$) (80). As a reasonable explanation, water is replaced from the active site during the substrate binding, and the binding of the natural substrate is more extensive than that of the non-natural enantiomer. Thermodynamically, this results in a more difficult binding for the natural enantiomer in hydrophobic solvents. In the case of the *Aspergillus oryzae* protease-catalysed propanolysis of N-acetylphenylalanine 2-chloroethyl ester, the partitioning of the phenyl group between the hydro-phobic binding pocket and the solvent seems to explain an almost complete inversion of enantioselectivity when going from hydrophilic to hydrophobic solvents (81).

For the transesterification of 1-phenylethanol with vinyl butyrate, subtilisin Carlsberg exerts a different type of solvent effect. In this case there is no correlation between enantioselectivity and log P of the solvent. Instead, there is a correlation with the permittivity and dipole moment (82). Clearly, if polar effects are considered alone, enantioselectivity should be higher (as was observed) the more rigid is the enzyme structure, i.e. in solvents of low permittivity. Namely, for the non-natural enantiomer to be reactive, the group R_{large} (e.g. Scheme 7.2a; in the case of subtilisin, the large pocket is on the left) should be bound into the sterically constrained smaller binding pocket. The observation that there is no correlation between enantioselectivity and hydro-phobicity or polar properties of the solvent for the subtilisin-catalysed acylation of racemic amines or for lipase-catalysed transesterifications (Table 7.6) seriously underlines the fact that a mechanistic explanation of solvent engineering is far from complete (24, 39, 70, 83). Efforts to find linear correla-tions by taking into account both hydrophobicity and polarity at the same time, or to use a two-parameter relationship which includes both permittivity and molar volume of the solvent, also led to poor linear correlations (84, 85).

Finally, the variation of *the structure of the co-substrate* may have an influence on enzymatic enantioselectivity. The chain length of an achiral acyl donor may

Table 7.6 Solvent effects on enantioselectivity for the lipase-catalysed transesterifications A (39), B (39, 70) and C (24)

Reaction A

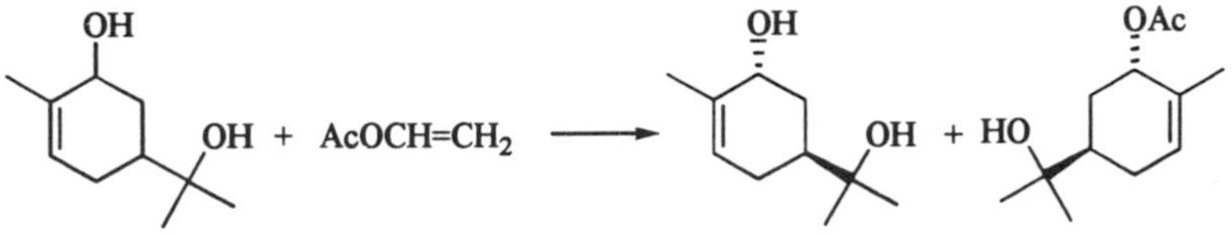

18   (1R,5S)-18   19

Reaction B

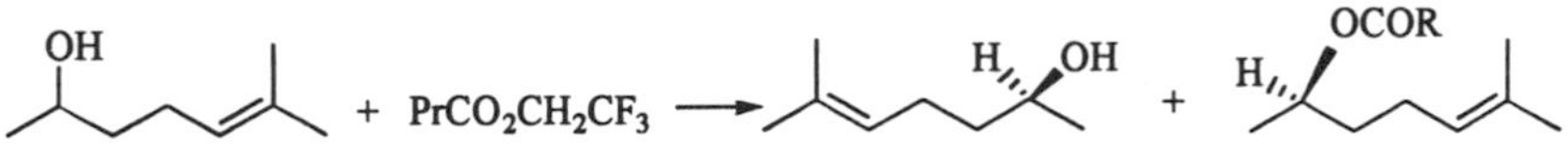

20   S-20   21

Reaction C

1   + $PrCO_2CH_2CF_3$ ⟶   (R)-1   +   (R)-3

Solvent	log P	ε	Lipase AK	Lipase PS		LPL		PPL
			C	A	B	A	B	B
Dioxane	−1.1	2.2		178	23	221	22	42
Acetone	−0.23	20.6		142	40	163	42	34
THF	0.49	7.6	6.9	69	27	112		40
3-Pentanone	0.8	17.0		212	47	249	45	22
Et$_2$O	0.85	4.2	7.5					
tert-Amyl alcohol	1.4	5.8	6.5	518	20	580	18	20
Pr$^i_2$O	1.9	3.9	10					
3-Methyl-3-pentanol	2.0	4.3		16		610		17
Benzene	2.0	2.3		32				62
Toluene	2.5	2.4	8.4	34			37	39
Bu$_2$O	2.9	1.1	5.0	22				20
Cyclohexane	3.1	2.0		13			14	26
Dodecane	6.6			21				23

affect enantioselectivity, although the mechanistic interpretation of the effect is unclear. Usually in the case of porcine pancreatic lipase, enantioselectivity depends only slightly on the chain length of the acyl donor. Systematic studies of structural effects for the porcine pancreatic lipase-catalysed resolution of epoxy alcohol (22) support this conclusion (Scheme 7.12) (86). Similarly, for the acylation of sulcatol (20) with 2,2,2-trifluoroethyl butyrate up to hexadecanoate in dodecane and up to octanoate in diethyl ether gave E values ranging from 19 to 27 (70, 72). Surprisingly, for the same reaction with 2,2,2-trifluoroethyl laurate in diethyl ether a fourfold increase in E was reported, although under similar conditions 2,2,2-trifluoroethyl butyrate and laurate resolved racemic

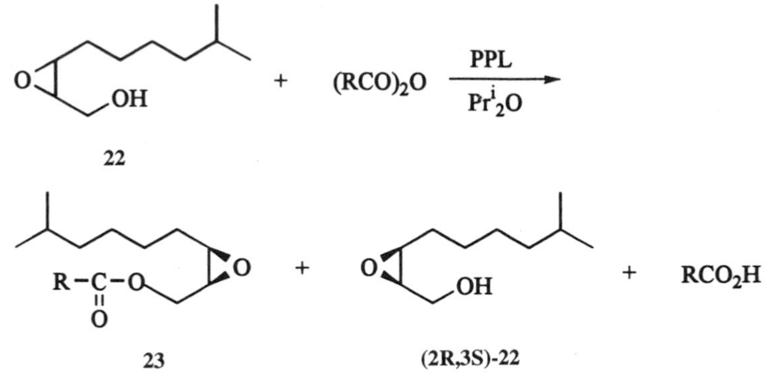

22

$$\text{epoxide–OH} \quad + \quad (RCO)_2O \quad \xrightarrow[\;Pr^i_2O\;]{PPL}$$

23      $R-\overset{O}{\underset{\parallel}{C}}-O$ $+$ (2R,3S)-22 $+$ RCO_2H

R	Me	Et	Pr	$CH_3(CH_2)_4$	Pr^i	Ph	Succinic Anhydride
E	13	9	10	11	8	11	7

Scheme 7.12

ethyl 7-hydroxyoctanoate and 9-hydroxy-2-decenoate with almost equal enantio-selectivity (E 60–70) (72, 87).

For the *Pseudomonas* sp. lipase-catalysed acylation of solketal (**1**) with 2,2,2-trifluoroethyl carboxylates, enantioselectivity increased from acetate ($E = 4.0$) through butyrate ($E = 8.4$) to hexanoate ($E = 8.3$) (24). It was proposed that this reflects increased enantioselectivity with increasing steric hindrance in the acyl enzyme intermediate. There is a tremendous increase in enantioselectivity for the *Pseudomonas fluorescens* lipase-catalysed acylation of tricarbonylchromium complex (**24**) with vinyl carboxylates (Scheme 7.13) (88). In the case of *Pseudomonas cepacia* lipase, the maximal enantioselectivity (E 7–8) was

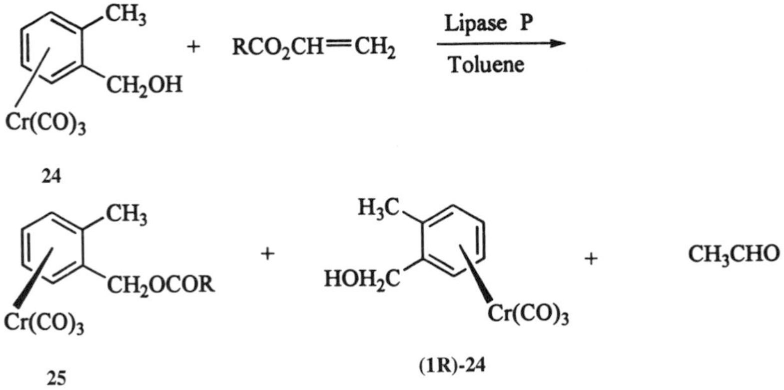

$$\text{(arene)}-CH_3, -CH_2OH, Cr(CO)_3 \quad + \quad RCO_2CH=CH_2 \quad \xrightarrow[\;Toluene\;]{Lipase\ P}$$

24

25    $CH_3, CH_2OCOR, Cr(CO)_3$ $+$ $H_3C, HOH_2C, Cr(CO)_3$ (1R)-24 $+$ CH_3CHO

R =	Me	$CH_3(CH_2)_6$	$CH_3(CH_2)_{14}$	Ph
E =	39	67	>221	38

Scheme 7.13

observed when vinyl hexanoate up to octanoate was used as an acyl donor for the 3-methyl isomer **24** (89). For the highly selective resolutions of 2-halo-1-arylethanols and *para*-substituted methyl mandelates, the acyl chain length of different isopropenyl or vinyl carboxylates did not play a role in enantio-selectivity (46, 89). Surprisingly, vinyl chloroacetate as an acyl donor for mandelates inverted the stereochemical preference of *Pseudomonas* AK lipase from *S* to *R* (90).

Candida cylindracea lipase effectively recognizes chirality in carboxylic acids (9). In accordance with this, the acyl moiety is very critical for *Candida cylindracea* lipase-catalysed enantioselective transesterifications. Thus, (*R*)-menthol was effectively acylated with straight-chain vinyl carboxylates, enantio-selectivity going through a minimum at $n = 2$–4 in $CH_3(CH_2)_nCO_2CH=CH_2$ (9). It is worth mentioning that in the *Candida cylindracea* lipase-catalysed hydrolysis of racemic 1-phenylethyl carboxylates from acetate to laurate, there is a maximum in enantioselectivity for valeric and caproic acid esters (91).

In the resolution of a racemic alcohol through the lipase-catalysed deacylation of its ester, enantiodiscrimination takes place in the formation of the acyl-enzyme intermediate (Scheme 7.1). As a natural consequence, the structure of an achiral alcohol (Nu_2H) should not have an effect on enantioselectivity. On the other hand, the reactivity increases with increasing carbon chain of an alcohol (92).

7.4 Practical resolution of racemic mixtures: transesterification

The synthesis of optically active natural products and pharmaceuticals usually starts with small optically active synthons. Functionally, many such compounds are alcohols, amines, carboxylic acids and their derivatives. That is why hydrolase-catalysed asymmetric acyl transfers in organic media have gained more and more attention. As shown in the previous chapter, it is worth while carefully to consider what is wanted and what are the ways to achieve the goal when the conditions for a kinetic enzymatic resolution are planned. Using enzyme screening it is possible to find the biocatalysts which are catalytically active and show at least some enantioselectivity under the reaction conditions. The improvement of enantioselectivity is the next step. The lack of clear corre-lation between the solvent parameters and enantioselectivity complicates the validation of the solvent. Usually ethers or toluene are good first choices, keeping in mind that the substrates and products must be soluble under the reaction conditions. Although dry enzyme powders are often used, stabilization of the catalyst through immobilization (see elsewhere in this book) is advantageous, especially when the reuse of an enzyme is important. While choosing between acylation and deacylation one should keep in mind that the two reactions with the same enzyme produce the product with the same enantio-selectivity. Moreover, a kinetic enzymatic resolution gives the less reactive enantiomer with highest possible optical purity (Figure 7.2). The hydrolase-catalysed deacylation of secondary alcohol esters is recommended especially when the less reactive enantiomer as an ester is preferred. That is the case, for example, when the alcohol itself is liable to racemization. Deacylation of thioesters is also known. Most of the published examples correspond to the O-acylations of racemic primary and secondary alcohols, N-acylations being comparatively less frequent. As to irreversible acyl transfers, esterification with acid anhydrides is inseparable from transesterification and as such is included in this section.

7.4.1 Resolution of primary alcohols

Primary alcohols (**26–43**) represent important compounds with a chiral center adjacent to the CH_2OH group (Scheme 7.14). Enantioselectivity for the enzyme-catalysed acylation of such compounds often stays low, owing to the remote position of the chiral centre from the reaction site. The reverse enzymatic reaction tends further to reduce optical purities, except when enol esters or acid anhydrides can be used as acyl donors. In spite of these defects, several primary alcohols have been successfully resolved by lipase catalysis in organic solvents.

Using *Pseudomonas fluorescens* lipase, the acetylation of compounds **26** ($R = CH_2SPh$, CH_2SePh, CH_2SO_2Ph, CH_2OCH_2Ph, Ph or $CH=CH–CH_3$) with vinyl acetate in chloroform proceeded with excellent enantioselectivity, leading to enantiopure (R)-(**26**) and the corresponding (S)-acetates (93). Enantio-selectivity was lower ($E = 12$) for the acetylation of propanol (**26**; R = Ph) in *tert*-butyl methyl ether (94). The *Pseudomonas* sp. lipase-catalysed acetylation of the primary hydroxyl group of compound **27** ($R = CH_2Ph$) proceeded with excellent (R)-selectivity ($E > 200$), whereas the E value was only 19 and 16 for $R = CH_2CH=CH_2$ and $CH_2CH=CHCH_3$, respectively, and there was no enantioselectivity with R = Et or Pr (95). The tertiary hydroxyl group of diol **27** was not acylated by lipase catalysis. On the other hand, for the resolution of propanediol (**28**; $R = o$-Me-$C_6H_4OCH_2$) with *Pseudomonas cepacia* lipase, enantioselectivity was associated with the secondary hydroxyl group at the stereogenic centre itself, sequential acetylation with vinyl acetate in THF–diethylamine affording the corresponding (R)-mono- and (S)-diacetates with ee 89–94% and 79–81%, respectively (96).

In the cases of compounds **29** ($R = R_1 = H$) and **30** ($R = H$, $R_1 = R_2 = Me$), the symmetrical shape and small size of the molecule explain the low enantio-selectivity ($E = 1–7$) usually observed (49, 97, 98), although the two enantiomers of solketal (Example 7.1) were previously prepared enantiopure with almost 40% theoretical yields (24). In the presence of porcine pancreatic lipase, aliphatic R_1 substituents enhanced enantioselectivity for the resolution of compound **29** (27, 97). This enhancement proceeded in favour of *cis* over *trans* stereochemistry, which is opposite to the selectivity found for the Sharpless epoxidation of allylic alcohols and for the porcine pancreatic lipase-catalysed

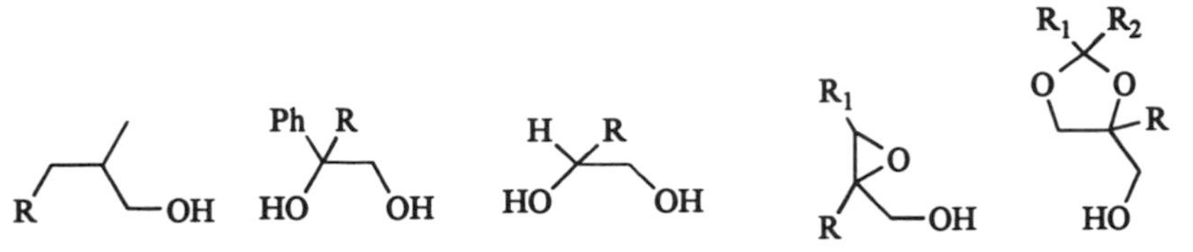

26 27 28 29 30

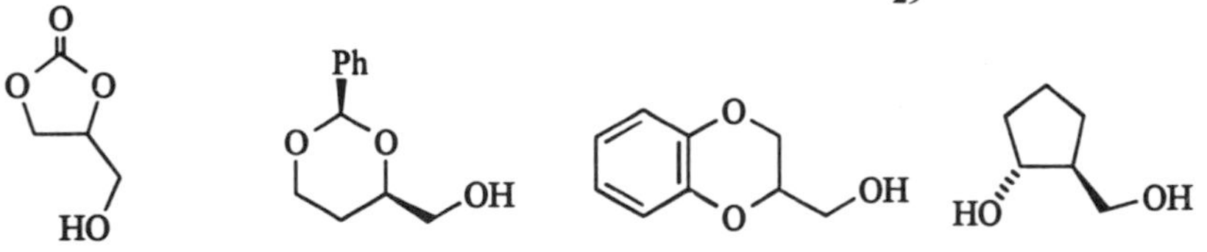

31 32 33 34

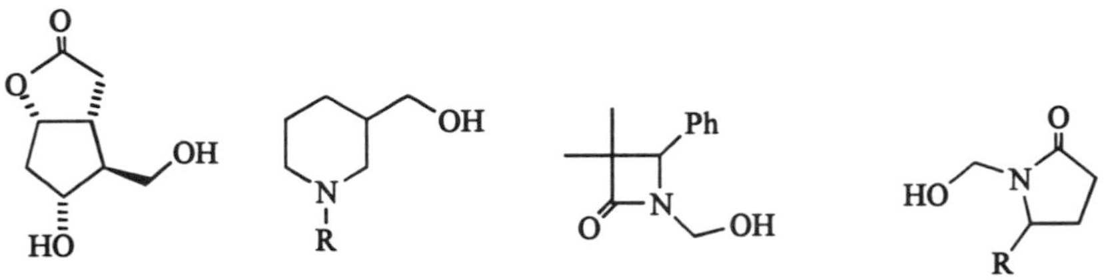

35 36 37 38

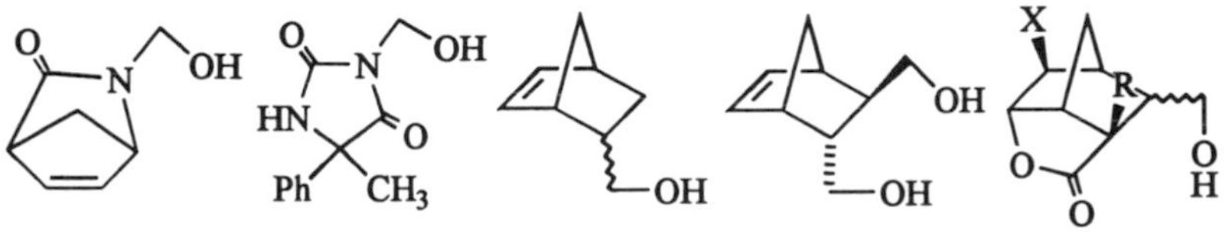

39 40 41 42 43

Scheme 7.14

hydrolysis of the esters of 3-substituted 2,3-epoxyalcohols (99, 100). The optically pure enantiomers of compound **29** (R_1 = H and R = $PhCH_2$, n-C_9H_{11} or $(CH_2)_3CH{=}CH_2$) prepared in the presence of *Pseudomonas fluorescens* lipase well describe the structural expectations for the substrate by the enzyme (93, 101). The nature of the substituents ($R_1 = R_2$ = Et, Pr^i, Ph or $PhCH_2$) on compound **30** clearly affected enantioselectivity (98). The *Pseudomonas fluorescens* lipase-catalysed kinetic resolution of carbonate (**31**) with vinyl acetate in chloroform followed by the enzymatic propanolysis of the acetate produced afforded (*R*)- and (*S*)-alcohols at over 90% and 96% ee, respectively (102).

Good enantioselectivity was observed for the *Pseudomonas fluorescens* lipase-catalysed acetylations of racemic alcohol **32** in THF ($E = 44$) (85, 103) and **33** in 1,4-dioxan (*E* of the order of 22) (104) with vinyl acetate as an acyl donor. In the first-mentioned case, the (S,S)-1,3-dioxan was preferentially recognized by the enzyme. In the latter case, (R)-1,4-benzodioxan (33) was mainly acylated. In other solvents and with other lipases, enantioselectivity was usually much lower. The same lipase facilitated the preparation of the less reactive (R)-alcohol (36; R = *tert*-butoxycarbonyl, Boc), the E value ranging between 23 and 35 in different solvents (105). N-Acetyl protection resulted in lower enantioselectivity and there was no selectivity for R = H in ethyl acetate (106). Notably, the 3-aminomethyl analogue was not resolvable. The *Pseudomonas cepacia* lipase-catalysed acetylation of the primary hydroxyl group in diol 35 with vinyl acetate led to the enantiomerically pure less reactive (1R,5S,6S,7S)-diol at 50% conversion (107). Diacylation was not detected, although sequential acetylation with vinyl acetate worked effectively in the case of diol 34 (108).

Optically pure azetidinones, the less reactive (R)-alcohol (37) (ee > 0.99) and the corresponding (S)-acetate (ee = 0.98), were produced by acetylating with vinyl acetate in the presence of *Pseudomonas cepacia* lipase in dichloromethane (109). Similarly, the acetylation of pyrrolidinones (38; R = Me, Et, Pr, Bu or pentyl) produced (S)-acetate with good enantioselectivity, the E value ranging from 27 to 97 according to the nature of the group R and the solvent (110). The change in priority order around the chiral centre resulted in (R)-acetate with R = Pr^i. *Pseudomonas cepacia* lipase immobilized on Hyflo Super Cel was a better biocatalyst than the free enzyme. The two enantiomers of the hydroxy-methyl derivative 39 were also prepared by using *Pseudomonas cepacia* lipase-mediated acetylation with vinyl acetate (111). *Pseudomonas* sp. lipase as a catalyst and vinyl acetate as an acyl donor in acetonitrile gave the recovered (S)-hydantoin 40 and the corresponding (R)-acetate with 56% (ee 0.43) and 31% (ee 0.80) chemical yields, respectively (112).

Porcine pancreatic lipase-catalysed transesterification in methyl acetate afforded *endo*-norbornenylmethanol enantiomers of 41 with moderate optical purities, the (2S)-isomer reacting preferentially (28). *Exo*-Methanol was better accepted by the lipase, but the acetylation proceeded with low selectivity. Similarly, the enzymatic acylation of alcohol 42 in alkyl carboxylates allowed the preparation of the less reactive (2S,3S)-diol with close to 90% ee at 60–70% conversion. Depending on the nature of the group X (H, Br, I) the porcine pancreatic lipase-catalysed resolution of *endo*-lactones 43 (R = H or Me) proceeded from good (E = 20) to excellent (E ≫ 100) enantioselectivity.

7.4.2 Resolution of secondary alcohols

The resolution of secondary alcohols, such as compounds 44–68 of Scheme 7.15, by hydrolase-catalysed acylation is often successful when the alcohol has one small (R$_{small}$) and one relatively large (R$_{large}$) substituent at the CHOH centre (Scheme 7.2a). Especially, the acylations of octan-2-ol (44; R = CH$_3$(CH$_2$)$_5$, R$_1$ = Me) and 1-phenylethanol (52; R = Me, X = H) by trans-esterification have been extensively studied (29, 35, 36, 41, 46, 49, 54, 63, 87,

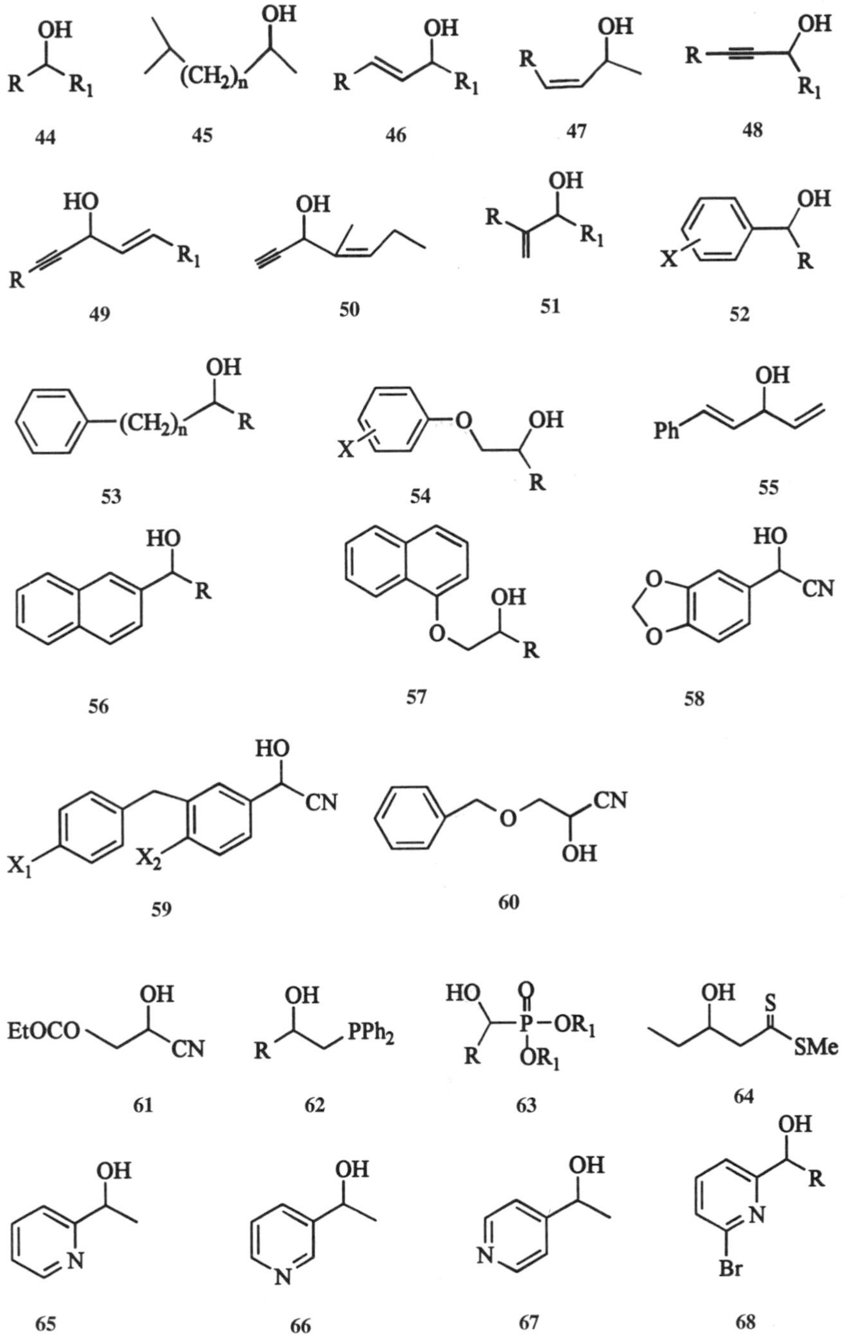

44   **45**   **46**   **47**   **48**

49   **50**   **51**   **52**

53   **54**   **55**

56   **57**   **58**

59   **60**

61   **62**   **63**   **64**

65   **66**   **67**   **68**

Scheme 7.15

113–118). Of the commonly used lipases, *Candida cylindracea* lipase is rather ineffective in the acylation of secondary alcohols. On the other hand, it has been successfully utilized for the deacylations of the corresponding esters. In addition to lipases, *Mucor* esterase showed good to excellent enantioselectivity for the resolution of various aliphatic and aromatic secondary alcohols in alkyl carboxylates (29).

Both crude and purified porcine pancreatic lipase resolved sulcatol (**20**; Table 7.6) excellently (39). This confirmed that it is the lipase (not esterases or proteases present in the commercial crude enzyme) which effectively acylated the secondary alcohol (10). Linear alkan-2-ols (**44**; R = Pr or longer and R_1 = Me) were acylated with excellent (R)-selectivity (E from 52 to >100) in diethyl ether in the presence of 2,2,2-trifluoroethyl laurate (**87**). Due to the similarities of the groups attached at the chiral centre, the acylation of butan-2-ol proceeded without selectivity. The enantiomers of alkan-2-ols with branched alkyl chains (**45**; n = 0–3) and 1-cyclohexyl and cyclohexenyl ethanols can also be prepared optically pure (39, 87, 118). Moreover, porcine pancreatic lipase was used to resolve racemic α-hydroxystannanes (**44**; R = $SnMe_3$ and R_1 = Me, Et or Pr; R = $SnEt_3$ and R_1 = Me or Et) with 2,2,2-trifluoroethyl valerate in diethyl ether (119). The acylated (S)-isomer usually formed with $E \gg 100$. With increasing sizes of the substituents R and R_1 the acylation rates, however, became too slow to be of any practical synthetic value.

(E)-Alkenyl (**46**; R_1 = Me) and alkynyl (**48**; R_1 = Me) groups adjacent to the reaction centre increased rigidity of the nucleophile and accordingly led to improved enantioselectivity of porcine pancreatic lipase (118). In the case of (Z)-allylic alcohols (**47**; R = n-C_7H_{15} or Ph) enantioselectivity was always poor (E = 6 or 2.5, respectively). Double bonds more remote from the reaction site lowered enantioselectivity, the effect being less pronounced the further away the double bond was situated on the chain. Thus, although E = 15 was obtained for the acylation of 4-dodecen-2-ol, porcine pancreatic lipase-mediated resolution of 6-hepten-2-ol with trifluoroethyl butyrate in diethyl ether gave the (R)-butyrate and (S)-alcohol with high chemical and optical yields (E = 60), affording the preparation of 97% enantiopure spiroketal-containing insect pheromone (118, 120).

The enantioselectivity lowering effect of the increasing size of the group R_{small} at the asymmetric centre is evident for the porcine pancreatic lipase-catalysed acylation of racemic 1-phenylalcohols, the E value decreasing from 160 to 25 when the group R in **52** (X = H) was progressively changed from methyl to pentyl (29). The resolution of carbinol **52** (R = $CH=CH_2$, X = H) with vinyl laurate in toluene at 50% conversion gave the enantiomers with high enantio-purity according to the values $[\alpha]_D^{28}$ –1.85 and +2.2 for the less reactive (S)-alcohol and for the (R)-alcohol obtained after hydrolysis of the produced ester, respectively (121). The loss of rigidity adjacent to the chiral centre resulted in a drop of selectivity (E = 15–30) in the case of phenyl substituted alkanols **53** (n = 1 or 2) (118).

Pseudomonas lipases are among the most practical biocatalysts for the resolution of secondary alcohols (7, 122), acylation usually proceeding with preferred (*R*)-selectivity in accordance with the model 'a' in Scheme 7.2. *Pseudomonas* lipases also allow the use of acid anhydrides as acyl donors. The use of succinic anhydride was reported to be advantageous because of an easy separation of the less reactive (*S*)-enantiomer from the half ester produced by extraction. A kilogram-scale resolution of compound **52** (R = Me and X = H) was performed by using this method in *tert*-butyl methyl ether (117).

Chiral haloaryl alcohols are important intermediates for the synthesis of different kinds of optically active compounds. In the presence of *Pseudomonas fluorescens* lipase, the kinetic resolutions of 2-halo-1-arylethanols (**52**; R = CH_2Cl and X = H or *m,p*-di-MeO; R = CH_2Br and X = *p*-Cl or *p*-MeO) and (**56**; R = CH_2Br) with enol esters in di-isopropyl ether afforded the (*R*)-alcohol and the corresponding (*S*)-ester with an ee value of >90% (48). The lipase also displayed high (*S*)-selectivity (usually *E* >100) for the acetylation of 3-chloro-propan-2-ols (**54**; R = CH_2Cl and X = H, *o*-, *m*- or *p*-Me, *m*- or *p*-OMe, 2,3-C_4H_4, *o*-$CH_2CH=CH_2$, 2-cyclo-C_5H_{11}, *p*-CH_2CN or *o*-$OCH_2CH=CH_2$) with vinyl acetate (123). The lipase was further utilized in the resolution of alcohols (**52**; R = *N*-benzyloxycarbonylpiperidine or CH_2Cl and X = *p*-F) with ee ≥ 0.97 for the (*R*)- and (*S*)-acetates produced as well as for the corresponding less reactive (*S*)- and (*R*)-alcohols, respectively, in etheral solutions of vinyl and isopropenyl acetates (124). The enantiomers of 6-(3-chloro-2-hydroxypropyl)-1,3-dioxin-4-one (ee ≥ 98%), which in turn can be converted to optically active 5,6-epoxyhexanoates, were prepared by resolving the racemate with vinyl acetate in the presence of *Pseudomonas cepacia* lipase (125). The (*S*)-selective acetylation of alcohols (**54**; R = CH_2Cl and X = *p*-CH_2CO_2Bu) and (**57**; R = CH_2Cl) in neat vinyl acetate (*E* > 100) or with acetic anhydride in di-isopropyl ether (*E* ≥ 48) by the same lipase made the enantioseparation possible (126, 127). The deacylation of the corresponding acetates by butan-1-ol proceeded with high enantioselectivity (*E* > 100) as well. Porcine pancreatic and *Candida cylindracea* lipases, on the other hand, showed low enantioselectivity in the same deacylation reactions. High enantioselectivity (*E* = 55–64) was further observed for the acetylation of ethanol (**56**; R = CF_3) with vinyl acetate in hexane in the presence of *Pseudomonas cepacia* lipase (128). The acetylation of the corresponding α-naphthyl ethanol proceeded smoothly but without selectivity.

As in the case of porcine pancreatic lipase, multiple bonds adjacent to the reaction centre are advantageous for resolution by *Pseudomonas* lipases. Thus, racemic (*E*)-1-buten-3-ol derivatives (**46**; R = $SiMe_3$, $SiBu^tMe_2$, $SiPhMe_2$ and R_1 = Me) were kinetically resolved with vinyl acetate to the less reactive (*S*)-alcohol and the (*R*)-acetate produced by using *Pseudomonas* AK lipase in pentane (129). For the acetylations of compounds **48** (R = Bu, Ph or $SiMe_3$ and R_1 = Me; R = Bu and R_1 = Et), **49** (R = H and R_1 = Ph; R = Ph and R_1 = H) and **51** (R = Ph, COBu, $CO(CH_2)_2Ph$, CO_2Bu, $CO_2CH_2CH=CH_2$, $CO_2CH_2C≡CH$,

$CO_2(CH_2)_2SPh$, etc., and R_1 = Me) in hexane, the E values ranged from >20 to >56 (130–132). For the *Pseudomonas cepacia* lipase-catalysed resolution of phenyl sulphones (**46**; R = SO_2Ph and R_1 = Me, Et, cyclohexyl, $C_{10}H_{21}$, Pr^i, Bu^t, $CH_2CH_2OCH_2Ph$ or $CH_2CH_2SO_2Ph$) with vinyl acetate, the (*R*)-isomer was always the fast-reacting enantiomer with an E value ≥ 45 in di-isopropyl ether (133, 134). The acylations of acrylonitriles (**51**; R = CN and R_1 = Me, Et, Pr, cyclohexyl or Ph), on the other hand, were extremely slow and showed only moderate enantioselectivities (135). The (*S*)-enantiomer of alcohol (**50**) is the structural component of the synthetic pyrethroid insecticides. In the presence of *Pseudomonas fluorescens* lipase in hexane, the (*S*)-alcohol of 99% ee was obtained at 68.6% conversion (136).

Optically active cyanohydrins (**44**; R = alkyl or aryl, R_1 = CN) are rather labile compounds. Accordingly, deacylation of racemic cyanohydrin esters is a reasonable choice for the enzymatic resolution because it leaves the important enantiomer as a more stable ester. By this reasoning, the *Candida cylindracea* lipase-catalysed deacylation of acylated cyanohydrins (**52**; R = CN) in di-isopropyl ether was used to produce the less reactive (*S*)-cyanohydrin esters (60, 62). Enantioselectivity was quite sensitive to the nature of the substituents X at the aromatic ring. Considerably better results were obtained by using *Pseudomonas cepacia* lipase as a catalyst. In this case, deacylation of racemic cyanohydrin propionates with hexan-1-ol produced the less reactive (*R*)-cyanohydrin esters (**52**; R = CN and X = *p*-CF_3, *p*-Br, H, *p*-Me, *p*-MeO or *m,p*-di-MeO) enantiopure at 50–60% conversion (62). The same lipase also catalysed highly enantioselective acetylations of the above-mentioned cyano-hydrins as well as the acetylations of compounds **56** (R = CN), **57** (R = CN), **58**, and **59** (X_1 = H or F and X_2 = F or H). When the reaction mixture at the beginning contained an aromatic aldehyde as a starting material, acetone cyanohydrin as a source of hydrogen cyanide, isopropenyl acetate, a basic resin and the lipase in di-isopropyl ether, it was possible to convert the aldehyde to the corresponding (*S*)-cyanohydrin acetate with almost 100% chemical yield and with more than 90% ee (62, 137). Lipoprotein lipase from *Pseudomonas* sp. was used to resolve cyanohydrins **53** (R = CN, *n* = 2), **57** (R = CN) and **60** by acetylation with isopropenyl acetate in dichloromethane (138). The resolution of aliphatic cyanohydrins (**44**; R = alkyl and R_1 = CN) was most successfully performed using (*S*)-selective acylation with vinyl butyrate in the presence of porcine pancreatic lipase or (*R*)-selective alcoholysis of the cyanohydrin ester with *Candida cylindracea* lipase, the E value ranging from 15 to >30 (63). In accordance with this, *Candida cylindracea* lipase-catalysed butanolysis of acetylated cyanohydrin **61** afforded (*R*)-cyanohydrin and (*S*)-enriched starting material at 40% conversion (139). The porcine pancreatic lipase-catalysed deacylation of the latter yielded (*S*)-cyanohydrin. The two cyanohydrin enantiomers were finally converted to enantiopure (*R*)- and (*S*)-4-amino-3-hydroxybutanoic acids.

Pseudomonas fluorescens lipase-catalysed acylations with vinyl acetate in *tert*-butyl methyl ether resolved dimethyl acetals **46** (R = CH(OMe)$_2$ and R$_1$ = Et, C$_5$H$_{11}$ or C$_7$H$_{15}$) into (*S*)-alcohols and (*R*)-acetates at 50% conversion, allowing the preparation of the corresponding hydroxy aldehydes with ee > 95% (140). Diethylacetals (**48**; R = CH(OEt)$_2$ and R$_1$ = Me, C$_5$H$_{11}$, C$_7$H$_{15}$, C$_{10}$H$_{21}$, C$_3$H$_7$CH=CH or Ph) were efficiently resolved as well (141). The direct acylation of aldehyde **46** (R = CHO and R$_1$ = C$_5$H$_{11}$) produced almost racemic products with porcine pancreatic (*E* < 1) and *Pseudomonas fluorescens* (*E* = 6) lipases (72, 140). Schiff base formation was supposed to be responsible for this. Hydroxycarboxylic acid esters, on the other hand, were accepted as substrates (142).

Rabbit gastric lipase acylated hydroxyphosphines (**62**; R = Me, Et, Pr, Pr^i, CH$_2$OMe, Ph or CH$_2$Ph) with isopropenyl acetate in toluene (143). The less reactive (*S*)-alcohols were obtained enantiopure (ee > 95%) at 60–70% conversion (*E* = 10–20). Lipase from *Pseudomonas fluorescens* catalysed a trans-esterification between vinyl acetate and the alcohol group of phosphonate (**63**; R = Et and R$_1$ = CH$_2$Ph), the reaction stopping spontaneously at 50% acetylation yielding (–)-acetate and (+)-alcohol with ee > 99% (144). Similarly, *Pseudomonas cepacia* lipase catalysed the acetylation of (*S*)-phosphonate (**63**; R = NHCO$_2$CH$_2$Ph and R$_1$ = Et) with complete selectivity (ee close to 100%) (145). Enantioselective (*E* = 12–90) acetylation of the (*R*)-alcohol of racemic (**64**) in isopropenyl acetate was also catalysed by *Pseudomonas lipases* (146). For the resolutions of β-, γ- and δ-hydroxy sulphones, the enantio-selectivities with few exceptions were good (*E* from 13 to >50) in the presence of *Pseudomonas cepacia* lipase, whereas porcine pancreatic lipase exerted lower enantioselectivities (147, 148).

The enantiomers of 1-pyridinylethanols (**65–68**; R = Me) are important chiral auxiliaries, the preparations of which were conveniently performed by using *Candida antarctica* lipase-catalysed acylation of racemic alcohols in *S*-ethyl thiooctanoate (149). At 50% conversion the less reactive (*S*)-alcohols and the (*R*)-octanoates produced were optically pure (ee ≥ 0.97 and ≥ 0.99, respectively). Evidently for steric reasons, there was no acylation of compound **68** (R = Bu^t). Also, *Pseudomonas fluorescens* lipase acylated ethanols **65–67** (*R*)-selectively (*E* ≥ 100) when vinyl acetate was used as an acyl donor and the reaction was performed in *tert*-butyl methyl ether (150).

Ever since it was found that the *Candida cylindracea* lipase-catalysed resolution of menthol (**74**) in organic solvents proceeded with higher enantio-selectivity than under emulsified conditions (ee > 95% and 70–85%, respectively) (151), non-aqueous enzymology has been widely used to resolve cyclic alcohols such as **69–81** (Scheme 7.16). Again, *Pseudomonas* lipases have served as highly valuable catalysts. Accordingly, *trans*-2-substituted cyclopentanols (**69**; R = Ph or CH$_2$Ph) were resolved with vinyl acetate to (1*S*,2*R*)-alcohol and (1*R*,2*S*)-acetate (*E* ≥ 100) in *tert*-butyl methyl ether (152). In the case of R = OPh or OCH$_2$Ph, the (1*S*,2*S*)-isomers were less reactive in accordance with the change of the priority order at the carbon 2. In the presence of *Pseudomonas*

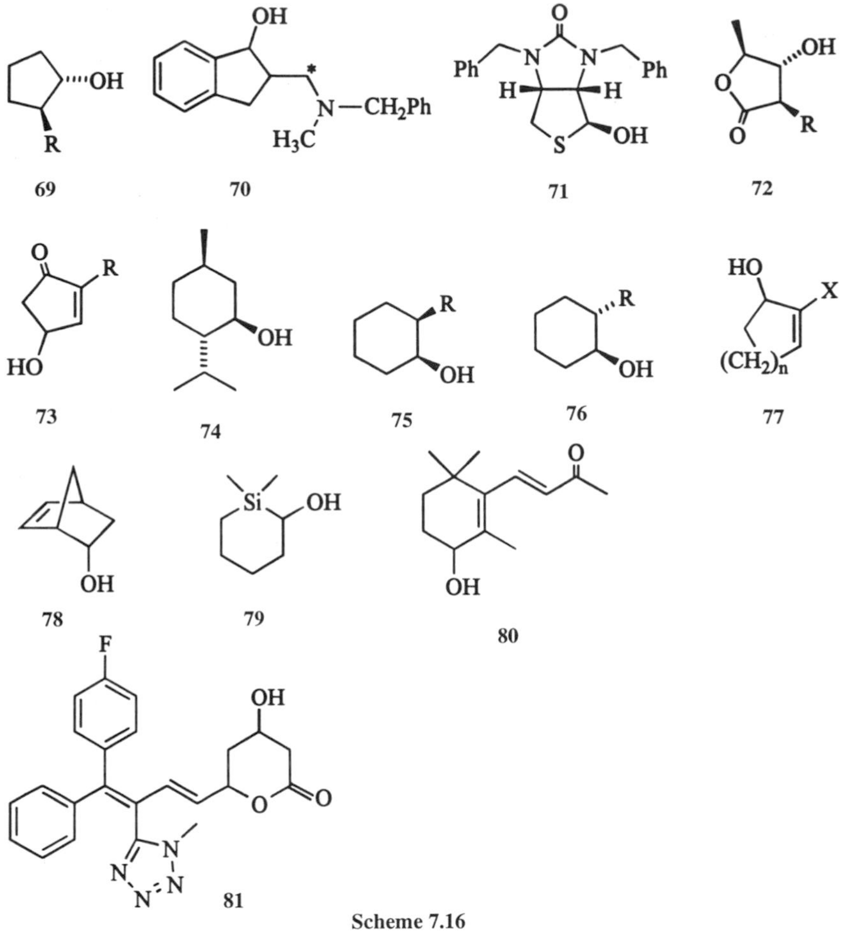

Scheme 7.16

fluorescens lipase, the resolution of radiolabelled compound **70** with vinyl acetate proceeded with almost complete selectivity, enabling the preparation of the two enantiomers of ^{14}C-labelled serotonin uptake inhibitor (153). Lipoprotein lipase from *Pseudomonas aeruginosa* adsorbed on Hyflo Super-Cel and vinyl acetate in toluene afforded the less reactive (4S)-D-biotin intermediate (**71**) in excellent chemical and optical yields (ee = 99.8% after recrystallization) (154). For the reaction of 2,2,2-trichloroethyl octanoate with pentanolide (**72**; R = C$_6$H$_{13}$) in benzene, porcine pancreatic lipase was the best biocatalyst tested (155). When the enantiomerically enriched ester product was subsequently hydrolysed with porcine pancreatic lipase, enantiopure (2S,3S,4R)-alcohol (**72**) was prepared. For the preparation of the (4R)-prostaglandin synthon (**73**; R = (CH$_2$)$_2$CH=CH(CH$_2$)$_2$CO$_2$Me), resolution of the racemate was performed on a 100 g scale with porcine pancreatic lipase in vinyl acetate (156).

The chemical yield was increased by inversion of the less reactive (4S)-enantiomer via a Mitsunobu reaction.

In a continuous column reactor, *Pseudomonas fluorescens* lipase on kieselguhr catalysed highly (1R)-specific ($E \geq 100$) acetylations of cyclohexanols **75** (R = Ph, CH_2Ph, OPh or OCH_2Ph) and **76** (R = OCH_2Ph) with vinyl acetate in *tert*-butyl methyl ether (157). Compared to the normal one-batch reaction, greatly enhanced rates of transformations were achieved. The pure enantiomers of five- to seven-membered cycloalkenols **77** (X = I or Br) were prepared by acetylating the (R)-enantiomer in the mixture of isopropenyl acetate and hexane (1:4) in the presence of *Candida antarctica* B lipase (158). Similarly, the *Pseudomonas cepacia* or *Mucor miehei* lipase-catalysed acetylations of various cyclohexenol derivatives such as sobrerol (**18**; Table 7.6) proceeded with high (R)-selectivity in vinyl acetate (39). In the presence of porcine pancreatic or *Candida cylindracea* lipases, only moderate (R)-selectivities were observed for the acetylation of alcohols **75** and **76** with R = $CH_2C_6H_4OMe$-p (159). In the case of the saturated analogue of compound **78**, on the other hand, the less reactive (1R,2S,4S)-alcohol (ee = 87.2%) and the (1S,2R,4R)-butyrate produced (ee = 87.6%) were obtained at 50% conversion in the porcine pancreatic lipase-catalysed acylation with 2,2,2-trichloroethyl butyrate (160). Interestingly, the corresponding reaction of the *exo*-isomer proceeded to 100% transformation without any selectivity. For the resolution of *endo*-alcohols (**12**; Scheme 7.9) and **78**, *Candida cylindracea* lipase provided moderate to excellent enantio-selectivity (51, 59). The lipase also afforded the preparation of (S)- (95% ee) and (R)-alcohols (96% ee) from racemic **79** provided that the enantiomerically enriched (S)-acetate obtained was further subjected to hydrolysis in another enzymatic transformation (161). A double resolution technique also worked for the preparation of the enantiomers of cyclohexenol **80** (162). In that case, the enzymatically produced (R)-acetate was further hydrolysed using *Chromobacterium viscosum* lipase as a catalyst in an acetylation–hydrolysis sequence. The ability of *Pseudomonas cepacia* lipase to catalyse the resolutions of rather complicated reagents is clearly seen in the preparation of the less reactive (R)-enantiomer of compound **81** using isopropenyl acetate as an acyl donor in toluene (163).

7.4.3 Acylation of diols and resolution of other polyfunctional compounds

When a diol contains both primary and secondary hydroxyl groups in a molecule the first-mentioned group regioselectively reacts first (164). With respect to enantioselectivity, it is often more advantageous, however, to direct the reaction to the secondary function at the chiral centre. Sequential acylation is one possibility to do this. Another way is the protection of the primary function. Suitable sterically hindered protecting groups often further increase the enantio-selectivity of an enzymatic reaction.

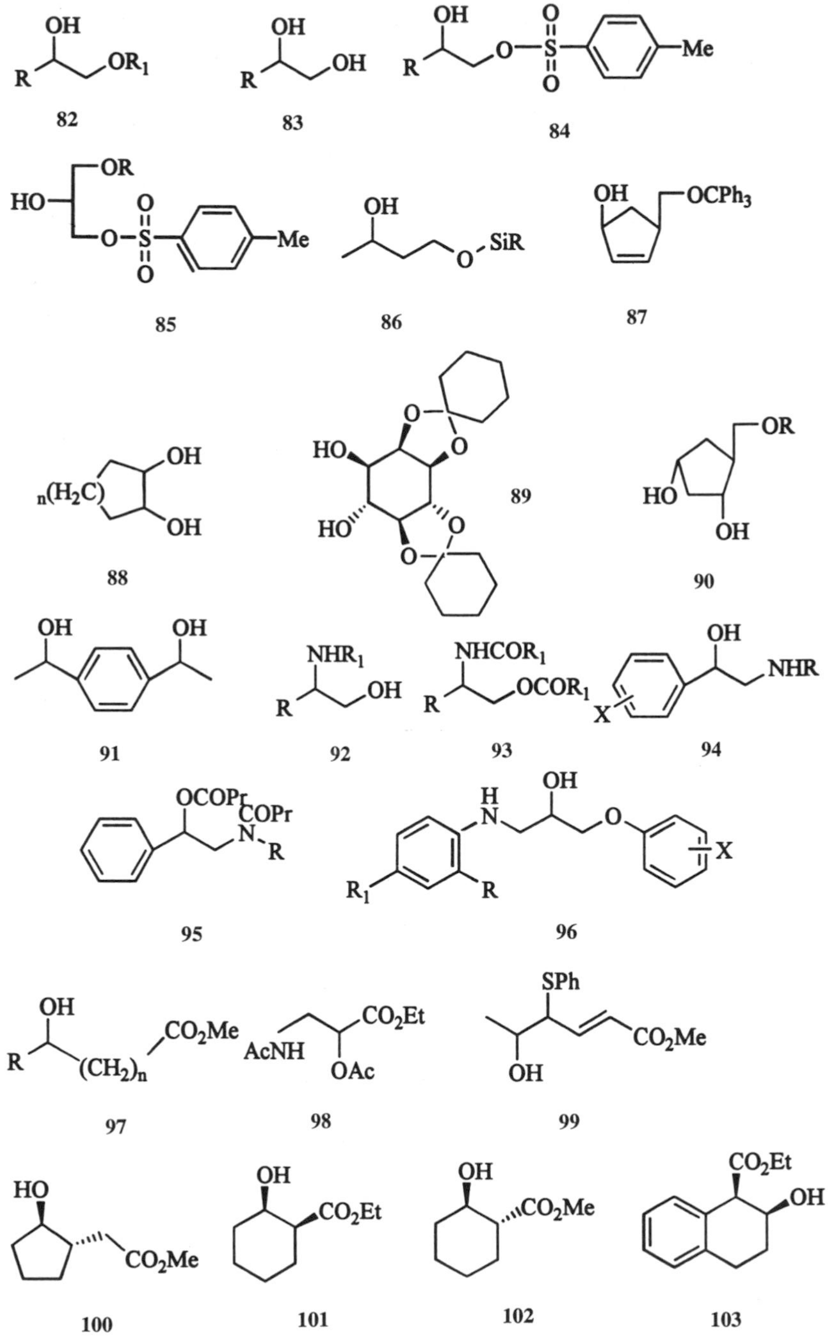
OH
OR1
R
82
OH
OH
R
83
OH
O
O
S
O
Me
R
84
OR
HO
O
O
S
O
Me
85
OH
O
SiR
86
OH
OCPh3
87
OH
n(H2C)
OH
88
O
O
HO
HO
O
O
89
OR
HO
OH
90
OH
OH
91
NHR1
R
OH
92
NHCOR1
R
OCOR1
93
OH
NHR
X
94
OCOPr
COPr
N
R
95
H
N
OH
O
X
R1
R
96
OH
CO2Me
R
(CH2)n
97
CO2Et
AcNH
OAc
98
SPh
CO2Me
OH
99
HO
CO2Me
100
OH
CO2Et
101
OH
CO2Me
102
CO2Et
OH
103

Scheme 7.17

In the presence of *Pseudomonas cepacia* and *fluorescens* lipases, pure (*R*)-acetates of *O*-trityl- (ee > 98%) and *O*-*tert*-butyl-1,2-diols (*E* > 100) were prepared by acetylation of racemic alcohols (**82**; R = Me or Et and R_1 = CPh_3 or Bu^t) with vinyl acetate (Scheme 7.17) (165, 166). In the case of R = CH_2Cl, the new product was the pure (*S*)-acetate due to the priority order of the substituents. For tosylate **84** (R = CH_2Cl), resolution with isopropenyl acetate in hexane similarly produced (*S*)-acetate with high enantioselectivity (*E* > 100) (167). With increasing alkyl chain length of R, enantioselectivity decreased (165, 168). The same trends were seen for the deacylation of the corresponding racemic butyrate with butan-1-ol. For the sequential acetylation of free diol **83** (R = Et) to (*R*)-mono- and (*S*)-diacetates, *E* = 53 was obtained under similar conditions (169, 170). Naturally, the nature of the group R in compound **83** strongly affected enantioselectivity, *E* varying from 12 for R = 1-naphthyl-OCH_2 to >100 for R = *o*- or *m*-$C_6H_4OCH_2$. The highly enantioselective acylations of (*R*)-glycerol derivatives (**85**; R = $C_{16}H_{33}$, $C_{10}H_{21}$ or C_4H_9) and (*R*)-alcohols (**86**; R = Me_2Bu^t or $Me_2(CMe_2Pr^i)$) further express the usability of *Pseudomonas fluorescens* lipase as an excellent resolution catalyst (171, 172). Moreover, asymmetric acylation of trityl-protected diol **87** with vinyl acetate yielded the two enantiomers enantiopure (ee > 95%) (173).

For the acylation of diols where both of the functional groups are identical the product distribution (diol and mono- and diacylated products) depends on the enzymatic enantioselectivity and acylation rates of each step. According to the principles shown in Scheme 7.3, *Pseudomonas cepacia* lipase catalysed the enantioselective diacylation of racemic butene-2,3-diol in vinyl acetate, facilitating the preparation of the diacetate ester of (*R*)-butanediol with 96% ee and the less reactive (*S*)-butanediol with 99% ee, both at reasonably good chemical yields (30 and 23%, respectively) (174). In the case of *trans*-cyclo-alkanediols (**88**; *n* = 1–3), *Pseudomonas* sp. lipase-catalysed acetylation with vinyl acetate in *tert*-butyl methyl ether afforded the corresponding (*R*,*R*)-diacetates (ee ≥ 98%) and (*S*,*S*)-diols (ee ≥ 95%) enantiopure (175). The disadvantage of the long reaction times was avoided by performing the reactions in a continuous column reactor. *Candida cylindracea* lipase showed excellent regio- and enantioselectivity for the acylations of racemic *myo*-inositol (**89**) and its isomeric 1,2:3,4-di-*O*-cyclohexylidene-*myo*-inositol, acylation in diethyl ether proceeding exclusively at the C-4 and C-5 positions of the (*R*)-enantiomers, respectively (176). *Pseudomonas fluorescens* lipase was active only in the case of compound **89**. It preferentially acylated the hydroxyl group at the C-3 position of the (*R*)-enantiomer. *Pseudomonas fluorescens* lipase was also highly regio- and enantioselective for the acetylation of *cis*-diol **90** (R = CPh_3) in vinyl acetate (ee ≥ 98% for (1*S*)- and (3*S*)-monoacetate), enantioselectivity decreasing considerably with decreasing size of the substituent R (177).

When two asymmetric centres with identical functional groups are far away from each other, the stereogenic centres behave independently. Accordingly, lipoprotein lipase from *Pseudomonas* sp. preferentially acetylated the (R)-stereogenic centres of diol **91** with acetic anhydride in benzene, resulting in the mixture of unreacted (S,S)-diol, (R,S)-monoacetate and (R,R)-diacetate (178).

In the case of amino alcohols, there are two different kinds of functional groups that can undergo enzymatic acylation. Although the resolution of racemic 2-aminobutan-1-ol and 1-aminopropan-2-ol by direct acylation has been described (179), it has been impossible to repeat the results afterwards. Moreover, there is a possibility of an $O \rightarrow N$ acyl migration, when the functional groups are close enough to each other, the amide product being thermo-dynamically more stable. Thus, the protection of the amino group was necessary to achieve enzymatic resolution. Porcine pancreatic lipase-catalysed O-acylation of amino alcohols **92** (R = Me or Et and $R_1 = CO_2Et$) in ethyl acetate as well as the O-deacylation of compound **93** (R = Et and R_1 = Me) with butan-1-ol in di-isopropyl ether afforded the (R)-esters and (R)-alcohol as the new reaction products, respectively (180, 181). The asymmetric O-acylation of racemic propan-2-ol derivatives **96** (R = COMe or COPh, R_1 = H or Cl and X = H, p-Cl, o- or p-Ac, o-PhCO or α-naphthyl) with 2,2,2-trichloroacetic anhydride in dioxan further illustrates the potential of porcine pancreatic lipase as an (R)-selective biocatalyst (182). *Pseudomonas cepacia* lipase, on the other hand, worked with high (S)-selectivity for the acylation of 2-aminoethanols **94** (R = CO_2Et or CO_2CH_2Ph and X = H; R = Ac and X = H, Cl, Br, Me, OMe) with 2,2,2-trifluoroethyl butyrate in toluene and with vinyl acetate in hexane–THF (3:2) (92, 183). The O-deacylation of ester **95** (R = H) in the presence of the same lipase was practically enantiospecific, the reaction stopping at 50% conversion. At this point the unreacted (R)-ester and the (S)-alcohol produced were enantiopure (92). Moreover, *Candida cylindracea* lipase (S)-selectively O-deacylated propionates **95** (R = H, Me or Pr^i).

As to enzymatic enantioselectivity, the acylation of the secondary alcohol function in hydroxy carboxylic acids leads to the best result. Further, to be able to avoid the harmful effect of acids on enzymes it is advantageous to resolve hydroxy alkyl carboxylates rather than free acids. The *Pseudomonas cepacia* lipase-catalysed resolution of propionates **10** (Table 7.5) is an elegant example of such a reaction (57, 58). In different solvents and with various acyl donors the resolution smoothly proceeded to 50% conversion, resulting in the enantiopure $(2S,3S)$-alcohol and $(2R,3R)$-ester with almost quantitative yields. Similarly, the highly selective butyrylation of methyl mandelate (**8**) with butyric anhydride was achieved in toluene (55). In the acetylation of butyl mandelate the less reactive alcohol was enantiomerically pure at 57% conversion (56).

Pseudomonas fluorescens lipase was used to resolve aliphatic alcohol **97** (R = $C_{14}H_{29}$ and $n = 0$) in vinyl acetate (184). The less reactive (R)-alcohol was obtained in 34% yield with enantiopurity > 98%. The less reactive (R)-alcohol (87% ee at 60% conversion) was also obtained by acetylating hydroxy methyl ester **97** (R = $C_{11}H_{23}$ and $n = 1$) in the presence of *Geotrichum candidum* lipase in water-saturated vinyl acetate (185). *Candida cylindracea* lipase was ineffective in these acylations, although it (R)-selectively catalysed the deacylation of isoserine derivative **98** with butan-1-ol as well as methanolysis of acetylated methyl mandelate (**8**) in di-isopropylether (60, 186). The enzymatic acylation of γ-hydroxy esters **46** ($R = CO_2Me$ and $R_1 = Me$, Et or Pr) with isopropenyl acetate also proceeded with good (R)-selectivity ($E > 30, > 150$ and > 20, respectively) in the presence of *Pseudomonas* K-10 lipase in hexane (142). Surprisingly, the enantioselectivity was reversed when R_1 was branched-chain Pr^i, Pr^iCH_2, cyclohexyl-CH_2 or $Me_2CHC(Me_2)SiO(CH_2)_2$. The treatment of δ-hydroxy ester **99** in the presence of isopropenyl acetate or phenylthio-acetate and *Alcaligenes* sp. lipase provided quantitatively the (4S,5R)-5-acetoxy ester (98–99% ee) and (4R,5S)-5-hydroxy ester (97–99% ee) (187).

The *Pseudomonas cepacia* lipase catalysed kinetic resolution of *trans*-alcohol **100** in vinyl acetate furnished the less reactive (1R,2S)-alcohol and (1S,2R)-acetate in 44 and 53% chemical yields, respectively (188). The oxidation of the free alcohol enantiomer provided an easy access to (R)-methyl 2-oxocyclo-pentane-1-acetate of 96% ee. *Pseudomonas cepacia* lipase was also highly enantioselective for the acylations of ethyl hydroxy carboxylates **101–103**, the two enantiomers being enantiopure (close to 100% ee) at 50% conversion (55). However, the transesterifications between compound **101** or **102** as a substrate and 6-aminohexan-1-ol as a nucleophile resulted in moderate optical purities (*c.* 50% ee for the two enantiomers at 50% conversion) with high *O*-selectivity (189).

7.4.4 Resolution of tertiary alcohols

For steric reasons, hydrolases do not usually catalyse transformations of tertiary hydroxyl groups in organic solvents. Such reactions are rare in aqueous solutions as well. However, in the presence of immobilized *Mucor miehei* lipase the propanolysis of chloroacetate **104** ($R = H$, Me, Et, Bu or C_6H_{13}) proceeded enantioselectively in di-isopropyl ether, the E value ranging from 13 to 38 (Scheme 7.18) (190). The cyclopropanol derivative produced had the (1S,6S) absolute configuration.

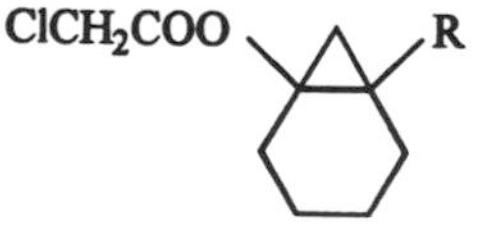

$ClCH_2COO$ R

104

Scheme 7.18

7.4.5 Resolution of amines and thiols

As already mentioned, hydrolase-catalysed direct acylations of amine groups are rare. The resolution of amines **105** ($R = Ph$, 1-naphthyl, CH_2CH_2Ph, 3-indolyl-CH_2, C_6H_{13}, Bu^i and cyclohexyl; Scheme 7.19), phenylalaninamide and 1,2,3,4-tetrahydro-1-naphthylamine were performed with subtilisin Carlsberg in

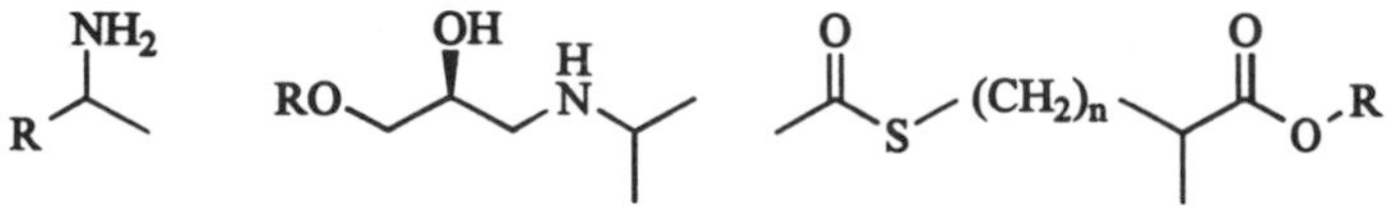

105   106   107

Scheme 7.19

3-methyl-3-pentanol (83). An interesting case among lipase-catalysed asymmetric transformations is the porcine pancreatic lipase-catalysed ring opening of 2,3-epoxypropan-1-ol derivatives with 2-propylamine (191). In toluene, (S)-propanol amines **106** (R = Ph, p-ClC$_6$H$_4$, p-MeCONHC$_6$H$_4$ or o-MeOC$_6$H$_4$) were prepared with moderate to good enantiopurities between 65 and 99% ee. In ethyl acetate or hexane the enantioselectivity was worse.

Little attention has also been paid to the enzymatic resolution of thiols. The porcine pancreatic and *Pseudomonas cepacia* lipase-catalysed deacylations of thioesters **107** (R = Me or Et and n = 0 or 1) with propan-1-ol in hexane took place with high chemo- and stereoselectivity, the thiotransesterification proceeding without affecting the ester moiety and producing the corresponding thiol with the (R) absolute configuration (192).

7.4.6 Resolution of carboxylic acids through transesterification

Carboxylic acids in organic media are most commonly resolved using hydrolase-catalysed esterification between an acid and an alcohol. Transesterifications between the corresponding ester (Scheme 7.20) and an achiral nucleophile is more advantageous, however, because water is not produced in these reactions. As a disadvantage, the two resolution products (the less reactive enantiomer and the new product) are esters, the separation of which presumes that their structures differ considerably. An elegant technique for the separation of the resolved enantiomers was achieved by using poly(ethyleneglycol) as a nucleophile for the alcoholysis of epoxybutanoate **108** in the presence of porcine pancreatic lipase in di-isopropyl ether (193). After filtration of the acylated polymeric alcohol, the less reactive (R)-epoxybutanoate was obtained in 96% ee at 50% conversion.

Mucor miehei lipase in a mixture of isobutyl alcohol and hexane (1:1) afforded (2S,3R)-isopropyl glycidate (ee = 97% at 36% conversion) from racemic methyl ester **109** (X = H) at < 40% conversion (194). Also, the alcoholysis of racemic *trans*-glycidate **109** (X = MeO) proceeded smoothly in

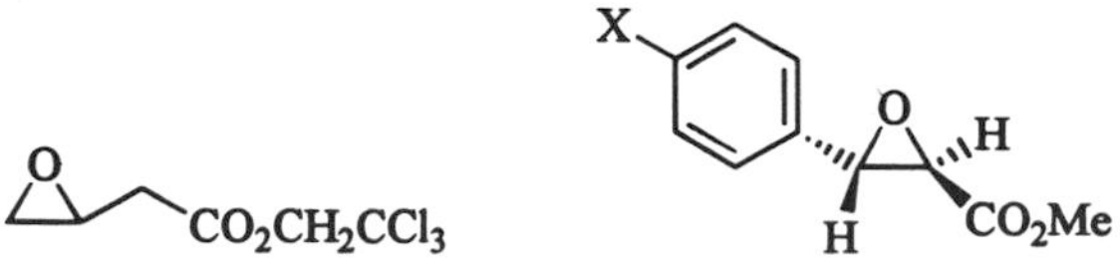

108   109

Scheme 7.20

the presence of various lipases and α-chymotrypsin (57, 195). In the case of *Candida cylindracea* lipase, the less reactive (2*R*,3*S*)-enantiomer was obtained at close to 90% ee and 60% conversion in *tert*-amyl alcohol (57). Enzymatic ester hydrolysis disturbed the reaction. This work was more successfully repeated by carefully drying the solvent and using *Candida cylindracea* lipase from Enzymatix (196).

Example 7.3 Resolution of methyl *trans*-3-(4-methoxyphenyl)glycidate (**109**; X = OMe) using octanolysis in *tert*-amyl alcohol (196). Methyl *trans*-3-(4-methoxyphenyl)glycidate (0.21 g, 1 mmol) was dissolved in 10 ml of *tert*-amyl alcohol (fractionally distilled from calcium oxide and stored over 3 Å molecular sieves) in a conical flask. Distilled octan-1-ol (0.26 g, 2 mmol) was added, followed by *Candida cylindracea* lipase (1 g, Enzymatix). The mixture was shaken in an orbital shaker at 30°C and 200 r.p.m. The reaction was monitored by chiral HPLC (25 cm Chiralcel OD column) using hexane/Pr^{i}OH (95:5) as eluent. The progress of the reaction was followed by observing the disappearance of methyl and appearance of octyl esters. The reaction was terminated after 5 h at 50% conversion by filtering off the enzyme. The solvent was removed, and the residue was purified by flash chromatography on silica gel, eluting with light petroleum–ethyl acetate (10:1 then 4:1), to give initially octyl (2*S*,3*R*)-*trans*-3-(4-methoxyphenyl)glycidate (24 mg, 16%, 99% ee) and finally methyl (2*R*,3*S*)-*trans*-3-(4-methoxyphenyl)glycidate (80 mg, 38%, 99% ee), both as colourless oils.

In the *Candida cylindracea* lipase-catalysed acidolysis of racemic ethyl 2-hydroxyhexanoate with octanoic acid, optically enriched (*S*)-2-hydroxy-hexanoic acid and ethyl (*R*)-2-hydroxyhexanoate were produced with moderate enantioselectivity (*E* = 12) (197). Esterification between the hydroxyl group of the ester substrate and octanoic acid became important only when the hydroxyl group was situated more remotely from the ester function.

Aminolysis of racemic amines (isobutylamine, 2-aminoheptane and α-methyl-benzylamine) with racemic ethyl 2-chloropropionate afforded amides with two chiral centres, one in the acid and the other in the amine moiety (198). *Candida cylindracea* lipase preferentially utilized the (*S*)-enantiomer of the ester and subtilisin the (*S*)-enantiomer of the amine.

7.4.7 Resolution of hydroperoxides

An interesting case of hydrolase-catalysed transformations in organic solvents is the preferred acylation of (*R*)-hydroperoxides of racemic **110** (R = Me and R$_1$ = Pr, Ph or 2-naphthyl; R = Et and R$_1$ = Ph) and **111** by lipoprotein lipase from *Pseudomonas fluorescens* (Scheme 7.21) (199, 200). Acylated hydroperoxide disintegrated spontaneously to carboxylic acid and ketone. Using cyclohexane as a solvent and isopropenyl acetate as an acyl donor,

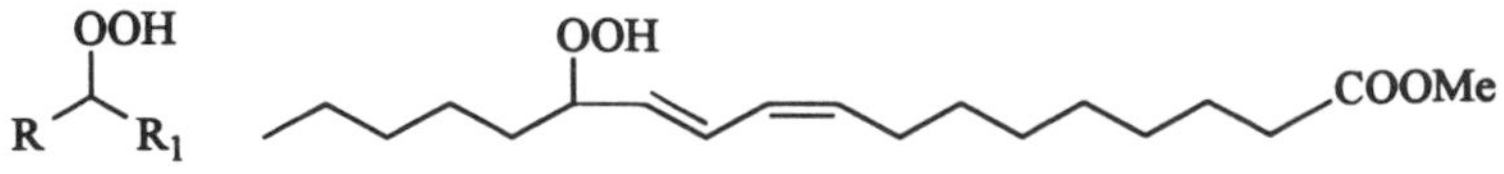

$$\underset{\underset{\textbf{110}}{R \quad R_1}}{\overset{\text{OOH}}{\wedge}} \qquad \underset{\textbf{111}}{\overset{\text{OOH}}{\wedge\wedge\wedge}} \text{COOMe}$$

Scheme 7.21

(S)-1-phenylethyl hydroperoxide was prepared enantiopure (100% ee) at 60% conversion (199). In other cases the optical purities, however, remained rather low.

7.5 Resolution of racemates without a chiral carbon centre and ferrocene-containing substrates

Generally, hydrolases have been exploited for the resolution of centrally chiral compounds with one or more stereogenic carbon centres. In addition to that, hydrolases have the potential to recognize other types of chiral centres in a molecule (Scheme 7.22). Thus, silylmethanol derivatives (**112**; R = Ph, p-MeC$_6$H$_4$ or p-FC$_6$H$_4$ and R$_1$ = Et or Pr) were resolved by papain-catalysed esterification with 5-phenylpentanoic acid in water-saturated 2,2,4-trimethyl-pentane with E values in the range 15–28 (201). It is also noteworthy that many lipases catalysed the esterification but did it without selectivity. *Pseudomonas* K-10 lipase, on the other hand, afforded the enantiomers (the less reactive (R)-methyl ester and the new (S)-butyl ester) of sulphinylalkanoates (**113**; R = p-ClC$_6$H$_4$ or 2-naphthyl and n = 1; R = p-ClC$_6$H$_4$ and n = 2) through butanolysis in hexane (202). The lipase-catalysed acylation of sulphinyl alcohols with vinyl acetate proceeded with low enantioselectivity (E < 10).

   Lipases in organic solvents also recognize axial and planar chirality. Axially chiral primary allenic alcohols of the type **114** were first asymmetrically esterified, although with low enantioselectivity, with lauric acid in hexane in the presence of *Candida cylindracea* lipase (203). Similarly, for the lipase-catalysed acylation of alcohols **115** and **116** with succinic anhydride and for the trans-esterification of vinyl acetate with primary or secondary hydroxyl groups of racemic spiro compounds, only moderate enantioselectivity was usually detected (204, 205). The acetylation of alcohol **117** at close to 50% conversion, however, afforded the less reactive (1S,5R,6S)-diol and the (1R,5S,6R)-mono-acetate counterpart at 89 and 62% ee values, respectively, in the presence of Amano lipase YS (205). In the presence of porcine pancreatic lipase, the acetylation proceeded with the same stereochemistry. The enantiomers of axially chiral [1,1'-binaphthyl]-2,2'-diol (**118**; R = H) have been prepared by *Pseudomonas* sp. lipase-catalysed acylation with vinyl esters as well as by the deacylation of monoacetate (**118**; R = Ac) with alkan-1-ols in di-isopropyl ether (206).

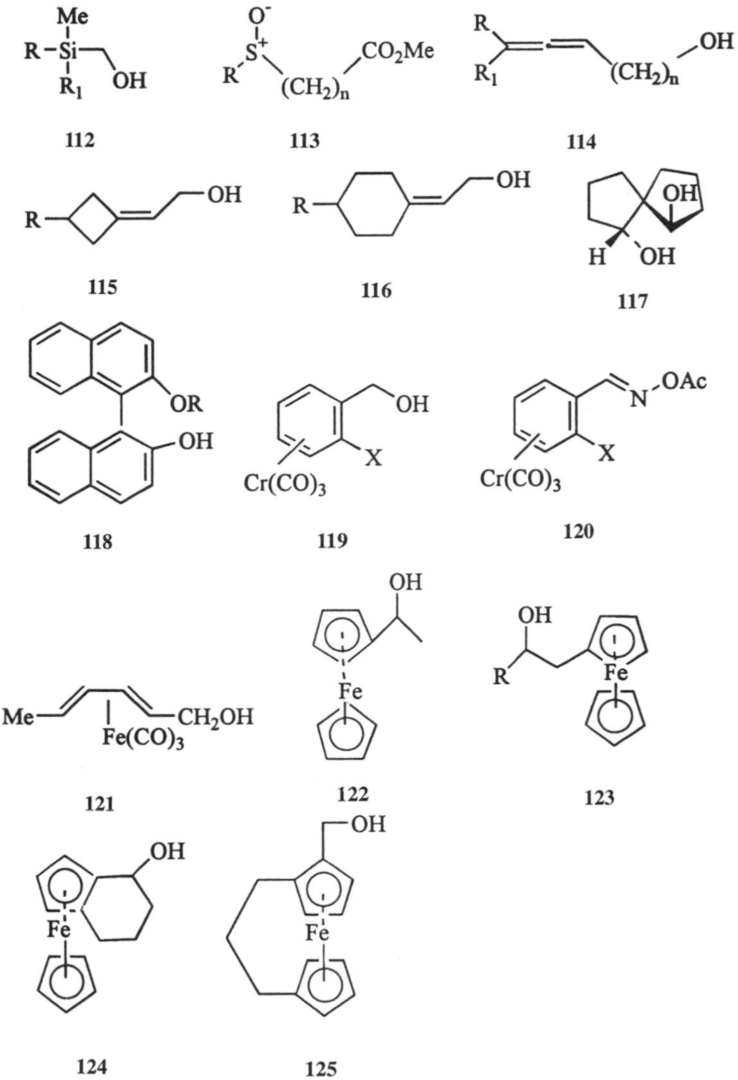

Scheme 7.22

Planar chiral organometallic compounds which belong to (arene)Cr(CO)$_3$ and (diene)Fe(CO)$_3$ series and bear CH$_2$OH substituents were resolved by lipase-catalysed transesterifications. Thus, the acylation of complexes **119** (X = Me, MeO or SiMe$_3$) with vinyl or isopropenyl carboxylates in the presence of *Pseudomonas* sp. lipases gave the corresponding (S)-acetate and the remaining (R)-alcohol, the optical purities of the products depending on the nature of the acyl group of the acyl donor (Scheme 7.13) (88, 207). The *Pseudomonas cepacia* lipase-catalysed deacylation of oxime acetate **120** with butan-1-ol in

3-pentanone proceeded with moderate enantioselectivity ($E = 12.3$), allowing the resolution of the racemic mixture (208). The acetylation of the free oxime proceeded with lower enantioselectivity. On the other hand, complex **121** in the presence of porcine pancreatic lipase afforded the recovered (S)-alcohol and the new (R)-acetate at 76 and 70% ee, respectively, at 53% conversion (209).

Pseudomonas fluorescens lipase allowed the resolution of ferrocene derivatives **122**, **123** (R = Me or Et) and **124** with vinyl esters and butyric anhydride in etheral solutions with high (R)-selectivity (53, 210, 211). High optical purities for (4R,5S)-acetate and recovered (4S,5R)-alcohol (68 and 92% ee at *c.* 50% conversion, respectively) were obtained in the Lipozyme-catalysed acetylation of complex **125** (212).

7.6 Hydrolases in other transformations

Chemo- and regioselectivities are applicable features of enzymatic reactions. Thus, porcine pancreatic and *Pseudomonas cepacia* lipases exclusively catalysed the thiotransesterification of thioesters **107** with propan-1-ol in hexane (192). Surprisingly, porcine pancreatic lipase chemospecifically resulted in the *O*-acylation of 2-mercaptoethanol with ethyl carboxylates $CH_3(CH_2)_nCO_2Et$ ($n = 0$–10) (213). Similarly, Lipozyme and *Candida cylindracea* lipase gave the *O*-acylated product in a fair yield, whereas *Pseudomonas* sp. lipase was ineffective in the acylation.

The formation of ω-aminoalkyl esters proceeded chemoselectively through the *Pseudomonas cepacia* lipase-catalysed transesterification of 2,2,2-trifluoroethyl and ethyl butyrates as well as of ethyl esters **101** and **102** with ω-aminoalkan-1-ols ($HO(CH_2)_nNH_2$, $n = 5$ or 6) in *tert*-amyl alcohol (189). With decreasing n the [ester]:[amide] ratio decreased. *N*-Acylated products were obtained in the acylation of 3-aminopropan-1-ol evidently because of the $O \rightarrow N$ acyl migration. Chemoselectivity for the acylation of ω-aminoalkan-1-ols was strongly dependent on the enzyme, acyl donor and solvent (214, 215). The preferred *O*-acylations of *N*-α-benzoyl-L-lysinol (**126**; Scheme 7.23) proceeded in the presence of subtilisin and *Pseudomonas* sp. lipoprotein, *Chromobacterium viscosum* and *Aspergillus niger* lipases whereas *N*-acylations were preferred in the cases of *Mucor miehei* and porcine pancreatic lipases (215). Moreover, for the acylation of amino alcohol **126** with 2,2,2-trifluoroethyl butyrate by *Pseudomonas* sp. lipoprotein lipase, the initial rate ratio ($v_{O\text{-acylation}}/v_{N\text{-acylation}}$) had the value 21 in 1,2-dichloroethane and 1.1 in *tert*-butyl alcohol. The corresponding values in the case of *Mucor miehei* lipase were 0.33 and 0.018, respectively.

The excellent regioselectivity of lipases towards primary hydroxyl groups in polyhydroxylated molecules was utilized in the acylations of chloramphenicol (**127**; R = H and X = NO_2) and its structural analogue thiamphenicol (**127**; R = H

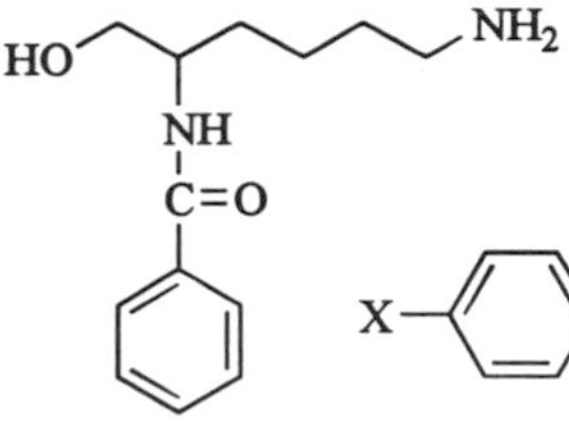

HO NH$_2$

NH

C=O

126

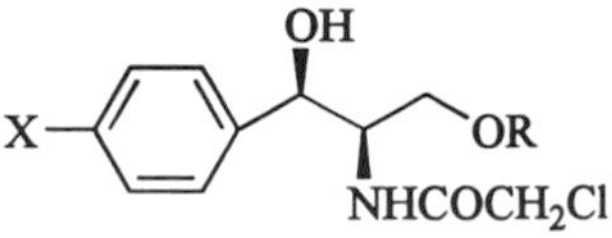

X— OH OR NHCOCH$_2$Cl

127

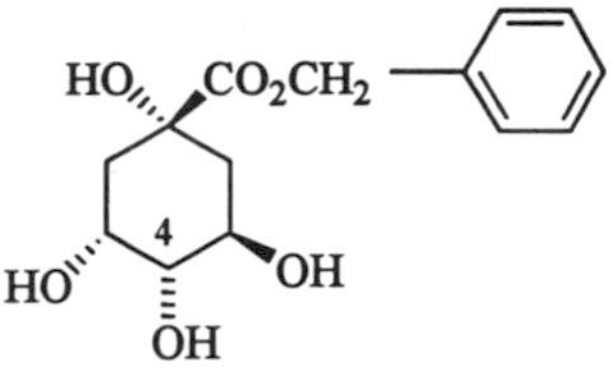

HO CO$_2$CH$_2$— HO 4 OH OH

128

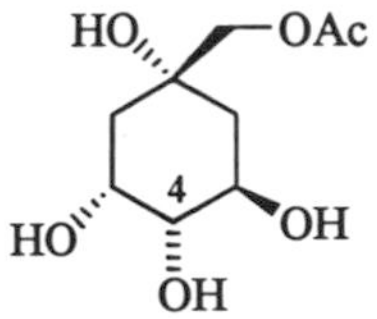

HO —OAc HO 4 OH OH

129

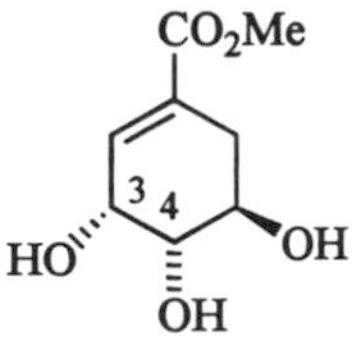

CO$_2$Me HO 3 4 OH OH

130

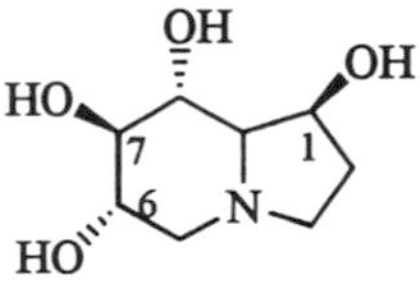

OH HO OH 7 1 6 HO N

131

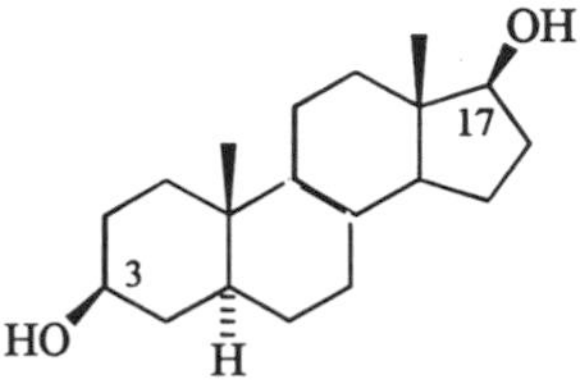

OH 17 HO 3 H

132

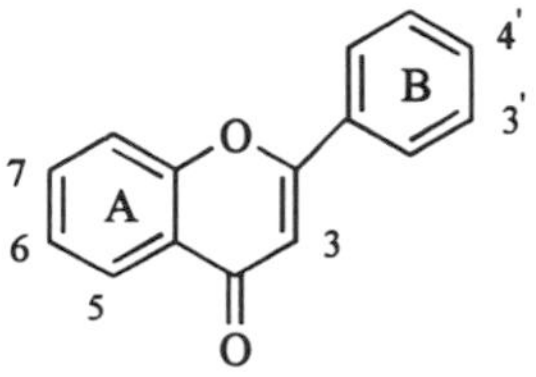

4' B 3' 7 O 6 A 3 5 O

133

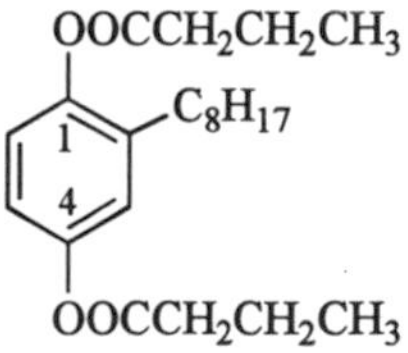

OOCCH$_2$CH$_2$CH$_3$ C$_8$H$_{17}$ 1 4 OOCCH$_2$CH$_2$CH$_3$

134

RO—C—CH$_2$SiR$^1_2$R^2 ‖ O

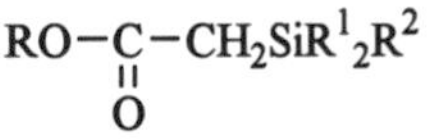

135

Scheme 7.23

and X = MeSO$_2$) (216). Acylations, especially with *Penicillium cyclopium*, *Pseudomonas fluorescens* and *Chromobacterium viscosum* lipases in methyl carboxylates or with trifluoroethyl esters in acetone, allowed the preparation of the esters of compounds **127** (R = COPr, CO(CH$_2$)$_{14}$CH$_3$, COCH=CHPh or COCH$_2$CH$_2$COOH) with almost theoretical yields. *Chromobacterium viscosum* lipase in vinyl acetate acetylated the hydroxyl group at position 4 of compound **128** in close to 90% isolated yield (217). In the presence of the same lipase vinyl acetate in pyridine yielded monoacetate **129** with approximately 50% yield, and the further acetylation in vinyl acetate again took place at position 4 of the molecule. In the case of triol **130** 3-*O*- and 4-*O*-acetyl derivatives were formed in the ratio 2.5:1.

Preparative-scale synthesis of a wide variety of 1-*O*-acyl derivatives of castanospermine (**131**) was performed by using subtilisin catalysis in pyridine (218). Further acylations in tetrahydrofuran preferred the acylation at position 6 by subtilisin and 7 by porcine pancreatic and *Chromobacterium viscosum* lipases. Interestingly, the subtilisin-catalysed hydrolysis of 1,7-diacylated product allowed the preparation of 7-*O*-acylated compound. Regioselective hydrolase-catalysed acylations and deacylations in organic solvents are also important tools in steroid chemistry. Thus, *Chromobacterium viscosum* lipase in acetone acylated the hydroxyl group at position 3 and subtilisin at position 17 of steroid **132** (219). *Candida cylindracea* lipase selectively acylated the 3α-hydroxyl group on several bile acid derivatives in benzene (220). The deacylation of steroid esters by the same lipase in di-isopropyl ether converted 3-esters into the corresponding alcohols, but 3α-, 17β- and 19-esters were not affected (221). 3,17α-Diacetoxy estradiol gave 60% dihydroxy product and 25% 17α-acetoxy-3-hydroxy-estradiol.

Vinyl acetate acetylated *meta*- and *para*-substituted phenols in cyclohexane–tetrahydrofuran (95:5) in the presence of *Chromobacterium viscosum* lipase on Hyflo Super Cel (222). The *Pseudomonas cepacia* lipase-catalysed acetylations of aromatic dihydroxyaldehydes and ketones, on the other hand, regioselectively proceeded into positions other than *ortho* to the carbonyl group in cyclohexane-*tert*-amyl alcohol (90:10) (223). These findings enabled regioselective protection–deprotection for polyhydric flavones in the presence of lipases. Thus, *Pseudomonas cepacia* lipase in tetrahydrofuran was used in regio-selective alcoholyses of various flavone acetates (224, 225). The acetoxy group at position 3 of flavone **133** was not affected. The acetoxy groups at positions 4′ and 3′ (ring B) were primarily converted to hydroxyl groups and the effectiveness of the alcoholysis at ring A followed the positional order 6, 7 and 5. The importance of solvent effects on regioselectivity was clearly shown for the butanolysis of dibutyrate **134** in the presence of *Pseudomonas* lipases (226). Regioselectivity when measured as the ratio of specificity constants, $(k_{cat}/K_M)_1/(k_{cat}/K_M)_4$, increased from the value of 0.5 to 2.5 with increasing hydrophobicity of the solvent from acetonitrile to cyclohexane. In the presence of lipases from *Chromobacterium viscosum*, *Candida cylindracea* and *Aspergillus niger*, however, the deacylation from position 1 took place at least 10 times faster than from position 4 independent of the solvent.

Finally, the lipase from *Rhizopus japonicus* was able to acylate various primary, secondary and tertiary alcohols with trimethylsilylketene in organic media giving trimethylsilylacetyl esters **135** (227). Acrylate monoesters, on the other hand, were synthesized on a preparative scale by the regioselective trans-esterification of a range of diols dissolved in ethyl acrylate in the presence of *Chromobacterium viscosum* lipase (228). The rates for the *Candida cylindracea* lipase-catalysed transesterification of vinyl acrylate with various alcohols followed the order 2-phenylethanol ≫ benzyl alcohol > hexan-1-ol > octan-1-ol > butan-1-ol in iso-octane (229).

References

1. Klibanov, A.M. (1989) Enzymatic catalysis in anhydrous organic solvents. *Trends Biochem.*, **14**, 141–144.
2. Mori, S., Nakata, Y. and Endo, H. (1992) Biosynthesis of cholesterol linoleate by polyethylene glycol-modified cholesterol esterase in organic solvents. *Biotechnol. Appl. Biochem.*, **15**, 278–282.
3. Mori, S., Nakata, Y. and Endo, H. (1992) Enzymatic properties of polyethylene glycol-modified cholesterol esterase in organic solvents. *Biotechnol. Appl. Biochem.*, **16**, 101–105.
4. International Union of Biochemistry and Molecular Biology (1992) *Enzyme Nomenclature 1992 IUBMB*, Academic Press, Inc., San Diego.
5. Brieva, R., Crich, J.Z. and Sih, C.J. (1993) Chemoenzymatic synthesis of the C-13 side chain of taxol: optically active 3-hydroxy-4-phenyl β-lactam derivatives. *J. Org. Chem.*, **58**, 1068–1075.
6. Fersht, A. (1984) *Enzyme Structure and Mechanism*, 2nd edn, Freeman, New York, pp. 405–426.
7. Kazlauskas, R.J., Weissfloch, A.N.E., Rappaport, A.T. and Cuccia, L.A. (1991) A rule to predict which enantiomer of a secondary alcohol reacts faster in reactions catalyzed by cholesterol esterase, lipase from *Pseudomonas cepacia*, and lipase from *Candida rugosa*. *J. Org. Chem.*, **56**, 2656–2665.
8. Cygler, M., Grochulski, P., Kazlauskas, R.J., Schrag, J.D., Bouthillier, F., Rubin, B., Serreqi, A.N. and Gupta, A.K. (1994) A structural basis for the chiral preferences of lipases. *J. Amer. Chem. Soc.*, **116**, 3180–3186.
9. Parida, S. and Dordick, J.S. (1993) Tailoring lipase specificity by solvent and substrate chemistries. *J. Org. Chem.*, **58**, 3238–3244.
10. Hultin, P.G. and Jones, J.B. (1992) Dilemma regarding an active site model for porcine pancreatic lipase. *Tetrahedron Lett.*, **33**, 1399–1402.
11. Toone, E.J., Werth, M.J. and Jones, J.B. (1990) Active-site model for interpreting and predicting the specificity of pig liver esterase. *J. Amer. Chem. Soc.*, **112**, 4946–4952.
12. Provencher, L., Wynn, H. and Jones, J.B. (1993) Enzymes in organic synthesis 51. Probing the dimensions of the large hydrophobic pocket of the active site of pig liver esterase. *Tetrahedron Asymmetry*, **4**, 2025–2040.
13. Chen, C.-S., Fujimoto, Y., Girdaukas, G. and Sih, C.J. (1982) Quantitative analyses of biochemical kinetic resolutions of enantiomers. *J. Amer. Chem. Soc.*, **104**, 7294–7299.
14. Chen, C.-S. and Sih, C.J. (1989) General aspects and optimization of enantioselective biocatalysis in organic solvents: The use of lipases. *Angew. Chem. Int. Ed. Engl.*, **28**, 695–707.
15. Jongejan, J.A., van Tol, J.B.A., Geerlof, A. and Duine, J.A. (1991) Enantioselective enzymatic catalysis 1. A novel method to determine the enantiomeric ratio. *Recl. Trav. Chim. Pays-Bas.*, **110**, 247–254.
16. van Tol, J.B.A., Jongejan, J.A., Geerlof, A. and Duine, J.A. (1991) Enantioselective enzymatic catalysis 2. Applicability of methods for enantiomeric ratio determinations. *Recl. Trav. Chim. Pays-Bas.*, **110**, 255–262.
17. Rakels, J.L.L., Straathof, A.J.J. and Heijnen, J.J. (1993) A simple method to determine the enantiomeric ratio in enantioselective biocatalysis. *Enzyme Microb. Technol.*, **15**, 1051–1056.
18. Guo, Z.-W., Wu, S.-H., Chen, C.-S., Girdaukas, G. and Sih, C.J. (1990) Sequential bio-catalytic kinetic resolutions. *J. Amer. Chem. Soc.*, **112**, 4942–4945.
19. Ozegowski, R., Kunath, A. and Schick, H. (1994) Lipase-catalyzed conversion of (±)-2-methyl-glutaric anhydride into (S)-2-methyl- and (R)-4-methyl-δ-valerolactone via a regio- and enantioselective sequential esterification. *Liebigs Ann. Chem.*, 1019–1023.
20. Macfarlane, E.L.A., Roberts, S.M. and Turner, N.J. (1990) Enzyme-catalysed inter-esterification procedure for the preparation of esters of a chiral secondary alcohol in high enantiomeric purity. *J. Chem. Soc., Chem. Commun.*, 569–571.
21. Fowler, P.W., Macfarlane, E.L.A. and Roberts, S.M. (1991) Highly diastereoselective inter-esterification reactions involving a racemic acetate and a racemic carboxylic acid catalysed by lipase enzymes. *J. Chem. Soc., Chem. Commun.*, 453–455.
22. Macfarlane, E.L.A., Rebolledo, F., Roberts, S.M. and Turner, N.J. (1991) Some inter-esterification reactions involving *Mucor miehei* lipase. *Biocatalysis*, **5**, 13–19.

23. Chen, C.-S. and Liu, Y.-C. (1991) Amplification of enantioselectivity in biocatalyzed kinetic resolution of racemic alcohols. *J. Org. Chem.*, **56**, 1966–1968.
24. Vänttinen, E. and Kanerva, L.T. (1994) Lipase-catalysed transesterification in the preparation of optically active solketal. *J. Chem. Soc., Perkin Trans. 1*, 3459–3463.
25. Chen, C.S., Wu, S.-H., Girdaukas, G. and Sih, C.J. (1987) Quantitative analyses of biochemical kinetic resolution of enantiomers. 2. Enzyme-catalyzed esterifications in water–organic solvent biphasic systems. *J. Amer. Chem. Soc.*, **109**, 2812–2817.
26. Cambou, B. and Klibanov, A.M. (1984) Preparative production of optically active esters and alcohols using esterase-catalyzed stereospecific transesterification in organic media. *J. Amer. Chem. Soc.*, **106**, 2687–2692.
27. Bianchi, D., Cabri, W., Cesti, P., Francalanci, F. and Rama, F. (1988) Enzymatic resolution of 2,3-epoxyalcohols, intermediates in the synthesis of the Gypsy moth sex pheromone. *Tetrahedron Lett.*, **29**, 2455–2458.
28. Janssen, A.J.M., Klunder, A.J.H. and Zwanenburg, B. (1991) Enzymatic resolution of norbor(ne)nylmethanols in organic media and an application to the synthesis of (+)- and (–)-*endo*-norbornene lactone. *Tetrahedron*, **47**, 5513–5538.
29. Janssen, A.J.M., Klunder, A.J.H. and Zwanenburg, B. (1991) Resolution of secondary alcohols by enzyme-catalyzed transesterification in alkyl carboxylates as the solvent. *Tetrahedron*, **47**, 7645–7662.
30. Janssen, A.J.M., Klunder, A.J.H. and Zwanenburg, B. (1991) PPl-catalyzed resolution of 1,2- and 1,3-diols in methyl propionate as solvent. An application of the tandem use of enzymes. *Tetrahedron*, **47**, 7409–7416.
31. Bianchi, D., Cesti, P., Golini, P., Spezia, S., Garavaglia, C. and Mirenna, L. (1992) Enzymatic preparation of optically active fungicide intermediates in aqueous and in organic media. *Pure Appl. Chem.*, **64**, 1073–1078.
32. Francalanci, F., Cesti, P., Cabri, W., Bianchi, D., Martinengo, T. and Foà, M. (1987) Lipase-catalyzed resolution of chiral 2-amino 1-alcohols. *J. Org. Chem.*, **52**, 5079–5082.
33. Gotor, V., Brieva, R. and Rebolledo, F. (1988) Enantioselective acylation of amino alcohols by porcine pancreatic lipase. *J. Chem. Soc., Chem. Commun.*, 957–958.
34. Fernández, S., Brieva, R., Rebolledo, F. and Gotor, V. (1992) Lipase-catalysed enantioselective acylation of *N*-protected or unprotected 2-aminoalkan-1-ols. *J. Chem. Soc., Perkin Trans. 1*, 2885–2889.
35. Öhrner, N., Martinelle, M., Mattson, A., Norin, T. and Hult, K. (1992) Displacement of the equilibrium in lipase catalysed transesterification in ethyl octanoate by continuous evaporation of ethanol. *Biotechnol. Lett.*, **14**, 263–268.
36. Hult, K. and Norin, T. (1992) Enantioselectivity of some lipases: Control and prediction. *Pure Appl. Chem.*, **64**, 1129–1134.
37. Frykman, H., Öhrner, N., Norin, T. and Hult, K. (1993) *S*-Ethyl thiooctanoate as acyl donor in lipase catalysed resolution of secondary alcohols. *Tetrahedron Lett.*, **34**, 1367–1370.
38. Ottolina, G., Carrea, G. and Riva, S. (1990) Synthesis of ester derivatives of chloramphenicol by lipase-catalyzed transesterification in organic solvents. *J. Org. Chem.*, **55**, 2366–2369.
39. Carrea, G., Ottolina, G., Riva, S. and Secundo, F. (1992) Effect of reaction conditions on the activity and enantioselectivity of lipases in organic solvents. In *Biocatalysis in Non-Conventional Media* (eds J. Tramper *et al.*), Elsevier, Amsterdam, 111–119.
40. Gerlach, D. and Schreier, P. (1989) Esterification in organic media for preparation of optically active secondary alcohols: effects of reaction conditions. *Biocatalysis*, **2**, 257–263.
41. Kirchner, G., Scollar, M.P. and Klibanov, A.M. (1985) Resolution of racemic mixtures via lipase catalysis in organic solvents. *J. Amer. Chem. Soc.*, **107**, 7072–7076.
42. Miyazawa, T., Mio, M., Watanabe, Y., Yamada, T. and Kuwata, S. (1992) Lipase-catalyzed transesterification procedure for the resolution of non-protein amino acids. *Biotechnol. Lett.*, **14**, 789–794.
43. Miyazawa, T., Kurita, S., Ueji, S., Yamada, T. and Kuwata, S. (1992) Resolution of racemic carboxylic acids via the lipase-catalyzed irreversible transesterification using vinyl esters; effects of alcohols as nucleophiles and organic solvents on enantioselectivity. *Biotechnol. Lett.*, **14**, 941–946.
44. Nishio, T., Kamimura, M., Murata, M., Terao, Y. and Achiwa, K. (1988) Enzymatic transesterification with the lipase from *Pseudomonas fragi* 22.39 B in a non-aqueous reaction system. *J. Biochem.*, **104**, 681–682.

45. Naemura, K., Ida, H. and Fukuda, R. (1993) Lipase YS-catalyzed enantioselective trans-esterification of alcohols of bicarbocyclic compounds. *Bull. Chem. Soc. Jpn*, **66**, 573–577.
46. Kanerva, L.T., Vihanto, J., Pajunen, E. and Euranto, E.K. (1990) Structural effects in the enzymatic resolution of 2-octanol. *Acta Chem. Scand.*, **44**, 489–491.
47. Degueil-Castaing, M., de Jeso, B., Drouillard, S. and Maillard, B. (1987) Enzymatic reactions in organic synthesis: 2-ester interchange of vinyl esters. *Tetrahedron Lett.*, **28**, 953–954.
48. Hiratake, J., Inagaki, M., Nishioka, T. and Oda, J. (1988) Irreversible and highly enantio-selective acylation of 2-halo-1-arylethanols in organic solvents catalyzed by a lipase from *Pseudomonas fluorescens*. *J. Org. Chem.*, **53**, 6130–6133.
49. Wang, Y.-F., Lalonde, J.J., Momongan, M., Bergbreiter, D.E. and Wong, C.-H. (1988) Lipase-catalyzed irreversible transesterifications using enol esters as acylating reagents: preparative enantio- and regioselective synthese of alcohols, glycerol derivatives, sugars and organometallics. *J. Amer. Chem. Soc.*, **110**, 7200–7205.
50. Ghogare, A. and Kumar, G.S. (1989) Oxime esters as novel irreversible acyl transfer agents for lipase catalysis in organic media. *J. Chem. Soc., Chem. Commun.*, 1533–1535.
51. Mischitz, M., Pöschl, U. and Faber, K. (1991) Limitations of enzymatic acylation using oxime esters: cosubstrate inhibition and the reversibility of the reaction. *Biotechnol. Lett.*, **13**, 653–656.
52. Berger, B. and Faber, K. (1991) 'Immunization' of lipase against acetaldehyde emerging in acyl transfer reactions from vinyl acetate. *J. Chem. Soc., Chem. Commun.*, 1198–1200.
53. Izumi, T., Tamura, F. and Sasaki, K. (1992) Enzymatic kinetic resolution of [4](1,2)ferro-cenophane derivatives. *Bull. Chem. Soc. Jpn*, **65**, 2784–2788.
54. Bianchi, D., Cesti, P. and Battistel, E. (1988) Anhydrides as acylating agents in lipase-catalyzed stereoselective esterification of racemic alcohols. *J. Org. Chem.*, **53**, 5531–5534.
55. Kanerva, L.T. and Sundholm, O. (1993) Enzymatic preparation of optically active α- and β-hydroxy carboxylic acid derivatives. *Acta Chem. Scand.* **47**, 823–825.
56. Ebert, C., Ferluga, G., Gardossi, L., Gianferrara, T. and Linda, P. (1992) Improved lipase-mediated resolution of mandelic acid esters by multivariate investigation of experimental factors. *Tetrahedron Asymmetry*, **3**, 903–912.
57. Kanerva, L.T. and Sundholm, O. (1993) Lipase catalysis in the resolution of racemic intermediates of diltiazem synthesis in organic solvents. *J. Chem. Soc., Perkin Trans. 1*, 1385–1389.
58. Kanerva, L.T. and Sundholm, O. (1993) Enzymatic acylation in the resolution of methyl *threo*-2-hydroxy-3-(4-methoxyphenyl)-3-(2-X-phenylthio)propionates in organic solvents. *J. Chem. Soc., Perkin Trans. 1*, 2407–2410.
59. Berger, B., Rabiller, C.G., Königsberger, K., Faber, K. and Griengl, H. (1990) Enzymatic acylation using acid anhydrides: crucial removal of acid. *Tetrahedron Asymmetry*, **1**, 541–546.
60. Bevinakatti, H.S., Banerji, A.A. and Newadkar, R.V. (1989) Resolution of secondary alcohols using lipase in di-isopropyl ether. *J. Org. Chem.*, **54**, 2453–2455.
61. Bevinakatti, H.S. and Newadkar, R.V. (1989) Lipase catalysis in organic solvents. Transesterification of *O*-formyl esters of secondary alcohols. *Biotechnol. Lett.*, **11**, 785–788.
62. Kanerva, L.T., Rahiala, K. and Sundholm, O. (1994) Optically active cyanohydrins and enzyme catalysis. *Biocatalysis*, **10**, 169–180.
63. Kanerva, L.T., Kiljunen, E. and Huuhtanen, T.T. (1993) Enzymatic resolution of optically active aliphatic cyanohydrins. *Tetrahedron Asymmetry*, **4**, 2355–2361.
64. Lin, G., Liu, S.-H., Wu, F.-C. and Jen, W.-J. (1993) Novel acyl donors for enzyme-catalyzed transacylation reactions. *Synthetic Comm.*, **23**, 2135–2138.
65. Keumi, T., Hiraoka, Y., Ban, T., Takahashi, I. and Kitajima, H. (1991) Utility of acyloxy-pyridines and acylazoles for the lipase-catalyzed enantioselective acylation of 1-phenylethanol. *Chem. Lett.*, 1989–1992.
66. Pozo, M., Pulido, R. and Gotor, V. (1992) Vinyl carbonates as novel alkoxycarbonylation reagents in enzymatic synthesis of carbonates. *Tetrahedron*, **48**, 6477–6484.
67. García, M.J., Rebolledo, F. and Gotor, V. (1992) Enzymatic synthesis of 3-hydroxybutyramides and their conversion to optically active 1,3-aminoalcohols. *Tetrahedron Asymmetry*, **3**, 1519–1522.
68. Ottolina, G., Carrea, G., Riva, S., Sartore, L. and Veronese, F.M. (1992) Effect of the enzyme form on the activity, stability and enantioselectivity of lipoprotein lipase in toluene. *Biotechnol. Lett.*, **14**, 947–952.

69. Holmberg, E. and Hult, K. (1991) Temperature as an enantioselective parameter in enzymatic resolutions of racemic mixtures. *Biotechnol. Lett.*, **13**, 323–326.
70. Secundo, F., Riva, S. and Carrea, G. (1992) Effects of medium and of reaction conditions on the enantioselectivity of lipases in organic solvents and possible rationales. *Tetrahedron Asymmetry*, **3**, 267–280.
71. Zaks, A., Klibanov, A.M. (1985) Enzyme-catalyzed processes in organic solvents. *Proc. Natl. Acad. Sci. USA*, **82**, 3192–3196.
72. Stokes, T.M. and Oehlschlager, A.C. (1987) Enzyme reactions in apolar solvents: The resolution of (±)-sulcatol with porcine pancreatic lipase. *Tetrahedron Lett.*, **28**, 2091–2094.
73. Kvittingen, L., Sjursnes, B., Anthonsen, T. and Halling, P. (1992) Use of salt hydrates to buffer optimal water level during lipase catalysed synthesis in organic media: a practical procedure for organic chemists. *Tetrahedron*, **48**, 2793–2802.
74. Högberg, H.-E., Edlund, H., Berglund, P. and Hedenström, E. (1993) Water activity influences enantioselectivity in a lipase-catalysed resolution by esterification in an organic solvent. *Tetrahedron Asymmetry*, **4**, 2123–2126.
75. Nordin, O., Hedenström, E. and Högberg, H.-E. (1994) Enantioselective transesterifications of 2-methyl-1-alcohols catalysed by lipases from *Pseudomonas*. *Tetrahedron Asymmetry*, **5**, 785–788.
76. Kitaguchi, H., Itoh, I. and Ono, M. (1990) Effects of water and water-mimicking solvents on the lipase-catalyzed esterification in an apolar solvent. *Chem. Lett.*, 1203–1206.
77. Gubicza, L. and Kelemen-Horvàth, I. (1993) Effect of water-mimicking additives on the synthetic activity and enantioselectivity in organic solvents of a *Candida cylindracea* lipase. *J. Mol. Catal.*, **84**, L27–L32.
78. Wescott, C.R. and Klibanov, A.M. (1994) The solvent dependence of enzyme specificity. *Biochim. Biophys. Acta*, **1206**, 1–9.
79. Laane, C., Boeren, S., Vos, K. and Veeger, C. (1987) Rules for optimization of biocatalysis in organic solvents. *Biotechnol. Bioeng.*, **30**, 81–87.
80. Sakurai, T., Margolin, A.L., Russell, A.J. and Klibanov, A.M. (1988) Control of enzyme enantioselectivity by the reaction medium. *J. Amer. Chem. Soc.*, **110**, 7236–7237.
81. Tawaki, S. and Klibanov, A.M. (1992) Inversion of enzyme enantioselectivity mediated by the solvent. *J. Amer. Chem. Soc.*, **114**, 1882–1884.
82. Fitzpatrick, P.A. and Klibanov, A.M. (1991) How can the solvent affect enzyme enantioselectivity? *J. Amer. Chem. Soc.*, **113**, 3166–3171.
83. Kitaguchi, H., Fitzpatrick, P.A., Huber, J.E. and Klibanov, A.M. (1989) Enzymatic resolution of racemic amines: crucial roles of the solvent. *J. Amer. Chem. Soc.*, **111**, 3094–3095.
84. Nakamura, K., Kinoshita, M. and Ohno, A. (1994) Effect of solvent on lipase-catalyzed transesterification in organic media. *Tetrahedron*, **50**, 4681–4690.
85. Herradón, B. and Valverde, S. (1994) Biocatalytic synthesis of chiral polyoxygenated compounds: modulation of the selectivity upon changes in the experimental conditions. *Tetrahedron Asymmetry*, **5**, 1479–1500.
86. Fukusaki, E., Senda, S., Nakazono, Y., Yuasa, H. and Omata, T. (1992) Lipase-catalyzed kinetic resolution of 2,3-epoxy-8-methyl-1-nonanol, the key intermediate in the synthesis of the Gypsy moth pheromone. *J. Ferm. Bioeng.*, **73**, 280–283.
87. Morgan, B., Oehlschlager, A.C. and Stokes, T.M. (1991) Enzyme reactions in apolar solvents. The resolution of branched and unbranched 2-alkanols by porcine pancreatic lipase. *Tetrahedron*, **47**, 1611–1620.
88. Yamazaki, Y. and Hosono, K. (1990) Facile resolution of planar chiral organometallic alcohols with lipase in organic solvents. *Tetrahedron Lett.*, **31**, 3895–3896.
89. Yamazaki, Y., Morohashi, N. and Hosono, K. (1991) Lipase-mediated homotopic and heterotopic double resolutions of a planar chiral organometallic alcohol. *Biotechnol. Lett.*, **13**, 81–86.
90. Miyazawa, T., Kurita, S., Ueji, S., Yamada, T. and Kuwata, S. (1992) Resolution of mandelic acids by lipase-catalysed transesterifications in organic media: Inversion of enantioselectivity mediated by the acyl donor. *J. Chem. Soc., Perkin Trans. 1*, 2253–2255.
91. Holmberg, E., Szmulik, P., Norin, T. and Hult, K. (1989) Hydrolysis and esterification with lipase from *Candida cylindracea*. Influence of the reaction conditions and acid moiety on the enantiomeric excess. *Biocatalysis*, **2**, 217–223.
92. Kanerva, L.T., Rahiala, K. and Vänttinen, E. (1992) Lipase catalysis in the optical resolution of 2-amino-1-phenylethanol derivatives. *J. Chem. Soc., Perkin Trans. 1*, 1759–1762.

93. Santaniello, E., Ferraboschi, P., Grisenti, P. and Manzocchi, A. (1992) α-Substituted primary alcohols as substrates for enantioselective lipase-catalyzed transesterification in organic solvents. In *Biocatalysis in Non-Conventional Media* (eds J. Tramper *et al.*), Elsevier, Amsterdam, pp. 533–540.

94. Delinck, D.L. and Margolin, A.L. (1990) Synthesis of chiral building blocks for selective adenosine receptor agents. Lipase-catalyzed resolution of 2-benzylpropanol and 2-benzylpropionic acid. *Tetrahedron Lett.*, **31**, 6797–6798.

95. Hof, R.P. and Kellogg, R.M. (1994) Lipase catalyzed resolutions of some α,α-disubstituted 1,2-diols in organic solvents; near absolute regio and chiral recognition. *Tetrahedron Asymmetry*, **5**, 565–568.

96. Theil, F., Ballschuh, S., Kunath, A. and Schick, H. (1991) Kinetic resolution of *rac*-3-(2-methylphenoxy)propane-1,2-diol (mephenesin) by sequential lipase-catalyzed transesterification. *Tetrahedron Asymmetry*, **2**, 1031–1034.

97. Kanerva, L.T. and Vänttinen, E. (1993) Biocatalytic resolution of 2,3-epoxyalcohols in organic solvents. *Tetrahedron Asymmetry*, **4**, 85–90.

98. Terao, Y., Tsuji, K., Murata, M., Achiwa, K., Nishio, T., Watanabe, N. and Seto, K. (1989) Facile process for enzymatic resolution of racemic alcohols. *Chem. Pharm. Bull.*, **37**, 1653–1655.

99. Gao, Y., Hanson, R.M., Klunder, J.M., Ko, S.Y., Masamune, H. and Sharpless, K.B. (1987) Catalytic asymmetric epoxidation and kinetic resolution: modified procedures including *in situ* derivatization. *J. Amer. Chem. Soc.*, **109**, 5765–5780.

100. Ladner, W.E. and Whitesides, G.M. (1984) Lipase-catalyzed hydrolysis as a route to esters of chiral epoxy alcohols. *J. Amer. Chem. Soc.*, **106**, 7250–7251.

101. Ferraboschi, P., Casati, S., Grisenti, P. and Santaniello, E. (1993) A chemoenzymatic approach to enantiomerically pure (*R*)- and (*S*)-2,3-epoxy-2-(4-pentenyl)-propanol, a chiral building block for the synthesis of (*R*)- and (*S*)-frontalin. *Tetrahedron Asymmetry*, **4**, 9–12.

102. Pallavicini, M., Valoti, E., Villa, L. and Piccolo, O. (1994) Lipase-catalyzed resolution of glycerol 2,3-carbonate. *J. Org. Chem.*, **59**, 1751–1754.

103. Herradón, B. (1993) Biocatalytic synthesis of chiral polyoxygenated compounds: effect of the solvent on the enantioselectivity of lipase catalyzed transesterifications in organic solvents. *Synlett*, 108–110.

104. Antus, S., Gottsegen, A., Kajtár, J., Kovács, T., Tóth, T.S. and Wagner, H. (1993) Lipase-catalyzed kinetic resolution of (±)-2-hydroxymethyl-1,4-benzodioxane. *Tetrahedron Asymmetry*, **4**, 339–344.

105. Wirz, B. and Walther, W. (1992) Enzymatic preparation of chiral 3-(hydroxymethyl)piperidine derivatives. *Tetrahedron Asymmetry*, **3**, 1049–1054.

106. Asensio, G., Andreu, C. and Marco, J.A. (1991) Enzyme-mediated enantioselective acylation of secondary amines in organic solvents. *Tetrahedron Lett.*, **32**, 4197–4198.

107. Weidner, J., Theil, F., Kunath, A. and Schick, H. (1991) Synthesis of enantiomerically pure prostaglandin intermediates by enzymatic transesterification of (1*S**,5*R**,6*R**,7*R**)-(±)-7-hydroxy-6-hydroxymethyl-2-oxabicyclo[3.3.0]octan-3-one. *Liebigs Ann. Chem.*, 1301–1303.

108. Weidner, J., Theil, F. and Schick, H. (1994) Kinetic resolution of (1*RS*,2*SR*)-2-(hydroxymethyl)cyclopentanol by a biocatalytic transesterification using lipase PS. *Tetrahedron Asymmetry*, **5**, 751–754.

109. Nagai, H., Shiozawa, T., Achiwa, K. and Terao, Y. (1992) Facile enzymatic preparation of enantiomeric β-lactams. *Chem. Pharm. Bull.*, **40**, 2227–2229.

110. Jouglet, B. and Rousseau, G. (1993) Enzymatic resolution of *N*-hydroxymethyl γ-butyrolactams. An access to optically active γ-butyrolactams. *Tetrahedron Lett.*, **34**, 2307–2310.

111. Nakano, H., Okuyama, Y., Iwasa, K. and Hongo, H. (1994) A facile lipase-catalyzed resolution of 2-azabicyclo[2.2.1]hept-5-en-3-ones. *Tetrahedron Asymmetry*, **5**, 1155–1156.

112. Mizuguchi, E., Achiwa, K., Wakamatsu, H. and Terao, Y. (1994) Lipase-catalyzed resolution of 5,5-disubstituted hydantoins. *Tetrahedron Asymmetry*, **5**, 1407–1410.

113. Jeromin, G.E. and Scheidt, A. (1991) Aliphatic optically active alcohols by enzyme-aided syntheses. *Tetrahedron Lett.*, **32**, 7021–7024.

114. Yoshida, N., Morita, H., Oyama, K. and Lee, H.-H. (1989) Enzymatic syntheses of chiral *sec*-alcohols and their NMR chemical shifts induced by a chiral lanthanide shift reagent. *Chem. Express*, **4**, 721–724.

115. Nishio, T., Kamimura, M., Murata, M., Terao, Y. and Achiwa, K. (1989) Production of optically active esters and alcohols from racemic alcohols by lipase-catalyzed stereoselective transesterification in non-aqueous reaction system. *J. Biochem.*, **105**, 510–512.

116. Kanerva, L.T., Vihanto, J., Halme, M.H., Loponen, J.M. and Euranto, E.K. (1990) Solvent effects in lipase-catalyzed transesterification reactions. *Acta Chem. Scand.*, **44**, 1032–1035.

117. Gutman, A.L., Brenner, D. and Boltanski, A. (1993) Convenient practical resolution of racemic alkyl-aryl alcohols via enzymatic acylation with succinic anhydride in organic solvents. *Tetrahedron Asymmetry*, **4**, 839–844.

118. Morgan, B., Oehlschlager, A.C. and Stokes, T.M. (1992) Enzyme reactions in apolar solvent. 5. The effect of adjacent unsaturation on the PPL-catalyzed kinetic resolution of secondary alcohols. *J. Org. Chem.*, **57**, 3231–3236.

119. Chong, J.M. and Mar, E.K. (1991) Preparation of enantiomerically enriched α-hydroxy-stannanes via enzymatic resolution. *Tetrahedron Lett.*, **32**, 5683–5686.

120. Ramaswamy, S. and Oehlschlager, A.C. (1991) Chemico-enzymatic syntheses of racemic and chiral isomers of 7-methyl-1,6-dioxaspiro[4.5]decane. *Tetrahedron*, **47**, 1157–1162.

121. Mallavadhani, U.V. and Rao, Y.R. (1994) Enantioselective synthesis of cinnamyl-1-phenyl-2-propenyl ether: a metabolite of marine green algae species *Caulerpa racemosa*. *Tetrahedron Asymmetry*, **5**, 23–26.

122. Laumen, K., Breitgoff, D. and Schneider, M.P. (1988) Enzymatic preparation of enantiomerically pure secondary alcohols. Ester synthesis by irreversible acyl transfer using a highly selective ester hydrolase from *Pseudomonas* sp.; an attractive alternative to ester hydrolysis. *J. Chem. Soc., Chem. Commun.*, 1459–1461.

123. Ader, U. and Schneider, M.P. (1992) Enzyme assisted preparation of enantiomerically pure β-adrenergic blockers III. Optically active chlorohydrin derivatives and their conversion. *Tetrahedron Asymmetry*, **3**, 521–524.

124. Nieduzak, T.R. and Margolin, A.L. (1991) Multigram lipase-catalyzed enantioselective acylation in the synthesis of the four stereoisomers of a new biologically active α-aryl-4-piperidinemethanol derivative. *Tetrahedron Asymmetry*, **2**, 113–122.

125. Sakaki, J., Sakoda, H., Sugita, Y., Sato, M. and Kaneko, C. (1991) Lipase-catalyzed asymmetric synthesis of 6-(3-chloro-2-hydroxypropyl)-1,3-dioxin-4-ones and their conversion to chiral 5,6-epoxyhexanoates. *Tetrahedron Asymmetry*, **2**, 343–346.

126. Bevinakatti, H.S. and Banerji, A.A. (1992) Lipase catalysis in organic solvents. Application to the synthesis of (R)- and (S)-atenolol. *J. Org. Chem.*, **57**, 6003–6005.

127. Bevinakatti, H.S. and Banerji, A.A. (1991) Practical chemoenzymatic synthesis of both enantiomers of propranolol. *J. Org. Chem.*, **56**, 5372–5375.

128. Gaspar, J. and Guerrero, A. (1995) Lipase-catalysed enantioselective synthesis of naphthyl trifluoromethyl carbinols and their corresponding non-fluorinated counterparts. *Tetrahedron Asymmetry*, **6**, 231–238.

129. Sparks, M.A. and Panek, J.S. (1991) Lipase mediated resolution of chiral (E)-vinylsilanes: an improved procedure for the production of (R)- and (S)-(E)-1-trialkylsilyl-1-buten-3-ol derivatives. *Tetrahedron Lett.*, **32**, 4085–4088.

130. Burgess, K. and Jennings, L.D. (1990) Biocatalytic resolutions of α-methylene-β-hydroxy esters and ketones. *J. Org. Chem.*, **55**, 1138–1139.

131. Burgess, K. and Jennings, L.D. (1990). Kinetic resolutions of chiral unsaturated alcohols that cannot be resolved efficiently via enantioselective epoxidation. *J. Amer. Chem. Soc.*, **112**, 7434–7436.

132. Burgess, K. and Jennings, L.D. (1991) Enantioselective esterifications of unsaturated alcohols mediated by a lipase prepared from *Pseudomonas* sp. *J. Amer. Chem. Soc.*, **113**, 6129–6139.

133. Domínguez, E., Carretero, J.C., Fernández-Mayoralas, A. and Conde, S. (1991) An efficient preparation of optically active (E)-γ-hydroxy-α,β-unsaturated phenyl sulfones using lipase-mediated acylations. *Tetrahedron Lett.*, **32**, 5159–5162.

134. Carretero, J.C. and Domínguez, E. (1992) Lipase-catalyzed kinetic resolution of γ-hydroxy phenyl sulfones. *J. Org. Chem.*, **57**, 3867–3873.

135. Bornscheuer, U., Schapöhler, S., Scheper, T. and Schügerl, K. (1991) Influences of reaction conditions on the enantioselective transesterification using *Pseudomonas cepacia* lipase. *Tetrahedron Asymmetry*, **2**, 1011–1014.

136. Mitsuda, S. and Nabeshima, S. (1991) Enzymatic optical resolution of a synthetic pyrethroid

alcohol. Enantioselective transesterification by lipase in organic solvent. *Recl. Trav. Chim. Pays-Bas*, **110**, 151–154.

137. Inagaki, M., Hiratake, J., Nishioka, T. and Oda, J. (1992) One-pot synthesis of optically active cyanohydrin acetates from aldehydes via lipase-catalyzed kinetic resolution coupled with *in situ* formation and racemization of cyanohydrins. *J. Org. Chem.*, **57**, 5643–5649.

138. Wang, Y.-F., Chen, S.-T., Liu, K.K.-C. and Wong, C.-H. (1989) Lipase-catalyzed irreversible transesterification using enol esters: resolution of cyanohydrins and syntheses of ethyl (*R*)-2-hydroxy-4-phenylbutyrate and (*S*)-propranolol. *Tetrahedron Lett.*, **30**, 1917–1920.

139. Lu, Y., Miet, C., Kunesch, N. and Poisson, J. (1990) A simple total synthesis of both enantiomers of γ-amino-β-hydroxybutanoic acid (GABOB) by enzymatic kinetic resolution of cyanohydrin acetates. *Tetrahedron Asymmetry*, **1**, 707–710.

140. Allevi, P., Anastasia, M., Cajone, F., Ciuffreda, P. and Sanvito, A.M. (1993) Enzymatic resolution of (*R*)- and (*S*)-(*E*)-4-hydroxyalk-2-enals related to lipid peroxidation. *J. Org. Chem.*, **58**, 5000–5002.

141. Allevi, P., Anastasia, M., Cajone, F., Ciuffreda, P. and Sanvito, A.M. (1994) Enzymatic resolution of the ethyl acetals of (*R*)- and (*S*)-4-hydroxyalk-2-ynals. *Tetrahedron Asymmetry*, **5**, 13–16.

142. Burgess, K. and Henderson, I. (1990) Lipase-mediated resolutions of SPAC reaction products. *Tetrahedron Asymmetry*, **1**, 57–60.

143. Kagan, H.B., Tahar, M. and Fiaud, J.-C. (1991) 2-Hydroxyalkyl diphenylphosphines: biocatalytic resolution and use as ligands for transition-metal catalysts. *Tetrahedron Lett.*, **32**, 5959–5962.

144. Khushi, T., O'Toole, K.J. and Sime, J.T. (1993) Biotransformation of phosphonate esters. *Tetrahedron Lett.*, **34**, 2375–2378.

145. Heisler, A., Rabiller, C., Douillard, R., Goalou, N., Hägele, G. and Levayer, F. (1993) Enzyme catalysed resolution of aminophosphonic acids—I—Serin and isoserin analogues. *Tetrahedron Asymmetry*, **4**, 956–960.

146. Nakamura, K., Kondo, H., Metzner, P. and Ohno, A. (1992) Lipase-catalyzed asymmetric transesterification of 3-hydroxydithiopentanoate in an organic solvent. *Bull. Inst. Chem. Res., Kyoto Univ.*, **70**, 302–307.

147. Chincilla, R., Nájera, C., Pardo, J. and Yus, M. (1990) Lipase-catalysed resolution of hydroxy sulfones. *Tetrahedron Asymmetry*, **1**, 575–578.

148. Jacobs, H.K., Mueller, B.H. and Gopalan, A.S. (1992) Chiral γ and δ hydroxysulfones via lipase catalyzed resolutions — Synthesis of (*R*)(+)-4-hexanolide and (2*R*,5*S*)-2-methyl-5-hexanolide using intramolecular acylation. *Tetrahedron*, **48**, 8891–8898.

149. Orrenius, C., Mattson, A. and Norin, T. (1994) Preparation of 1-pyridinylethanols of high enantiomeric purity by lipase catalysed transesterifications. *Tetrahedron Asymmetry*, **5**, 1363–1366.

150. Seemayer, R. and Schneider, M.P. (1992) Preparation of optically pure pyridyl-1-ethanols. *Tetrahedron Asymmetry*, **3**, 827–830.

151. Langard, G., Baratti, J., Buono, G. and Triantaphylides, C. (1986) Lipase catalyzed reactions and strategy for alcohol resolution. *Tetrahedron Lett.*, **27**, 29–32.

152. Seemayer, R. and Schneider, M.P. (1991) Enzymatic hydrolysis and esterification. Routes to optically pure cyclopentanols. *Recl. Trav. Chim. Pays-Bas*, **110**, 171–174.

153. Cregge, R.J., Wagner, E.R., Freedman, J. and Margolin, A.L. (1990) Lipase-catalysed transesterification in the synthesis of a new chiral unlabeled and carbon-14 labeled serotonin uptake inhibitor. *J. Org. Chem.*, **55**, 4237–4238.

154. Tokuyama, S., Yamano, T., Aoki, I., Takanohashi, K. and Nakahama, K. (1993) Optical resolution of a D-biotin chiral intermediate by use of lipoprotein lipase. *Chem. Lett.*, 741–744.

155. Nishida, T., Nihira, T. and Yamada, Y. (1991) Facile synthesis of racemic 2-hexyl-3-hydroxy-4-pentanolide (NFX-2) and its optical resolution. *Tetrahedron*, **47**, 6623–6634.

156. Babiak, K.A., Ng, J.S., Dygos, J.H. and Weyker, C.L. (1990) Lipase-catalysed irreversible transesterification using enol esters: resolution of prostaglandin synthons 4-hydroxy-2-alkyl-2-cyclopentenones and inversion of the 4*S* enantiomer to the 4*R* enantiomer. *J. Org. Chem.*, **55**, 3377–3381.

157. Laumen, K., Seemayer, R. and Schneider, M.P. (1990) Enzyme preparation of enantiomerically pure cyclohexanols: ester synthesis by irreversible acyl transfer. *J. Chem. Soc., Chem. Commun.*, 49–51.

158. Johnson, C.R. and Sakaguchi, H. (1992) Enantioselective transesterifications using immobilized, recombinant *Candida antarctica* lipase B: Resolution of 2-iodo-2-cycloalken-1-ols. *Synlett*, 813–816.
159. Zarevúcka, M., Rejzek, M., Pavlik, M., Wimmer, Z., Zima, J. and Legoy, M.-D. (1994) Enzyme mediated resolution of alcohols. *Biotechnol. Lett.*, **16**, 807–812.
160. Meltz, M. and Saccomono, N.A. (1992) Lipase mediated optical resolution of bicyclic secondary carbinols. *Tetrahedron Lett.*, **33**, 1201–1202.
161. Fritsche, K., Syldatk, C., Wagner, F., Hengelsberg, H. and Tacke, R. (1989) Enzymatic resolution of *rac*-1,1-dimethyl-1-sila-cyclohexan-2-ol by ester hydrolysis or transesterification using a crude lipase preparation of *Candida cylindracea*. *Appl. Microbiol. Biotechnol.*, **31**, 107–111.
162. Kakeya, H., Sugai, T. and Ohta, H. (1991) Biochemical preparation of optically active 4-hydroxy-β-ionone and its transformation of (*S*)-6-hydroxy-α–ionone. *Agric. Biol. Chem.*, **55**, 1873–1876.
163. Patel, R.N., McNamee, C.M. and Szarka, L.J. (1992) Enantioselective enzymatic acetylation of racemic [4-[4α,6β(*E*)]]-6-[4,4-bis(4-fluorophenyl)-3-(1-methyl-1H-tetrazol-5-yl)-1,3-butadienyl]-tetrahydro-4-hydroxy-2H-pyran-2-one. *Appl. Microbiol. Biotechnol.*, **38**, 56–60.
164. Ramaswamy, S., Morgan, B. and Oehlschlager, A.C. (1990) Porcine pancreatic lipase mediated selective acylation of primary alcohols in organic solvents. *Tetrahedron Lett.*, **31**, 3405–3408.
165. Kim, M.-J. and Choi, Y.K. (1992) Lipase-catalyzed enantioselective transesterification of *O*-trityl 1,2-diols. Practical synthesis of (*R*)-tritylglycidol. *J. Org. Chem.*, **57**, 1605–1607.
166. Goergens, U. and Schneider, M.P. (1991) A facile chemoenzymatic route to enantiomerically pure oxiranes: building blocks for biologically active compounds. *J. Chem. Soc., Chem. Commun.*, 1064–1066.
167. Chen, C.-S., Liu, Y.-C. and Marsella, M. (1990) A convenient chemoenzymatic synthesis of (*R*)- and (*S*)-(chloromethyl)oxirane. *J. Chem. Soc., Perkin Trans. 1*, 2559–2561.
168. Chen, C.-S. and Liu, Y.-C. (1989) A chemoenzymatic access to optically active 1,2-epoxides. *Tetrahedron Lett.*, **30**, 7165–7168.
169. Theil, F., Weidner, J., Ballschuh, S., Kunath, A. and Schick, H. (1993) Kinetic resolution of aliphatic 1,2-diols by a lipase-catalyzed sequential acetylation. *Tetrahedron Lett.*, **34**, 305–306.
170. Theil, F., Weidner, J., Ballschuh, S., Kunath, A. and Schick, H. (1994) Kinetic resolution of acyclic 1,2-diols using a sequential lipase-catalyzed transesterification in organic solvents. *J. Org. Chem.*, **59**, 288–393.
171. Chênevert, R. and Gagnon, R. (1993) Lipase-catalyzed enantioselective esterification or hydrolysis of 1-*O*-alkyl-3-*O*-tolsylglycerol derivatives. Practical synthesis of (*S*)-(+)-1-*O*-hexadecyl-2,3-di-*O*-hexadecanoylglycerol, a marine natural product. *J. Org. Chem.*, **58**, 1054–1057.
172. Goergens, U. and Schneider, M.P. (1991) Enzymatic preparation of enantiomerically pure and selectively protected 1,2- and 1,3-diols. *J. Chem. Soc., Chem. Commun.*, 1066–1068.
173. Evans, C.T., Roberts, S.M., Shoberu, K.A. and Sutherland, A.G. (1992) Potential use of carbocyclic nucleosides for the treatment of AIDS: Chemoenzymatic syntheses of the enantiomers of carbovir. *J. Chem. Soc., Perkin Trans. 1*, 589–592.
174. Garon, G. and Kazlauskas, R.J. (1993) Sequential kinetic resolution of (±)-2,3-butanediol in organic solvents using lipase from *Pesudomonas cepacia*. *Tetrahedron Asymmetry*, **4**, 1995–2000.
175. Seemayer, R. and Schneider, M.P. (1991) Enzymatic preparation of optically pure *trans*-1,2-cycloalkanediols. *J. Chem. Soc., Chem. Commun.*, 49–50.
176. Ling, L., Li, X., Watanabe, Y., Akiyama, T. and Ozaki, S. (1993) Enzymatic resolution of racemic 1,2:5,6-di-*O*-cyclohexylidene and 1,2:3,4-di-*O*-cyclohexylidene-*myo*-inositol. *Bioorg. Med. Chem.*, **1**, 155–159.
177. Henly, R., Elie, C.J.J., Buser, H.P., Ramos, G. and Moser, H.E. (1993) *Tetrahedron Lett.*, **34**, 2923–2926.
178. Wallace, J.S., Baldwin, B.W. and Morrow, C.J. (1992) Separation of remote diol and triol stereoisomers by enzyme-catalyzed esterification in organic media or hydrolysis in aqueous media. *J. Org. Chem.*, **57**, 5231–5239.
179. Gotor, V., Brieva, R. and Rebolledo, F. (1988) Enantioselective acylation of amino alcohols by porcine pancreatic lipase. *J. Chem. Soc., Chem. Commun.*, 957–958.

180. Francalanci, F., Cesti, P., Cabri, W., Bianchi, D., Martinengo, T. and Foà, M. (1987) Lipase-catalyzed resolution of chiral 2-amino 1-alcohols. *J. Org. Chem.*, **52**, 5079–5082.
181. Bevinakatti, H.S. and Newadkar, R.V. (1990) Lipase-catalyzed kinetic resolution of *N,O*-diacetyl-2-amino-1-butanol in diisopropyl ether. *Tetrahedron Asymmetry*, **1**, 583–586.
182. Kamal, A. and Rao, M.V. (1991) Lipase-catalyzed resolution: enantioselective esterification of 2-propanol amines. *Tetrahedron Asymmetry*, **2**, 751–754.
183. Izumi, T. and Fukaya, K. (1993) Baker's yeast reduction of α-(acylamino)acetophenones and lipase catalyzed resolution of 2-acylamino-1-arylethanols. *Bull. Chem. Soc. Jpn*, **66**, 1216–1221.
184. Sugai, T. and Ohta, H. (1990) Lipase-catalyzed kinetic resolution of 2-hydroxyhexadecanoic acid and its esters. *Agric. Biol. Chem.*, **54**, 3337–3338.
185. Feichter, C., Faber, K. and Griengl, H. (1990) Chemoenzymatic preparation of optically active long-chain 3-hydroxyalkanoates. *Biocatalysis*, **3**, 145–158.
186. Lu, Y., Miet, C., Kunesch, N. and Poisson, J.E. (1993) A simple total synthesis of naturally occurring hydroxy-amino acids by enzymatic kinetic resolution. *Tetrahedron Asymmetry*, **4**, 893–902.
187. Akita, H., Umezawa, I., Takano, M., Matsukura, H. and Oishi, T. (1991) A facile chemo-enzymatic route to enantiomerically pure 4,5-disubstituted-2-hexenoate derivatives. *Chem. Pharm. Bull.*, **39**, 3094–3096.
188. Hashimoto, S., Miyazaki, Y. and Ikegami, S. (1992) A facile access to (*R*)-methyl-2-oxocyclo-pentane-1-acetate via lipase-catalyzed irreversible transesterification. *Synth. Commun.*, **22**, 2717–2722.
189. Kanerva, L.T., Kosonen, M., Vänttinen, E., Huuhtanen, T.T. and Dahlqvist, M. (1992) Studies on the chemo- and enantio-selectivity of enzymatic monoacylations of amino alcohols. *Acta Chem. Scand.*, **46**, 1101–1105.
190. Barnier, J.-P., Blanco, L., Rousseau, G., Guibé-Jampel, E. and Fresse, I. (1993) Enzymatic resolution of cyclopropanols. An easy access to optically active cyclohexanones possessing an α-quaternary chiral carbon. *J. Org. Chem.*, **58**, 1570–1574.
191. Kamal, A., Damayanthi, Y. and Rao, M.V. (1992) Stereoselective synthesis of (*S*)-propanol amines: lipase catalyzed opening of epoxides with 2-propylamine. *Tetrahedron Asymmetry*, **3**, 1361–1364.
192. Bianchi, D. and Cesti, P. (1990) Lipase-catalyzed stereoselective thiotransesterification of mercapto esters. *J. Org. Chem.*, **55**, 5657–5659.
193. Wallace, J.S., Reda, K.B., Williams, M.E. and Morrow, C.J. (1990) Resolution of a chiral ester by lipase-catalyzed transesterification with poly(ethylene glycol) in organic media. *J. Org. Chem.*, **55**, 3544–3546.
194. Gou, D.-M., Liu, Y.-C. and Chen, C.-S. (1993) A practical chemoenzymatic synthesis of the taxol C-13 side chain *N*-benzoyl-(2*R*,3*S*)-3-phenylisoserine. *J. Org. Chem.*, **58**, 1287–1289.
195. Gentile, A., Giordano, C., Fuganti, C., Ghirotto, L. and Servi, S. (1992) The enzymatic preparation of (2*R*,3*S*) phenyl glycidic acid esters. *J. Org. Chem.*, **57**, 6635–6637.
196. Roberts, S.M. and Casy, G. (eds) (1994) *Candida cylindracea* lipase enantioselective trans-esterification of methyl *trans*-3-(4-methoxyphenyl)glycidate with octan-1-ol. In *Preparative Biotransformations*, John Wiley & Sons Ltd, Chichester, Update 3, pp. 5:1.9–15.
197. Engel, K.-H., Bohnen, M. and Dobe, M. (1991) Lipase-catalysed reactions of chiral hydroxy-acid esters: competition of esterification and transesterification. *Enzyme Microb. Technol.*, **13**, 655–660.
198. Brieva, R., Rebolledo, F. and Gotor, V. (1990) Enzymatic synthesis of amides with two chiral centres. *J. Chem. Soc., Chem. Commun.*, 1386–1387.
199. Baba, N., Mimura, M., Hiratake, J., Uchida, K. and Oda, J. (1988) Enzymatic resolution of racemic hydroperoxides in organic solvent. *Agric. Biol. Chem.*, **52**, 2685–2687.
200. Baba, N., Tateno, K., Iwasa, J. and Oda, J. (1990) Lipase-catalysed kinetic resolution of racemic methyl 13-hydroperoxy-9Z,11E-octadecadienoate in an organic solvent. *Agric. Biol. Chem.*, **54**, 3349–3350.
201. Fukui, T., Kawamoto, T. and Tanaka, A. (1994) Enzymatic preparation of optically active silyl-methanol derivatives having a stereogenic silicon atom by hydrolase-catalyzed enantioselective esterification. *Tetrahedron Asymmetry*, **5**, 73–82.
202. Burgess, K., Henderson, I. and Ho, K.-K. (1992) Biocatalytic resolution of sulfinylalkanoates: a facile route to optically active sulfoxides. *J. Org. Chem.*, **57**, 1290–1295.

203. Gil, G., Ferre, E., Meou, A., Le Petit, J. and Triantaphylides, C. (1987) Lipase-catalyzed ester formation in organic solvents. Partial resolution of primary allenic alcohols. *Tetrahedron Lett.*, **28**, 1647–1648.

204. Fiaud, J.-C., Gil, R., Legros, J.-Y., Aribi-Zouioueche, L. and König, W.A. (1992) Kinetic resolution of 3-*t*-butyl and 3 phenyl cyclobutylidenethanols through lipase-catalyzed acylation with succinic anhydride. *Tetrahedron Lett.*, **33**, 6967–6970.

205. Naemura, K. and Furutani, A. (1991) Lipase-catalyzed asymmetric and enantioselective esterification of spiro[3.3]heptanes in organic solvents. *J. Chem. Soc., Perkin Trans. 1*, 2891–2892.

206. Inagaki, M., Hiratake, J., Nishioka, T. and Oda, J. (1989) Lipase-catalyzed stereoselective acylation of [1,1'-binaphthyl]-2,2'-diol and deacylation of its esters in an organic solvent. *Agric. Biol. Chem.*, **53**, 1879–1884.

207. Nakamura, K., Ishihara, K., Ohno, A., Uemura, M., Nishimura, H. and Hayashi, Y. (1990) Kinetic resolution of (η^6-arene)chromium complexes by a lipase. *Tetrahedron Lett.*, **31**, 3603–3604.

208. Baldoli, C., Maiorana, S., Carrea, G. and Riva, S. (1993) Studies on the enzymatic resolution of chiral tricarbonyl(benzaldehyde oxime)chromium complexes. *Tetrahedron Asymmetry*, **4**, 767–772.

209. Howell, J.A.S., Palin, M.G., Jaouen, G., Top, S., El Hafa, H. and Cense, J.M. (1993) Asymmetric biochemical reduction, acylation and hydrolysis in the (diene)Fe(CO)$_3$ series: experimental results and molecular modelling studies. *Tetrahedron Asymmetry*, **4**, 1241–1252.

210. Boaz, N.W. (1989) Enzymatic esterification of 1-ferrocenylethanol: an alternate approach to chiral ferrocenyl bis-phosphines. *Tetrahedron Lett.*, **30**, 2061–2064.

211. Kim, M.-J., Cho, H. and Choi, Y.K. (1991) Highly enantioselective transformations of ferrocene-containing substrates by bacterial lipases. *J. Chem. Soc., Perkin Trans. 1*, 2270–2272.

212. Izumi, T., Hino, T. and Ishihara, A. (1993) Enzymatic kinetic resolution of [3](1,1')-ferrocenophane derivatives. *J. Chem. Tech. Biotechnol.*, **56**, 45–49.

213. Baldessari, A., Iglesias, L.E. and Gros, E.G. (1992) Regiospecific acylation of 2-mercapto-ethanol by lipase-catalysed transesterification in organic solvents. *J. Chem. Research (S)*, 204–205.

214. Chinsky, N., Margolin, A.L. and Klibanov, A.M. (1989) Chemoselective enzymatic mono-acylation of bifunctional compounds. *J. Amer. Chem. Soc.*, **111**, 386–388.

215. Tawaki, S. and Klibanov, A.M. (1993) Chemoselectivity of enzymes in anhydrous media is strongly solvent dependent. *Biocatalysis*, **8**, 3–19.

216. Ottolina, G., Carrea, G. and Riva, S. (1990) Synthesis of ester derivatives of chloramphenicol by lipase-catalyzed transesterification in organic solvents. *J. Org. Chem.*, **55**, 2366–2369.

217. Danieli, B. and De Bellis, P. (1992) 104. Studies on the enzymatic acylation of quinic acid, shikimic acid, and their derivatives in organic solvents. *Helv. Chim. Acta*, **75**, 1297–1304.

218. Margolin, A.L., Delinck, D.L. and Whalon, M.R. (1990) Enzyme-catalyzed regioselective acylation of castanospermine. *J. Amer. Chem. Soc.*, **112**, 2849–2854.

219. Riva, S. and Klibanov, A.M. (1988) Enzymochemical regioselective oxidation of steroids without oxidoreductases. *J. Amer. Chem. Soc.*, **110**, 3291–3295.

220. Riva, S., Bovara, R., Ottolina, G., Secundo, F. and Carrea, G. (1989) Regioselective acylation of bile acid derivatives with *Candida cylindracea* lipase in anhydrous benzene. *J. Org. Chem.*, **54**, 3161–3164.

221. Njar, V.C.O. and Caspi, E. (1987) Enzymatic transesterification of steroid esters in organic solvents. *Tetrahedron Lett.*, **28**, 6549–6552.

222. Nicolosi, G., Piattelli, M. and Sanfilippo, C. (1992) Acetylation of phenols in organic solvent catalyzed by a lipase from *Chromobacterium viscosum*. *Tetrahedron*, **48**, 2477–2482.

223. Nicolosi, G., Piatteli, M. and Sanfilippo, C. (1993) Lipase-catalyzed regioselective protection of hydroxyl groups in aromatic dihydroxyaldehydes and ketones. *Tetrahedron*, **49**, 3143–3148.

224. Natoli, M., Nicolosi, G. and Piattelli, M. (1990) Enzyme-catalysed alcoholysis of flavone acetates in organic solvent. *Tetrahedron Lett.*, **31**, 7371–7374.

225. Natoli, M., Nicolosi, G. and Piattelli, M. (1992) Regioselective alcoholysis of flavonoid acetates with lipase in an organic solvent. *J. Org. Chem.*, **57**, 5776–5778.

226. Rubio, E., Fernandez-Mayorales, A. and Klibanov, A.M. (1991) Effect of the solvent on enzyme regioselectivity. *J. Amer. Chem. Soc.*, **113**, 695–696.

227. Yamamoto, Y., Ozasa, N. and Sawada, S. (1993) Acylation of alcohols with silylketenes catalyzed by lipase in organic media. *Chem. Express*, **8**, 305–308.
228. Hajjar, A.B., Nicks, P.F. and Knowles, C.J. (1990) Preparation of monomeric acrylic ester intermediates using lipase catalysed transesterifications in organic solvents. *Biotechnol. Lett.*, **12**, 825–830.
229. Ikeda, I., Tanaka, J. and Suzuki, K. (1991) Synthesis of acrylic esters by lipase. *Tetrahedron Lett.*, **32**, 6865–6866.

8 Peptide synthesis

H. KITAGUCHI

8.1 General aspects of protease-catalyzed peptide synthesis (1, 2)

Peptide bond formation by protease catalysis offers several advantages over chemical methods, including absence of racemization, minimal activation and side-chain protection requirement, and the enzyme's high chemo-, regio-, and enantiospecificity. For instance, Z-L-Asp-L-Phe–OMe, a precursor of an artificial sweetener aspartame, was synthesized from Z-L-Asp–OH and DL-Phe–OMe in the presence of a protease (thermolysin) (3). A chemical condensation (e.g. by dicyclohexylcarbodiimide) between the two substrates should give a mixture of four isomeric products (two regioisomers of Asp and two enantiomers of Phe), whereas only the desired product was obtained by the enzymatic reaction. Enzymatic methods are especially useful for peptide segment condensation, where side chain protection–deprotection and racemization through carboxyl terminus activation by chemical methods are serious problems (4). One of the most successful examples of protease-catalyzed peptide segment condensation can be found in a semi-synthesis of insulin (5).

In spite of these virtues, enzymatic methods are not routinely employed in a laboratory, partly because they suffer from some shortcomings, including:

(a) unfavorable thermodynamics in water,
(b) narrow specificity of enzymes, and
(c) secondary cleavage of the growing peptide chain.

Because the inherent function of a protease is to catalyze the hydrolysis of a peptide bond, (a) and (c) are unavoidable problems if the reaction is carried out in water. To be free from these problems, organic solvents should be a good alternative as a reaction medium.

As a synthetic catalyst, an enzyme's high specificity is double-edged; no protease can be a catalyst for all kinds of amino acid substrates because of its inherent substrate specificity. In organic media, however, it has recently been found that the specificity of enzymes can be relaxed dramatically (see Chapter 4), which enables a protease to be a more generally useful catalyst for peptide synthesis. Thus, the use of organic solvents is also very beneficial to overcome the problem (b).

There are two distinct mechanisms for protease-catalyzed peptide synthesis: thermodynamically controlled and kinetically controlled mechanisms. The former is the reverse reaction of protease-catalyzed peptide bond cleavage, whereas the latter is a transamidation reaction of amino acid esters.

$$R\text{-}COO^- + H_3N^+\text{-}R' \rightleftarrows R\text{-}COOH + H_2N\text{-}R'$$

$$\rightleftarrows R\text{-}CO\text{-}NH\text{-}R' + H_2O \tag{8.1}$$

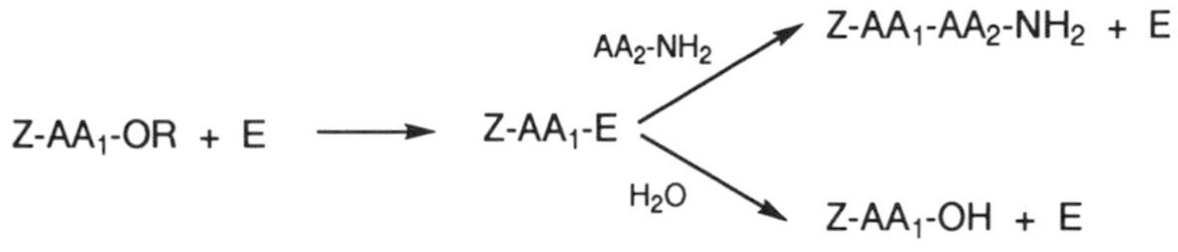

$$Z\text{-}AA_1\text{-}OR + E \longrightarrow Z\text{-}AA_1\text{-}E \begin{array}{l} \xrightarrow{AA_2\text{-}NH_2} Z\text{-}AA_1\text{-}AA_2\text{-}NH_2 + E \\ \xrightarrow{H_2O} Z\text{-}AA_1\text{-}OH + E \end{array} \tag{8.2}$$

Z-AA$_1$-OR : *N*-Z-amino acid ester

E : serine or thiol protease

Z-AA$_1$-E : acyl intermediate

AA$_2$-NH$_2$: amino acid amide

From equations (8.1) and (8.2), it is easily understood that to replace water with organic solvents as a reaction medium is beneficial for either mechanism.

In this chapter, effects of organic solvents on protease-catalyzed peptide bond formation in three reaction medium systems are discussed:

(i) aqueous solutions containing a fraction of water-miscible organic solvent,
(ii) organic–aqueous biphasic systems, and
(iii) organic solvents containing little or virtually no water.

Because a protease is dissolved in aqueous solution for (i) or (ii), the effects of the organic solvent on the catalytic activity can be rationalized by conventional enzyme chemistry. In contrast, enzyme powder is suspended in an organic solvent for (iii), and the reaction medium affects the catalyst in a completely different way from (i) and (ii). Thus, we need a new theory on how to control a protease-catalyzed peptide synthesis reaction in (iii).

8.2 Thermodynamically controlled synthesis

Thermodynamically controlled peptide synthesis is the reverse reaction of protease-catalyzed peptide bond cleavage, and the equilibrium is illustrated by equation (8.1). According to the principle of microscopic reversibility, both peptide bond formation and cleavage reactions proceed via the same reaction path (via the same intermediate). For instance, Ac–Phe–OH forms an acyl intermediate with [195]Ser of α-chymotrypsin, which is then deacylated with

Leu–NH$_2$, resulting in the formation of a Phe–Leu peptide bond (equation 8.3). The acyl intermediate formation is supported by the fact that the incubation of Bz-L-Phe–OH and chymotrypsin in H$_2$^{18}O leads to the formation of Bz-L-Phe–^{18}OH, although Bz-D-Phe–OH does not undergo oxygen exchange (6). Acyl intermediate formation from a carboxylic acid substrate is a very slow process, however, and it should be the rate-determining step of thermo-dynamically-controlled peptide synthesis.

$$(8.3)$$

In equation (8.1), it is obvious that the equilibrium is totally shifted to the left (peptide hydrolysis) in water, and to reduce the concentration of water in the reaction medium by replacement with organic solvents is a good way to shift the equilibrium to peptide bond formation. Moreover, it has been revealed that organic solvents also affect the equilibrium in several other ways, which are discussed in the following sections about three reaction medium systems.

8.2.1 Addition of water-miscible solvents

The first equilibrium of equation (8.1) is a non-enzymatic ionic equilibrium. Because only neutral species are involved in the next enzymatic equilibrium, the apparent total equilibrium constant (K_{ob}) of equation (8.1) is shown in equation (8.4).

$$K_{ob} = \frac{K}{(1 + 10^{pH-pK_1})(1 + 10^{pK_2-pH})}$$

pK$_1$: pKa for a carboxylic acid substrate

pK$_2$: pKa for an amine substrate

$$(8.4)$$

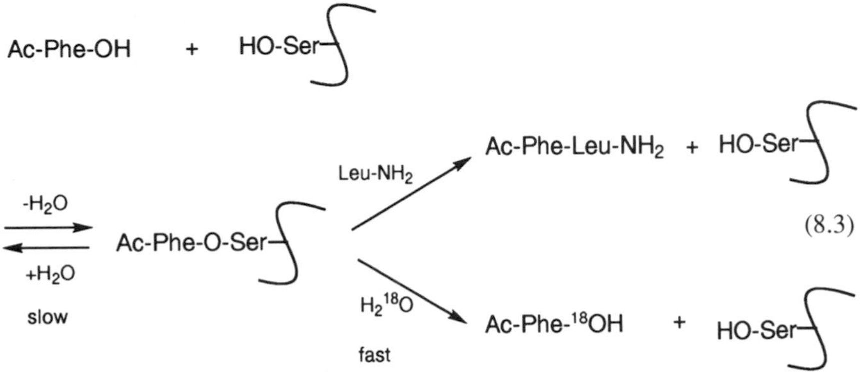

In water, K, pK$_1$, and pK$_2$ are constant, and the only way to maximize K_{ob} is to set the pH value of the reaction medium at (pK$_1$ + pK$_2$)/2. Even at this pH, however, the equilibrium of equation (8.1) is usually shifted to the left. For instance, assuming pK_1 is 4 and pK_2 is 8, it is calculated that the synthetic yield is only 8.4% at pH 6 with 1 M substrates.

The reason for the low yield in the above calculation is that pK_1 and pK_2 are different by four orders of magnitude, and 99% of the substrates are ionized in an aqueous solution whose pH is in the middle of them. It has been found, however, that addition of water-miscible organic solvents alters the apparent pK_a value dramatically and improves K_{ob} (Table 8.1) (7). For instance, the pK_a values of Ac–Gly–OH (pK_1) and Gly-NH_2 (pK_2) in water are 3.60 and 8.20, while 6.93 and 8.10 when dimethylsulfoxide (DMSO) is mixed at 80%. DMSO affects only the pK_a of the carboxyl substrate, resulting in reducing ΔpK_a from 4.6 to 1.2, and shifts the equilibrium to the peptide bond formation. For instance, K_{syn} for a chymotrypsin-catalyzed peptide synthesis (1 mM Z-Trp–OH + 100 mM Gly-NH_2) was only 0.45 M^{-1} in water, while it was improved to 2.12 M^{-1} in 60% glycerol, and 38 M^{-1} in 85% 1,4-butanediol. Optimal pH range for the chymotrypsin-catalyzed reaction in 60% triethylene glycol was higher than that in water by about 0.5 unit, as expected by the selective pK_a change of the carboxylic acid substrate (3.60 in water, 4.70 in 60% triethylene glycol). They also demonstrated a linear correlation between the synthetic equilibrium constant and $\Delta\Delta pK_a$ (7).

Similar effects of water-miscible solvents on the equilibrium shift have been found for other proteases. The yield of a dipeptide synthesis (50 mM Z-Arg–OH + 500 mM Leu-NH_2) catalyzed by papain or trypsin was enhanced by 4–5 times in the presence of 50% dimethylformamide (DMF) (8).

Although addition of water-miscible solvents is a simple and useful way to improve the synthetic yield, one associated problem is that high concentration of organic solvent often retards the amidase activity of a protease (2 and references cited therein). When the concentration of organic solvent exceeds 50%, the enzyme is usually deactivated and the synthetic yield decreases (8). An exception is a group of polyols, which are known as protein stabilizers (9). For instance, 1,4-butanediol does not deactivate proteases even at 85% concentration (7).

One of the most successful examples of a protease-catalyzed peptide synthesis via thermodynamically-controlled mechanism is a semisynthesis of

Table 8.1 Effect of cosolvent on pK_1 and pK_2 (7)

Cosolvent (80% v/v)	App. pK_1 of Ac–Gly–OH	App. pK_2 of Gly–NH_2	ΔpK_a
(Water)	3.60	8.20	4.60
Glycerol	4.48	8.20	3.72
Ethylene glycol	4.83	8.20	3.37
Acetonitrile	5.00	8.05	3.05
Acetone	5.49	8.00	2.51
Dioxan	5.53	8.00	2.47
DMSO	6.93	8.10	1.17
1,4-Butanediol	5.10	8.15	3.05

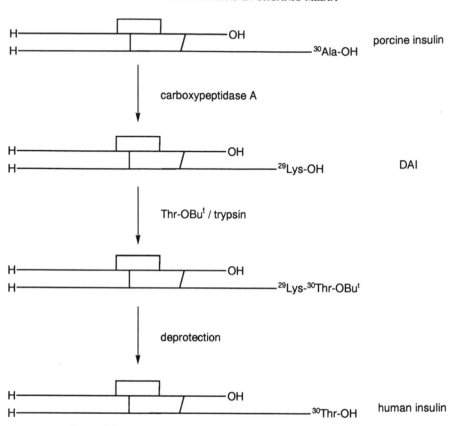

Figure 8.1 Enzymatic conversion from porcine insulin to human insulin.

insulin (5, 10, 11). Porcine insulin differs from human insulin only in the carboxyl-terminal residue of the B chain (porcine: Ala, human: Thr). Therefore, to eliminate Ala(B30) from porcine insulin (desalanine-B30-insulin; DAI) and to add the Thr residue is an attractive method for the production of human insulin (11) (Figure 8.1).

DAI was obtained by a limited hydrolysis of porcine insulin with carboxypeptidase A. Incubation of DAI (10 mM) with large excess of Thr–OBu^t (0.5 M) and trypsin (0.1 mM) in aqueous solution (pH 6.5) in the presence of 60% organic solvents (ethanol:DMF, 1:1) at 37°C for 20 h converts DAI to human insulin in 73% yield. One problem suspected under the above reaction conditions is trypsin-catalyzed cleavage of the Lys(B22)–Gly(B23) bond, but this was actually not observed. Partial deactivation of trypsin caused by high concentration of organic solvents seems to explain the unexpected stability of the Lys–Gly bond (11).

Direct transpeptidation between porcine insulin and Thr–OMe by using trypsin is an alternative way to synthesize human insulin (12). High concentration of organic solvent is a key factor for this reaction, too. When the concentration of DMF was less than 50%, the product yield was very low due to significant hydrolysis of Lys(B22)–Gly(B23). At the concentration of more than 67%, however, the hydrolysis was suppressed and the conversion reached 60% (13).

Lysine-specific *Achromobacter* protease was found to catalyze the elimination of Ala(B30) from porcine insulin and the addition of Thr–OBu^t (14). The yield of the condensation reaction between DAI (10 mM) and Thr–OBu^t (0.5 M) reached 85% when 60% organic solvent (ethanol:DMF, 1:1) was added to water (pH 6.5). Once again, high concentration organic solvent plays a key role.

8.2.2 Biphasic systems

Examination of equation (8.1) reveals that elimination of the product peptide from the reaction system makes the equilibrium shift to product formation (the law of mass action). For instance, if the product peptide precipitates while the substrates are soluble in aqueous solution, the synthetic yield is expected to increase. Indeed, it has been demonstrated that the yield depends strongly on the solubility of the product peptide (5 and references cited therein).

Biphasic reaction systems are an alternative method to use the above principle. Partition of the product peptide from the aqueous phase, where the enzymatic reaction occurs, to the organic phase is the driving force to shift the equilibrium to peptide bond formation.

Moreover, in biphasic systems, the apparent pK_1 and pK_2 are defined as in equations (8.5) and (8.6), and it is obvious that with higher values of α, pK_1 increases while pK_2 decreases. Thus, with higher portion of the organic phase, more fraction of the substrates is deionized, shifting the equilibrium to the peptide bond formation. At an optimized biphasic system, it was calculated that the equilibrium constant is enhanced by 10^5 times (15, 16).

$$pK_{1\,app} = pK_1 + \log(1 + \alpha \cdot P_1)$$

α : volume ratio (V_{org}/V_{aq}) (8.5)

P_1 : partition coefficient for a carboxylic acid substrate

$$pK_{2\,app} = pK_2 - \log(1 + \alpha \cdot P_2)$$

        (8.6)

α : volume ratio (V_{org}/V_{aq})

P_2 : partition coefficient for an amine substrate

α-Chymotrypsin was found to catalyze the condensation reaction between Ac–Trp–OH (0.4 mM) and Leu–NH$_2$ (0.4 mM) in 98% ethyl acetate (water phase: pH 7), giving the dipeptide Ac–Trp–Leu–NH$_2$ in 44% conversion. When

Table 8.2 Papain-catalyzed tripeptide synthesis in homogeneous and biphasic systems (17)

Solvent	Yield of tripeptide amide (%)	
	Ac–Leu–Phe–OH + Leu–NH$_2$	Boc–Leu–Phe–OH + Leu–NH$_2$
Methanol (15%) + water: homogeneous	25	30
Ethyl acetate (25%) + water: biphasic	49	61
Carbon tetrachloride (25%) + water: biphasic	60	78

[Carboxyl component] = 0.1 M, [Leu–NH$_2$] = 0.1 M, 30°C, 20 h.

one of the substrates was used in large excess, the conversion reached 100%. In contrast, the yield was only 0.01% in water under the same conditions. The pK_1 of Ac–Trp–OH in water is 3.2, which is enhanced by 3 orders of magnitude in a biphasic system, although the pK_2 for Leu–NH$_2$ (8.0) hardly changes upon the replacement of the reaction medium. This ΔpK_a reduction is supposed to be the reason for the high conversion in a biphasic system (16).

Papain was successfully used as a catalyst for tripeptide syntheses in a biphasic system. Product yields in biphasic systems (ethyl acetate:water, carbon tetrachloride:water) were found to be higher than those in a homogeneous system (methanol + water) (17) (Table 8.2).

The use of immobilized enzymes in organic media can be regarded as a biphasic system if the carrier contains sufficient amount of water. This system was successfully applied to the production of an artificial sweetener aspartame (Asp–Phe–OMe) (3).

Isowa *et al.* found that thermolysin catalyzes the condensation reaction between Z-L-Asp–OH and L-Phe–OMe in 96% yield in water (pH 6–8) (18). When racemic phenylalanine substrate was employed, only the L-isomer reacted with Z-L-Asp–OH, and the D-isomer formed the salt with the side-chain carboxylic acid group of the product dipeptide, which precipitated from water (92% yield). This precipitation should explain the high synthetic yield even though the reaction was carried out in water. For a continuous production system, however, the reaction mixture should be homogeneous, and an alternative trick for product precipitation is necessary to make the equilibrium shift towards product formation. Biphasic systems seem to be an ideal way to achieve the goal.

Oyama *et al.* immobilized thermolysin on ion-exchange resin (XAD-7, XAD-8) by physical adsorption and found that the water-wet immobilized enzyme effectively catalyzed the condensation between Z-L-Asp–OH and L-Phe–OMe in ethyl acetate at 40°C in 85–93% yield (19). Because the enzyme exists in the resin pores which are filled with water, this system is regarded as a biphasic system. Nakanishi *et al.* applied the above biphasic reaction to the continuous production system, and demonstrated that the yield could be maintained by 90% for more than 300 h under optimized conditions (20).

In biphasic systems, the product peptide usually partitions into the organic phase and is protected from further protease-catalyzed peptide bond cleavage. Thus this system is advantageous for peptide segment condensation, especially when a peptide bond susceptible to the protease employed is included in the substrate segments.

When thermolysin was used as a catalyst for a tetrapeptide synthesis (Z-Gly–Gly–OH + Phe–Leu–OEt) in water (pH 5.4, 40°C), the conversion was only 10% shortly after the reaction was started, and then dropped rapidly to almost 0% after 24 h (21). Moreover, two by-products, Z-Gly–Gly–(Phe)$_2$–Leu–OEt and Z-Gly–Gly–(Phe)$_3$–Leu–OEt, which have one or two more phenylalanine residues than the wanted tetrapeptide, were detected by HPLC during the reaction. Formation of these by-products is rationalized by equation (8.7).

Z-GlyGly-OH + PheLeu-OEt	⇌	Z-GlyGlyPheLeu-OEt (8.7a)
Z-GlyGlyPheLeu-OEt + H$_2$0	⇌	Z-GlyGlyPhe-OH + Leu-OEt (8.7b)
Z-GlyGlyPhe-OH + PheLeu-OEt	⇌	Z-GlyGlyPhePheLeu-OEt (8.7c)
Z-GlyGlyPhePheLeu-OEt + H$_2$0	⇌	Z-GlyGlyPhePhe-OH + Leu-OEt (8.7d)
Z-GlyGlyPhePhe-OH + PheLeu-OEt	⇌	Z-GlyGlyPhePhePheLeu-OEt (8.7e)

Thermolysin has been successfully utilized for the synthesis of various dipeptides in water (2). When a substrate segment contained a susceptible peptide bond for the protease such as Phe–Leu, however, unwanted hydrolysis of the product followed by miscouplings led to a complex mixture. In contrast, when the above tetrapeptide synthesis was carried out in a biphasic system (ethyl acetate:water), the authors observed greatly improved purity and yield of the product tetrapeptide. Under optimized condition (pH 4.5, $\alpha = 10$, 40°C, [Phe–Leu–OEt]:[Z-Gly–Gly–OH] = 3), the yield and purity are 91% and 97%, respectively, after 24 h.

Although biphasic systems are an attractive method, they also have several drawbacks and their utility seems to be limited:

(a) The reaction rate in biphasic systems is relatively slower than that in water, because interphase diffusion of substrates is presumed to be a rate-determining step.
(b) The enzyme tends to denaturate at the organic–water surface.
(c) In order not to be miscible with water, the organic layer should consist of rather hydrophobic solvents, in which oligopeptides are barely soluble.

To avoid these problems, the reaction medium should be homogeneous and contain as low a content of water as possible. These conditions are described in the next section.

8.2.3 Use of organic media containing a low content of water

Because water is the principal cause of almost all the drawbacks of protease-catalyzed peptide synthesis, it is obvious that non-aqueous organic solvents are ideal reaction media. Unfortunately, protease-catalyzed peptide synthesis by a thermodynamically controlled mechanism in non-aqueous media has never been demonstrated, while some proteases (especially thermolysin) have been successfully utilized as catalysts for the synthesis of oligopeptides in organic solvents containing at most a few percent of water. The difference between the present system and a biphasic system is that the reaction medium is homogeneous and the enzyme often functions as a solid for the former, while the medium is biphasic and enzyme is dissolved in the water phase for the latter.

Ooshima *et al.* reported that thermolysin effectively catalyzed the synthesis of Z-Asp–Phe–OMe when its powder was suspended in an organic medium containing 2% water (ethyl acetate:benzene:methanol:water = 50:29:19:2; note that the medium is homogeneous) (3). Assuming that the active portion of the catalyst is localized at the external surface of the solid, they calculated that the specific activity of solid thermolysin is roughly equal to that of solute enzyme in water.

Sakina *et al.* successfully applied thermolysin to peptide segment condensation. The carboxyl terminus of oxidized insulin B chain was elongated with Phe–NH$_2$ or Leu–NH$_2$ by thermolysin catalysis in an organic medium (DMF:ethanol, 1:1) containing 10% water (24). When an excess amount of Leu–NH$_2$ was used, the conversion reached 84%. The protease also catalyzed the synthesis of a pentapeptide (Fmoc–Asp–Tyr(SO$_3$Ba$_{1/2}$)–OH + Met–Gly–Trp–OMe) in acetonitrile containing 5% water in 66% yield (25).

Despite these successes, significant concentrations of water may still be the cause of unwanted peptide bond cleavage. To reduce water concentration further, however, usually results in the loss of the thermolysin's activity. Water activates thermolysin in organic media, presumably because it forms multiple hydrogen bonds with protein molecules and makes its conformation flexible enough to exhibit the catalytic activity. Based on this hypothesis, Kitaguchi and Klibanov introduced a new methodology, the water mimic method, to overcome the present dilemma (26).

Thermolysin did not exhibit any catalytic activity for a peptide bond formation (Z-Phe–OH + Phe–NH$_2$) in anhydrous *tert*-amyl alcohol, though the reaction became noticeable with 1% water and very fast with 4% water. Assuming that water activates the enzyme because of its ability to form multiple hydrogen bonds, several hydrogen-bond forming solvents were screened as possible enzyme activators. Table 8.3 confirms that water which is essential for the enzyme activity can be replaced, at least partially, by such a 'water mimic': the activity in *tert*-amyl alcohol containing 1% water and 9% formamide is comparable to that with 4% water (optimal concentration), and exceeds that with 1% water 200-fold. Similar activation effects were observed for other

Table 8.3 Effect of cosolvents on thermolysin's activity in *tert*-amyl alcohol (26)

Cosolvent (% v/v)	Relative activity[a]
None	0
1% Water	1
4% Water	184
1% Water + 9% formamide	200
1% Water + 9% ethylene glycol	79
1% Water + 9% glycerol	43
1% Water + 9% ethylene glycol monomethyl ether	7.4
1% Water + 9% ethylene glycol dimethyl ether	3.2
1% Water + 9% dimethylformamide	2.7

[a] Determined by the initial rates for a tetrapeptide synthesis (Z-Gly–Gly–Phe–OH + Leu–NH$_2$).

hydrogen-bond solvents, ethylene glycol and glycerol. When the acidic hydrogen atoms of formamide or ethylene glycol were substituted with methyl groups, the enzyme activating effect diminished dramatically, which strongly supports the hypothesis that water's activation capability comes from its capability to form multiple hydrogen bonds.

When thermolysin was used as a catalyst for a tetrapeptide synthesis Z-Gly–Gly–Phe–OH + Phe–NH$_2$) in *tert*-amyl alcohol containing 4% water, a significant amount of secondary cleavage of the Gly–Phe bond was observed. With 1% water and 9% formamide, however, virtually no by-product was detected by HPLC, while the reaction rate was almost the same as that with 4% water. Thus, this solvent system was successfully applied to peptide segment condensation reactions with thermolysin catalysis.

Even under such an unusual medium condition, thermolysin's specificities were completely conserved. L-Phe and L-Leu are favored substrates as amino group donors, while any D-amino acid substrate never participates in the reaction. A lysine substrate can be used without the protection of the side chain amino group.

Nishino *et al.* expanded the concept and developed an improved enzymatic reaction system (27, 28). Instead of the thermolysin + *tert*-amyl alcohol + formamide system, they proposed to use trypsin in a co-solvent of hexafluoro-isopropyl alcohol (HFIP) and DMF (1/1, v/v); 4% was found to be the optimal

Table 8.4 Effect of the solvent on the yield of trypsin-catalyzed tetrapeptide synthesis (26)

Solvent	Yield (%)[a]
HFIP: DMF (1:1) + 4% water	83
HFIP + 4%	0
DMF + 4% water	0

[a] Abz–Gly–Phe–Arg–OH (10 mM) + Leu–Nba (50 mM).

water concentration for trypsin's catalytic activity in this solvent system. As shown in Table 8.4, a mixed solvent of HFIP and DMF is much more advantageous than either HFIP or DMF alone (containing 4% water in all cases), presumably because the hydrogen-bonding capacity of HFIP is accepted by DMF (27). This novel solvent system is especially useful because of HFIP's high capability to dissolve poorly soluble oligopeptides, and it was successfully applied to the synthesis of α-MSH 13-peptide hormone (28).

Ac-Phe-Tyr-Met-Glu-His-Phe-Arg-OH + Trp(CHO)-Gly-Lys(ClZ)-Pro-Val-NH$_2$

$$\xrightarrow[\text{HFIP/DMF/H}_2\text{O}]{\text{trypsin}} \text{Ac-Phe-Tyr-Met-Glu-His-Phe-Arg-Trp(CHO)-Gly-Lys(ClZ)-Pro-Val-NH}_2 \quad (8.8)$$

$$\xrightarrow[\text{2) OH}^-]{\text{1) H}_2\text{ / Pd}} \text{Ac-Phe-Tyr-Met-Glu-His-Phe-Arg-Trp-Gly-Lys-Pro-Val-NH}_2$$

α-MSH

Although equation (8.8) is a sophisticated application of non-aqueous enzymology, it should be pointed out that it is still far from ideal. First, the side-chain amino group of Lys residue in the substrate was protected prior to the enzymatic condensation, because otherwise trypsin and 4% water would cause an unwanted peptide bond cleavage at this position. Second, the conversion was only about 30% when both substrates were used at the same concentration; the conversion reached to 95% only when the amino substrate was used in a five times excess.

In summary, although protease catalysis in organic media with a low content of water seems a promising method, there still remain several problems to overcome. Two points are especially important to be considered for practical application: the solubility of peptide segments and the secondary cleavage of peptide bonds. To replace water essential for the catalytic activity with other hydrogen-bonding solvents which dissolve oligopeptides well (formamide, HFIP) seems to be a way to satisfy these requirements, and further investigations on this approach should be carried out in due course.

8.3 Kinetically controlled synthesis

According to the reaction scheme of a kinetically controlled mechanism illustrated by equation (8.2), it is obvious that the catalyst should be a serine or thiol protease, which catalyzes the reaction via the acyl intermediate, while all sorts of proteases should, in principle, work for thermodynamically controlled mechanism. α-Chymotrypsin, subtilisins, and papain are the enzymes which are utilized most frequently. In general, kinetically controlled synthesis proceeds

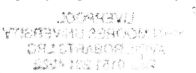

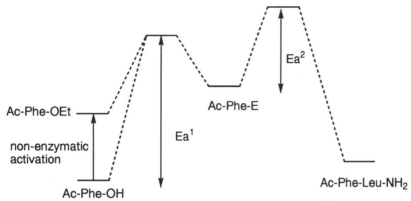

Figure 8.2 Energy diagram of chymotrypsin-catalyzed peptide synthesis. Ea^1: activation energy for thermodynamically controlled synthesis. Ea^2: activation energy for kinetically controlled synthesis.

faster and needs a smaller amount of enzyme than thermodynamically controlled synthesis, because the substrate for the former is an ester of amino acids, which is an activated form of carboxylic acid. Figure 8.2 illustrates the energy diagram of chymotrypsin-catalyzed synthesis of Ac–Phe–Leu–NH$_2$ via the two mechanisms.

According to equation (8.2), peptide synthesis is a competitive reaction with hydrolysis, and conditions of a high concentration of H$_2$N–R' and/or low concentration of water should be satisfied for effective peptide bond formation ($k_4[\text{H}_2\text{N–R}'] > k_3[\text{H}_2\text{O}]$). Fastrez and Fersht measured the deacylation rate of an acyl intermediate (Ac–Phe–chymotrypsin) with water, Gly–NH$_2$, and Ala–NH$_2$, and concluded that k_4/k_3 is 630 for Gly–NH$_2$ and 2400 for Ala–NH$_2$ (29). Thus, it is calculated that, in water ([H$_2$O] = 55 M), [Gly–NH$_2$] should be 87 mM to make the initial rate of the synthesis equal to that of hydrolysis ($k_4[\text{H}_2\text{N–R}'] = k_3[\text{H}_2\text{O}]$), and should be around 1 M to make the synthetic yield reasonably high. If water concentration decreases upon replacement with organic solvents, however, effective peptide synthesis can be achieved even if H$_2$N–R' is not used in large excess.

Another issue to be considered for kinetically controlled peptide synthesis is protease-catalyzed hydrolysis of the product peptide. Because the peptide bond cleavage catalyzed by a serine protease is a slower process than the peptide bond formation from the corresponding ester substrate, the yield of the product peptide becomes optimal in a short period, followed by gradual decomposition (2). Thus, the product peptide should be protected from further degradation to achieve high product yield.

In the following sections, the effects of organic solvents on kinetically controlled peptide synthesis are described in two reaction medium systems. (Biphasic systems are excluded from the following discussion because they seem not to be so useful for kinetically controlled synthesis. As described in the

LIVERPOOL
JOHN MOORES UNIVERSITY
AVRIL ROBARTS LRC
TEL. 0151 231 4022

previous section, biphasic systems have been successfully employed for the thermodynamically controlled process because ionized substrates tend to partition to the aqueous phase, while the product is usually transported to the organic phase. For the kinetically controlled process, however, neutral ester substrates are less likely to partition to the aqueous phase than ionized carboxylic acid substrates, implying that the substrate concentration for the enzymatic reaction should be low.)

8.3.1 Addition of water-miscible solvents

It is quite obvious that reducing the water concentration is beneficial to avoid the hydrolysis of the acyl intermediate and the product peptide. In addition to that, it has been demonstrated that the added organic solvent exhibits more fundamental effects on the protease's catalytic activity.

A serine protease catalyzes the hydrolysis of an ester substrate according to its esterase activity, while the catalyst functions as an amidase for peptide bond cleavage. Thus, the ratio between esterase and amidase activity should be maximized for the effective formation and protection of the product peptide. For the amide hydrolysis, formation of the acyl intermediate is rate-determining and pH independent (as long as the active site His is deprotonated), while deacylation, which is general-base catalyzed and pH dependent, is the rate-determining step for the ester hydrolysis (30). Thus, high pH is preferable to increase the esterase versus amidase ratio.

Wong and his research group have demonstrated that amidase activity of serine and thiol proteases (trypsin, chymotrypsin, papain, subtilisin) is destroyed significantly when water-miscible organic solvents (dioxan, DMF, DMSO, acetonitrile) are added at 50–60%, while there is virtually no loss of esterase activity (31–35). Thus, aqueous solution with high pH (around 9) containing about 50% of organic solvent was found to be a good reaction medium. In this solvent system, they synthesized various kinds of peptides, including dipeptides containing D-amino acids as an amino donor. An interesting application is a papain-catalyzed one-pot tripeptide synthesis illustrated by equation (8.9) (34).

$$\text{Z-AA}_1\text{-OR}_1 + \text{AA}_2\text{-OR}_2 + \text{D-AA}_3\text{-OR}_3 \xrightarrow{\text{papain}} \text{Z-AA}_1\text{-AA}_2\text{-D-AA}_3\text{-OR}_3 \quad (8.9)$$

$$2 - 69\ \%$$

Schellenberger et al. succeeded in the effective synthesis of a decapeptide, D-Phe(6)–GnRH, by a 3 + 7 segment condensation by using α-chymotrypsin in aqueous solution, pH 8, containing 48% DMF (36). The conversion reached 97.5% in 6.5 h, and 16.2 g of the product peptide was obtained after purification. Thus, this enzymatic reaction system is a practical method for large-scale peptide synthesis (equation 8.10).

p-Glu-His-Trp-OEt + Ser-Tyr-D-Phe-Leu-Arg-Pro-Gly-NH$_2$

$\xrightarrow[\substack{\text{DMF (48 \%) / H}_2\text{0}\\\text{pH 8}}]{\text{chymotrypsin}}$ p-Glu-His-Trp-Ser-Tyr-D-Phe-Leu-Arg-Pro-Gly-NH$_2$ (8.10)

D-Phe(6)-GnRH

It is not clear why high concentrations of water-miscible organic solvents selectively eliminate the amidase activity. It could be a conformation change of the enzyme, which damages its catalytic activity for a reaction with higher activation energy (amide hydrolysis) more significantly (35). Another possibility is an increase of local pH, which selectively accelerates general-base catalyzed acyl intermediate formation (ester exchange). In any case, the effect of organic solvents seems to be independent of the kind of protease or solvent, and the use of high concentrations of organic solvents should be generally applicable for kinetically controlled peptide synthesis.

It has been revealed that the amidase activity of a serine protease is also selectively eliminated by chemical modification. Thiosubtilisin, a subtilisin whose hydroxyl group on the Ser residue at the active site is chemically transformed to a thiol group, which is inactive for peptide hydrolysis while exhibiting 10–30% catalytic activity towards the hydrolysis of activated esters, was demonstrated to be a useful catalyst for peptide synthesis (37). Methylation of [57]His in chymotrypsin was also found to eliminate its amidase activity selectively (38, 39). These modified enzymes are especially useful for segment condensation. For instance, RNase T$_1$ (12–22) was synthesized by an 8 + 4 segment condensation in aqueous solution, pH 8.3, containing 54% DMF in the presence of thiosubtilisin (37) (equation 8.11).

Fmoc-Ser-Ser-Ser-Asp-Val-Ser-Thr-Ala-OC$_6$H$_4$Cl-p + Gln-Ala-Ala-Gly

$\xrightarrow[\substack{\text{DMF (54 \%) / H}_2\text{0}\\\text{pH 8.3}}]{\text{thiosubtilisin}}$ Fmoc-Ser-Ser-Ser-Asp-Val-Ser-Thr-Ala-Gln-Ala-Ala-Gly   (8.11)

RNase T$_1$ (12-23)

Although this is a very sophisticated method, one drawback is that activated esters (p-chlorophenyl ester for thiosubtilisin, cyanomethyl ester for methylchymotrypsin) have to be used as substrate because the esterase activity of these modified enzyme is also significantly lower than that of the native ones. The reaction would have to be carried out at high temperature (50°C) to use simple esters (e.g. methyl ester) as a substrate, although the enzyme tends to be inactivated under such a condition. To use a thermostable mutant of thiosubtilisin (such as thiosubtilisin 8397/C206Q (42, 43), see below) might offer a solution.

One problem associated with kinetically controlled synthesis in alkaline aqueous solutions containing high concentrations of water-miscible organic solvents is that a protease tends to be less stable under these conditions than in neutral aqueous solution without organic solvents. Met(O)192-α-chymotrypsin, in which ^{192}Met of chymotrypsin is oxidized to methionine sulfoxide, was found to be more stable than the native enzyme under the present synthetic conditions (32, 39). Wong and coworkers developed a series of site-specific mutants of subtilisin BPN′ (subtilisin 8350, 8397) which are more stable than the wild type both in water and in 50% DMF (40–43). For instance, subtilisin 8350, a mutant of subtilisin BPN′ by six site-specific mutations (Met50Phe, Gly169Ala, Asn76Asp, Gln206Cys, Tyr217Lys, and Asn218Ser), is 100 times more stable than the wild-type enzyme in aqueous solution at room temperature. By using the mutant enzyme in aqueous solution containing high concentrations (typically 50%) of DMF, they synthesized various oligopeptides in good yields. An intriguing example among them is a polymerization of methionine illustrated by equation (8.12). They found that the degree of polymerization could be controlled by the concentration of DMF: the higher the concentration of DMF, the higher the molecular weight of the polymer. The polymer obtained in 50% DMF contained more than 50 methionine residues, while 20–30 methionine residues were incorporated in 10% DMF. High-concentration DMF induced a high degree of polymerization presumably because it improved the solubility of the product polymer as well as selectively eliminating the amidase activity.

$$\text{Met-OMe} \xrightarrow[\substack{\text{DMF (x \%) / H}_2\text{O} \\ \text{pH 7.5}}]{\text{subtilisin 8350}} (\text{Met})_y\text{-OMe} \qquad (8.12)$$

$$\substack{x=50, y=50 \\ x=10, y=20\text{-}30}$$

8.3.2 Use of anhydrous organic solvents

Anhydrous organic solvents should be an ideal reaction medium for kinetically controlled synthesis because any unwanted hydrolytic side reaction, in principle, can be avoided. Klibanov and his research group demonstrated that enzymes are catalytically active in various anhydrous organic solvents and opened the door for this novel methodology. They have also carried out systematic investigations on how the nature of solvent affects the enzyme's catalytic activity and selectivity, and applied their findings to enantio-, regio-, and chemoselective peptide synthesis.

First, they revealed that the enantioselectivity of serine proteases is dramatically relaxed in organic solvents (44). For instance, subtilisin Carlsberg exhibited strict L-selectivity for the hydrolysis of Ac–Phe–OEtCl $((k_{cat}/K_m)_L/(k_{cat}/K_m)_D = 15\ 000)$ in water, while the ratio became only 5.4 for the transesterification reaction with propyl alcohol in anhydrous butyl ether. Similar selectivity relaxation was observed for other serine proteases (trypsin α-chymotrypsin, elastase, α-lytic protease, subtilisin BPN′) and seems to be a

Table 8.5 Subtilisin-catalyzed synthesis of D–L and D–D type peptides in *tert*-amyl alcohol (46)

Acyl donor	Nucleophile	Product	Yield (%)
Ac–D-Phe–OEtCl	L-Phe–NH$_2$	Ac–D-Phe–L-Phe–NH$_2$	67
F–D-Ala–OEtCl	D-Ala–NH$_2$	F–D-Ala–D-Ala–NH$_2$	65
F–D-Ala–OEtCl	L-Phe–L-Leu–NH$_2$	F–D-Ala–L-Phe–L-Leu–NH$_2$	51

general phenomenon. They also found the enantioselectivity is linearly correlated with the hydrophobicity of the solvent; the more hydrophobic, the less selective.

Their recent study demonstrated that protease's enantioselectivity is even inverted upon simple replacement of the reaction medium. The L/D selectivity of *Aspergillus oryzae* protease for Ac–Phe–OEtCl (defined above) is 6.6 in a hydrophilic solvent, acetonitrile, while it is 0.24 in a hydrophobic solvent, toluene (45). Once again, linear correlation between the protease's enantioselectivity and the solvent's hydrophobicity was observed.

These findings opened the possibility for the formation of D–L or D–D type peptide bonds by using a protease (Table 8.5). Margolin *et al.* succeeded in the synthesis of peptides with D-amino acid residues at their carboxyl side by utilizing subtilisin in anhydrous *tert*-amyl alcohol (46). Although synthesis of L–D type peptides was achieved by using high concentration of D-amino donors to compete with the hydrolysis of the acyl intermediate in aqueous solution with alkaline pH and containing high concentration of organic solvent (31–33), the D–L type has never been synthesized with protease catalysis due to the enzyme's strict selectivity. It became possible only in anhydrous organic solvents, where the enantioselectivity is dramatically relaxed (or even inverted).

The subtilisin–*tert*-amyl alcohol system was found to exhibit unique regio- and chemoselectivity as well as enantioselectivity. When lysine was used as an amino donor without protection of its side chain, subtilisin catalyzed peptide bond formation selectively on the ε-amino group (47).

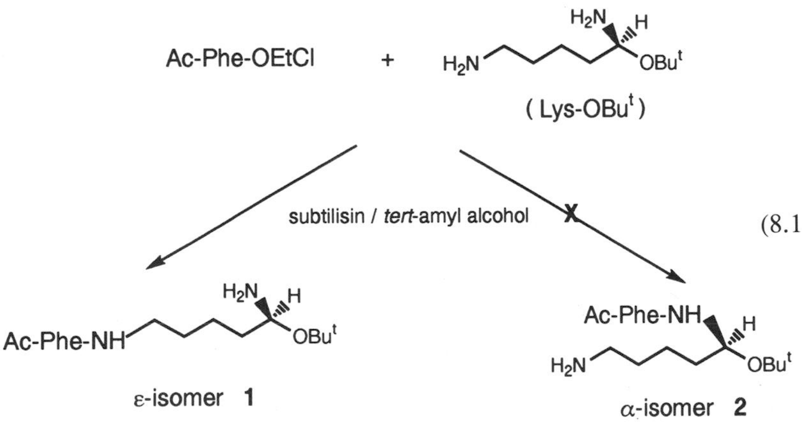

(8.13)

Similar ε-selectivity was found for other bacterial serine proteases, while the peptide with the α-amino group (**2** in equation 8.13) is a main product by using mammalian proteases (α-chymotrypsin, elastase) (48). Interestingly, in water, subtilisin readily catalyzed the hydrolysis of **2** while it was completely inactive to the ε-isomer **1**. Kinetically-controlled synthesis is not the reverse reaction of enzymatic peptide hydrolysis, and the mechanistic difference should be sufficient to explain the opposite selectivities in peptide bond formation and cleavage (see Figure 8.2).

In contrast to the preference for the ε-amino group of lysine, subtilisin is completely inactive to the side-chain hydroxyl group of serine. This chemo-selectivity was applied to the selective modification of amino alcohols. When 6-aminohexanol was employed as a nucleophile, the amide was a sole product with subtilisin, while only the ester was obtained with lipase catalysis (49) (equation 8.14).

Ac-Phe-OEtCl   + H_2N ⁀⁀⁀ OH $\xrightarrow[\text{\textit{tert}-amyl alcohol}]{\text{subtilisin}}$ Ac-Phe-NH ⁀⁀⁀ OH

74 %

$C_3H_7\overset{\overset{\displaystyle O}{\|}}{C}$-OEtCl + HO ⁀⁀⁀ NH_2 $\xrightarrow[\text{\textit{tert}-amyl alcohol}]{\text{lipase}}$ $C_3H_7\overset{\overset{\displaystyle O}{\|}}{C}$-O ⁀⁀⁀ NH_2

47 %

$$(8.14)$$

Even substrate specificity of a protease may be drastically altered or even inverted in anhydrous organic solvents (50). An aromatic hydrophobic amino acid substrate is known to be a specific substrate for α-chymotrypsin in water: the protease catalyzed the hydrolysis of Ac-L-Phe–OEt much more effectively than Ac-L-Ser–OMe ($(k_{cat}/K_m)_{Phe}/(k_{cat}/K_m)_{Ser} = 46\,000$). For the trans-esterification reaction with propyl alcohol in anhydrous octane, however, the ratio was 0.3: the selectivity was completely inverted. Similarly, the substrate specificity (as defined above) with subtilisin was 800 in water, while it was 0.4 in octane. Thus, the catalytic activity of a protease in anhydrous organic medium is not limited to its specific substrates in water, implying that a protease is a generally useful catalyst for peptide synthesis in anhydrous organic medium. This substrate specificity inversion was rationalized by thermodynamic considerations (51, 52).

Because the kinetically controlled mechanism depends on the esterase activity of serine or thiol proteases, other hydrolases with esterase activity also should work as a catalyst for peptide synthesis. Lipases are good candidates because they have rather broad substrate specificity and are widely used for organic synthesis. Margolin and Klibanov found that porcine pancreatic lipase catalyzed peptide bond formation in anhydrous toluene or THF (53). Because a lipase is not restricted to protease's specificities, it is a useful catalyst for the synthesis of unnatural peptides such as those containing D-amino acids, or those linked through the ε-amino group of lysine or the β-hydroxyl group of serine (54) (equation 8.15).

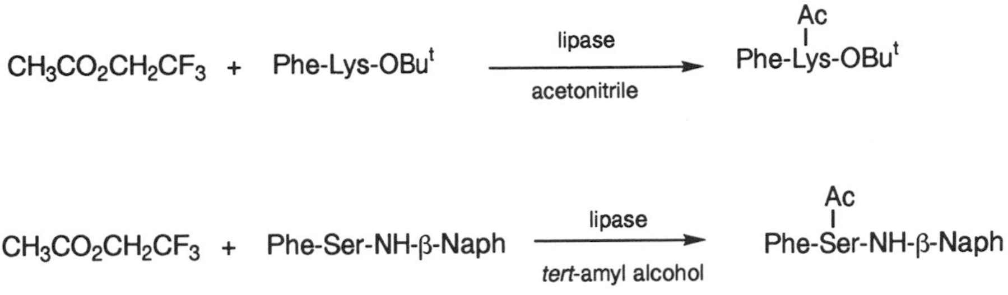

$$CH_3CO_2CH_2CF_3 \; + \; \text{Phe-Lys-OBu}^t \quad \xrightarrow[\text{acetonitrile}]{\text{lipase}} \quad \overset{\overset{\displaystyle Ac}{\displaystyle |}}{\text{Phe-Lys-OBu}^t}$$

$$CH_3CO_2CH_2CF_3 \; + \; \text{Phe-Ser-NH-}\beta\text{-Naph} \quad \xrightarrow[\textit{tert}\text{-amyl alcohol}]{\text{lipase}} \quad \overset{\overset{\displaystyle Ac}{\displaystyle |}}{\text{Phe-Ser-NH-}\beta\text{-Naph}}$$

$$(8.15)$$

As described in Chapter 3, organic solvents with high hydrophobicity are suitable as a reaction medium for enzymatic reactions. In this sense, subtilisin is somehow exceptional because it exhibits high catalytic activity even in rather hydrophilic solvents (*tert*-amyl alcohol, THF, acetonitrile, acetone, pyridine), while α-chymotrypsin shows the same level of activity only in very hydrophobic solvents (Table 8.6) (55). It makes subtilisin uniquely suitable as a catalyst for peptide synthesis because peptide substrates, especially long-chain peptide segments, are soluble only in hydrophilic solvents. Subtilisin is catalytically active even in anhydrous DMF, and Riva *et al.* successfully utilized the protease for selective modification of saccharides in DMF (56). The activity of subtilisin in anhydrous DMF, however, is significantly lower than in *tert*-amyl alcohol, and a new type of protease which retains most of its activity in anhydrous DMF has been explored. Site-specific mutants of subtilisin BPN′, which were found to be more stable than the wild type in aqueous solution, also exhibited improved stability in anhydrous DMF (40–43). For instance, the half-life in DMF at 25°C is 20–30 min for subtilisin BPN′, 25 h for mutant 8350, and 350 h for mutant 8397.

Table 8.6 Catalytic activity[a] of proteases in organic solvents (53)

Solvent	V/K_m for subtilisin[b]	V/K_m for chymotrypsin[b]
Octane	2000	1700
Toluene	150	120
tert-Amyl alcohol	2100	38
Pyridine	97	< 0.1
THF	120	7.2
Acetone	810	0.6
Acetonitrile	150	0.4

[a] For the transesterification reaction between Ac–Phe–OEt and propanol; [b] $\text{min}^{-1} \times 10^6$.

According to a recent result, subtilisin BPN' lyophilized from phosphate buffer solution was much more stable in anhydrous DMF than that from tris buffer (43). In general, enzymes lyophilized from inorganic buffers exhibited remarkable stability in DMF compared to those from organic buffers. Another intriguing point is that lyophilization from a higher-concentration phosphate buffer made the enzyme more stable in anhydrous DMF. These results are rationalized by considering that inorganic salts do not dissolve in DMF, and form a water-abundant protection layer around the enzyme molecule, whereas organic salts eventually dissolve and the enzyme is exposed directly to DMF. Mutants 8350 and 8397 are much more stable than the wild type under the same lyophilization conditions, and have been applied for regioselective acylation of sugars, nucleosides and ribosides.

In summary, protease catalysis in anhydrous organic solvents is a useful method for kinetically controlled synthesis, not only because unwanted competitive hydrolysis is avoidable, but also because, by choosing a suitable solvent, a protease can be converted to a new catalyst with unique activity and selectivity which are significantly different from those in water. This new methodology ('solvent engineering' (57)) should be complementary to another method which produces a new type of enzyme ('protein engineering'), and a combination between them makes enzymatic methods very attractive for peptide synthesis.

References

1. Kullmann, W. (1987) *Enzymatic Peptide Synthesis*. CRC Press, Boca Raton, FL, USA.
2. Jakubke, H.-D., Kuhl, P. and Konnecke, A. (1985) *Angew, Chem. Int. Edn Engl.*, **24**, 85.
3. Oyama, K. and Kihara, K. (1984) *Chemtech*, **14**, 100.
4. Bodanszky, M. (1984) *Principles of Peptide Synthesis*. Springer-Verlag, Berlin.
5. Morihara, K. (1987) *Tibtech*, 164.
6. Bender, M.L. and Kemp, K.C. (1957) *J. Amer. Chem. Soc.*, **79**, 116.
7. Homandberg, G.A., Mattis, J.A. and Laskowski, M. Jr (1978) *Biochemistry*, **17**, 5220.
8. Tsuzuki, H., Oka, T. and Morihara, K. (1980) *J. Biochem.*, **88**, 669.
9. Inoue, K., Watanabe, K., Tochino, Y., Kobayashi, M. and Shigeta, Y. (1981) *Biopolymers*, **20**, 1845.
10. Inoue, K., Watanabe, K., Morihara, K., Tochino, Y., Kanaya, T., Emura, J. and Sakakibara, S. (1979) *J. Amer. Chem. Soc.*, **101**, 751.
11. Morihara, K., Oka, T. and Tsuzuki, H. (1979) *Nature*, **280**, 412.
12. Markussen, J. (1982) US patent 4 343 898.
13. Jonczk, A. and Gattner, H.-G. (1981) *Hoppe-Seylers Z. Physiol. Chem.*, **362**, 1591.
14. Morihara, K., Oka, T., Tsuzuki, H., Tochino, Y. and Kanaya, T. (1980) *Biochem. Biophys. Res. Commun.*, **92**, 396.
15. Martinek, K., Semenov, A.N. and Berezin, I.V. (1981) *Biochim. Biophys. Acta*, **658**, 76.
16. Martinek, K. and Semenov, A.N. (1981) *Biochim. Biophys. Acta*, **658**, 90.
17. Kuhl, P., Könnecke, A., Döring, G., Däumer, H. and Jakubke, H.-D. (1980) *Tetrahedron Lett.*, **21**, 893.
18. Isowa, Y., Ohmori, M., Ichikawa, T., Mori, K., Nonaka, Y., Kihara, K., Oyama, K., Satoh, H. and Nishimura, S. (1979) *Tetrahedron Lett.*, **28**, 2611.
19. Oyama, K., Nishimura, S., Nonaka, Y., Kihara, K. and Hashimoto, T. (1981) *J. Org. Chem.*, **46**, 5241.

20. Nakanishi, K., Kamikubo, T. and Matsuno, R. (1985) *Bio/Technology*, **3**, 459.
21. Nakanishi, K., Kimura, Y. and Matsuno, R. (1986) *Bio/Technology*, **4**, 452.
22. Oka, T. and Morihara, K. (1980) *J. Biochem.*, **88**, 807.
23. Ooshima, H., Mori, H. and Harano, Y. (1985) *Biotechnol. Lett.*, **7**, 789.
24. Sakina, K., Kawazura, K., Morihara, K. and Yajima, H. (1988) *Chem. Pharm. Bull.*, **36**, 4345.
25. Sakina, K., Kawazura, K., Morihara, K. and Yajima, H. (1988) *Chem. Pharm. Bull.*, **36**, 3915.
26. Kitaguchi, H. and Klibanov, A.M. (1989) *J. Amer. Chem. Soc.*, **111**, 9272.
27. Nishino, N., Xu, M., Mihara, H. and Fujimoto, T. (1992) *Chem. Lett.*, 327.
28. Nishino, N., Xu, M., Mihara, H. and Fujimoto, T. (1992) *Tetrahedron Lett.*, **33**, 3137.
29. Fastrez, J. and Fersht, A.R. (1973) *Biochemistry*, **12**, 2025.
30. Fersht, A. (1985) *Enzymatic Structure and Mechanism*, 2nd Edn. W.H. Freeman, New York.
31. West, J.B. and Wong, C.-H. (1986) *J. Org. Chem.*, **51**, 2728.
32. West, J.B. and Wong, C.-H. (1986) *J. Chem. Soc., Chem. Commun.*, 417.
33. Barbas, C.F. III and Wong, C.-H. (1987) *J. Chem. Soc., Chem. Commun.*, 533.
34. Barbas, C.F. III and Wong, C.-H. (1988) *Tetrahedron Lett.*, **29**, 2907.
35. Barbas, C.F. III, Matos, J.R., West, J.B. and Wong, C.-H. (1988) *J. Amer. Chem. Soc.*, **110**, 5162.
36. Schellenberger, V., Schellenberger, U., Jakubke, H.-D., Hansicke, A., Bienert, M. and Krause, E. (1990) *Tetrahedron Lett.*, **31**, 7305.
37. Nakatsuka, T., Sasaki, T. and Kaiser, E.T. (1987) *J. Amer. Chem. Soc.*, **109**, 3808.
38. West, J.B., Scholten, J., Stolowich, N.J., Hogg, J.L., Scott, A.I. and Wong, C.-H. (1988) *J. Amer. Chem. Soc.*, **110**, 3709.
39. West, J.B., Hennen, W.J., Lalonde, J.L., Bibbs, J.A., Zhong, Z., Meyer, E.F. Jr. and Wong, C.-H. (1990) *J. Amer. Chem. Soc.*, **112**, 5313.
40. Wong, C.-H., Chen, S.-T., Hennen,, W.J., Bibbs, J.A., Wang, Y.-F., Liu, J.L.-C., Pantoliano, M.W., Whitlow, M. and Bryan, P.N. (1990) *J. Amer. Chem. Soc.*, **112**, 945.
41. Zhong, Z., Liu, J.L.-C., Dinterman, L.M., Finkelman, M.A.J., Mueller, W.T., Rollence, M.L., Whitlow, M. and Wong, C.-H. (1991) *J. Amer. Chem. Soc.*, **113**, 683.
42. Wong, C.-H., Schuster, M., Wang, P. and Sears, P. (1993) *J. Amer. Chem. Soc.*, **115**, 5893.
43. Sears, P., Schuster, M., Wang, P., Witte, K. and Wong, C.-H. (1994) *J. Amer. Chem. Soc.*, **116**, 6251.
44. Sakurai, T., Margolin, A.L., Russell, A.J. and Klibanov, A.M. (1988) *J. Amer. Chem. Soc.*, **110**, 7236.
45. Tawaki, S. and Klibanov, A.M. (1992) *J. Amer. Chem. Soc.*, **114**, 1882.
46. Margolin, A.L., Tai, D.-F. and Klibanov, A.M. (1987) *J. Amer. Chem. Soc.*, **109**, 7885.
47. Kitaguchi, H., Tai, D.-F. and Klibanov, A.M. (1988) *Tetrahedron Lett.*, **29**, 5487.
48. Kitaguchi, H., Ono, M., Itoh, I. and Klibanov, A.M. (1991) *Agric. Biol. Chem.*, **55**, 3067.
49. Chinsky, N., Margolin, A.L. and Klibanov, A.M. (1989) *J. Amer. Chem. Soc.*, **111**, 386.
50. Zaks, A. and Klibanov, A.M. (1986) *J. Amer. Chem. Soc.*, **108**, 2767.
51. Wescott, C.R. and Klibanov, A.M. (1993) *J. Amer. Chem. Soc.*, **115**, 1629.
52. Wescott, C.R. and Klibanov, A.M. (1993) *J. Amer. Chem. Soc.*, **115**, 10362.
53. Margolin, A.L and Klibanov, A.M. (1987) *J. Amer. Chem. Soc.*, **109**, 3802.
54. Gardossi, L., Bianchi, D. and Klibanov, A.M. (1991) *J. Amer. Chem. Soc.*, **113**, 6328.
55. Zaks, A. and Klibanov, A.M. (1988) *J. Biol. Chem.*, **263**, 3194.
56. Riva, S., Chopineau, J., Kieboom, A.P.G. and Klibanov, A.M. (1988) *J. Amer. Chem. Soc.*, **110**, 584.
57. Klibanov, A.M. (1989) *Trends Biochem. Sci.*, **14**, 141.

9 Productivity of enzymatic catalysis in non-aqueous media: New developments

E.N. VULFSON, I. GILL and D.B. SARNEY

9.1 Introduction

Enzymatic catalysis in non-aqueous solvents has gained considerable interest in recent years as an efficient approach to the synthesis of natural products, pharmaceuticals, fine chemicals and food ingredients (1–7). The high selectivity and mild reaction conditions associated with enzymatic transformations have made them very attractive for the synthesis of complex bioactive compounds, which are often difficult to obtain by standard chemical routes. Hence the regio- and stereospecificity of enzymes and, in particular, the dependence of these characteristics on the reaction conditions and the reaction media employed have been actively investigated in the last few years (see other chapters of this book for an up-to-date review).

In contrast, scant attention has been directed towards improving the productivity of enzymatic transformations. Indeed, enzyme-based processes often suffer from insufficient productivities from a technological standpoint, which has invariably led to a reduced competitive edge as compared to conventional chemical approaches. Although significant improvements have been achieved through the application of enzymes in organic solvents (3, 8–12), especially where the synthesis and modification of substances showing a poor solubility in water are concerned, a number of problems still exist in transferring these protocols from the laboratory to an industrial scale. In particular, enzymes typically display relatively low catalytic activities when used in non-aqueous media, thus necessitating the use of comparatively large amounts of biocatalyst. Since, in the majority of cases, the cost of enzymes remains too high to justify their disposal on completion of the synthesis, issues such as the operational stability of enzymes and their recovery and reuse become crucial in determining the commercial viability of the process. Hence, it is the productivity of enzymatic catalysis, in terms of kilograms of product obtained per kilogram of enzyme used, which frequently determines the feasibility of incorporating an enzymatic step into the overall technological cycle. Consequently, numerous research papers have addressed the issue of enzyme activity and stability in organic solvents, and substantial progress has been made in our understanding of the key factors involved (8–15). In practice, however, it is difficult to select a solvent which preserves a high level of enzyme activity and at the same time is capable of solubilizing high concentrations of dissimilar (hydrophilic and hydrophobic) substrates and, in addition, satisfies the overall technological requirements of the process.

In this chapter, some recent developments in enzymatic catalysis in non-aqueous media will be reviewed, with an emphasis placed on new methods that display enhanced productivities. In particular, enzymatic solvent-free syntheses, transformations in heterogeneous eutectic mixtures of substrates, and the design of continuous bioreactors are discussed in some detail. It should be stressed that it is not the intention of the authors to present a comprehensive survey of this important subject, but to analyse some emerging trends and to assess their impact on the future development and practical utilization of the technology.

9.2 Enzymatic solvent-free synthesis

'Low-water enzymology' has made substantial progress over the past two decades: This evolution began with the investigation of predominantly aqueous systems containing small quantities of water-miscible organic solvents, continued with the development of aqueous–organic two-phase reaction mixtures and the dispersion of an aqueous phase in a bulk organic medium by means of micro-emulsions, and finally led to the introduction and use of enzyme suspensions in almost anhydrous organic solvents (10). The latter development was especially significant because it conclusively showed that very little water is actually required for enzymes to maintain their catalytically active conformations, and that the existence of a defined aqueous phase (however small) is not a pre-requisite for efficient catalysis. However, if this is so, it would be techno-logically attractive to take a further step and attempt to dispense with the bulk solvent (or, in some cases, with organic solvents altogether) by performing the enzymatic reaction in a mixture of the substrates themselves. This approach, if feasible, would combine the precision of biological catalysis with the high levels of productivity achieved by the best conventional methods.

It should be stressed that the idea of using one of the substrates as the bulk solvent has been in the literature for some time. For example, Cambou and Klibanov (16) carried out the resolution of racemic alcohols dissolved in a 'matrix ester' (methyl propionate, tributyrin or ethyl acetate). Similarly, vinyl acetate (17) is often used as both a solvent and an acylating agent in enzymatic esterification and transesterification reactions. Primary alcohols have been utilized by Snijder et al. (18) as both reaction solvents and sacrificial substrates for the regeneration of NADH in the alcohol dehydrogenase-catalysed reduction of aldehydes. However, in all of these cases, one substrate (solvent) was used in a large excess over the other, thus offering very little practical advantage. Clearly this is rather different from the situation when an enzymatic reaction is carried out in a nearly stoichiometric mixture of the substrates in the presence of a small amount (if any) of added solvent.

Generally one can envisage two immediate difficulties with the implementation of a solvent-free synthesis: (i) the reaction mixture should be sufficiently liquid and remain so during the course of the transformation; and (ii) there should be a driving force for the reaction to proceed in the required direction until completion. Fortunately, neither of these problems arise in the transesterification of oils, and the synthesis of speciality triglycerides was the first enzymatic solvent-free process introduced into industry.

The potential for utilizing 1,3-specific lipases in the manufacture of high-value asymmetric triglycerides was recognized as early as the beginning of the 1980s, when Unilever (19), Kao Corporation (20) and Fuji Oil (21) filed their first patent applications concerned with the enzymatic synthesis of cocoa butter-like fats. Natural cocoa butter consists almost exclusively of 1,3-disaturated-2-oleoylglycerol, where palmitic, stearic and oleic acids account for 95% of the total fatty acid content. Thus, lipase-catalysed transesterification of the inexpensive 1,3-dipalmitoyl-2-oleoyl-glycerol (palm-oil mid-fraction) with 1,3-distearoyl-2-oleoyl-glycerol (palm-oil second stearin fraction) or 1,3-specific acyl exchange in the presence of stearic acid or methyl stearate should give a product with a triglyceride composition and properties similar to the natural product. Alternatively, a cocoa butter-equivalent fat can be obtained by blending the palm oil fractions with 1,3-distearoyl-2-oleoyl-glycerol, enzymatically prepared from high-oleic acid sunflower oil (triolein) and stearic acid (Figure 9.1). Although the initial studies were carried out using organic solvents, the companies opted for solvent-free syntheses which are currently performed on multitonne scales (22, 23). This has allowed a reduction in the production costs to the level where the enzymatically obtained products are competitive with natural cocoa butter, despite a significant fall in the price of the latter on the world market.

The development of enzymatic solvent-free transesterification as a generic technology has quickly led to the introduction of several new speciality triglycerides into the market place. For example, 1,3-dibehenoyl-2-oleoyl-glycerol (Fuji Oil Co.) is used in the food industry to prevent 'chocolate blooming' (surface crystallization of fats) in confectionery products (24), and 1,3-dioleoyl-2-palmitoyl-glycerol, enzymatically prepared by Unilever, has found applications as a human milk-fat substitute in baby foods (Figure 9.1) (25). It is clear that apart from the technological advantages offered by this methodology, the avoidance of organic solvents is particularly attractive in the context of the food industry, where stringent regulations over the use of solvents exist (26).

These examples of the successful application of lipase technology to the highly competitive area of low- to mid-value oil and fat products emphasize the critical advantages offered by solvent-free bioprocessing in this market niche. Similar enzymatic (trans)esterifications in neat mixtures of substrates are also being applied to low-volume, high-value food and pharmaceutical products,

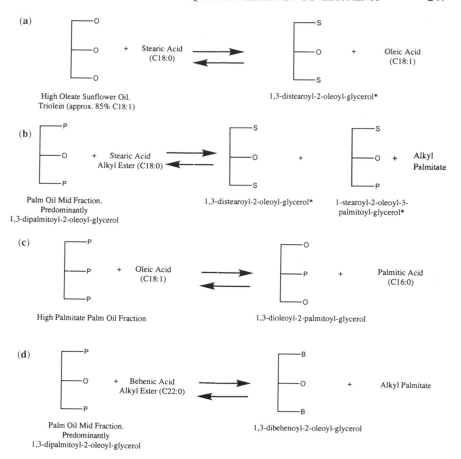

Figure 9.1 The industrial synthesis of structured triglycerides as exemplified by the enzymatic production of cocoa butter substitutes (a, b), human milk fat alternatives (c) and confectionery additives (d). O: oleoyl [$OCO(CH_2)_7CH=CH(CH_2)_7CH_3$]; S: stearoyl [$OCO(CH_2)_{16}CH_3$]; P: palmitoyl [$OCO(CH_2)_{14}CH_3$]; B: behenoyl [$OCO(CH_2)_{20}CH_3$]. The oils and fats used in these transformations are not pure triglycerides (as schematically represented) but contain the highest proportion of the triglycerides shown. *These products are blended with other vegetable fat fractions for the final cocoa butter-substitute formulation.

including simple alkyl and terpenyl flavour and fragrance esters (such as ethyl isovalerate, heptyl oleate, geranyl acetate and citronellyl acetate) (27–31), and polyunsaturated fatty acid (PUFA) enriched glycerides for biomedical applications (32–35). For example, esters of C2–C8 alcohols such as isopropyl myristate, isopropyl palmitate and 2-ethylhexyl palmitate are currently produced by Unichema on a scale of several thousand tonnes per annum for applications in cosmetics and personal care products (27).

The reactions above may be considered as rather convenient examples where the substrates and products are miscible liquids at the reaction temperature and, therefore, this approach would be difficult to implement with other bio-transformations. However, recent research has demonstrated that this is not necessarily the case, and that the concept of enzymatic solvent-free synthesis can be utilized efficiently using a wide range of substrates. In particular, it has been successfully implemented for the regioselective preparation of sugar fatty acid esters.

The regioselective acylation of sugars is a difficult task due to the presence of several hydroxyl groups of similar reactivity. Recently, carbohydrate chemists have benefited from the application of esterases, which has allowed them to circumvent (at least partially) the tedious introduction and removal of protecting groups normally required for achieving the desired regioselectivity. Since the first demonstration of the utility of these enzymes in the selective acylation of monosaccharides by Klibanov and coworkers (36, 37), numerous elegant syntheses have been described in the literature (see refs 38, 39 and references cited therein). However, this methodology, although undoubtedly suitable for the preparation of sugar derivatives on a laboratory scale, has certain drawbacks where its application on a larger scale is concerned. For instance, there are only a few solvents (such as dimethyl sulphoxide (DMSO), dimethylformamide and pyridine) which can solubilize sugars at high enough concentrations, and not many enzymes retain their catalytic activity in these media. Additionally, even those enzymes that remain active (subtilisin and several lipases) usually display very low catalytic activities. As a result, a large excess of the biocatalyst is required and even then the reaction takes days rather than hours. In addition, the operational stability of enzymes in these highly polar solvents is low and a significant loss of catalytic activity occurs during the course of the reaction. These limitations become especially apparent in the synthesis of bulk products such as sugar fatty acid esters, where concerns over the health and environmental safety of currently used products have invoked great interest in their preparation by enzymatic routes (40).

Although the synthesis of sugar fatty acid esters has been successfully accomplished in anhydrous organic solvents by several groups (36, 41–45), a low productivity and the reduced operational stability of enzymes in DMSO and pyridine have prevented the commercialization of this approach. An alternative method discussed here relies upon the prior modification ('hydrophobization') of the sugar moiety followed by solvent-free esterification with molten fatty acids (46–50). Although this synthetic scheme appears to be more complex than the direct one-step acylation procedure, the reaction kinetics and overall productivity are good. For this reason, this newer methodology appears to be technologically more attractive.

Adelhorst et al. (46) performed the regioselective, solvent-free esterification of simple 1-O-alkyl-glycosides using a slight molar excess of molten fatty acids (Figure 9.2a). The rate of enzymatic esterification was found to be markedly dependent on the length of the alkyl group. Thus, only 20% yield was obtained with glucose and 1-O-methyl-glucoside after 1 and 21 days of incubation respec-

(a)

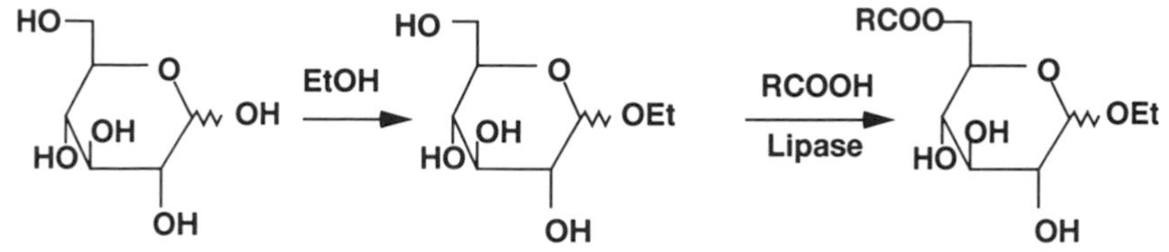

(b)

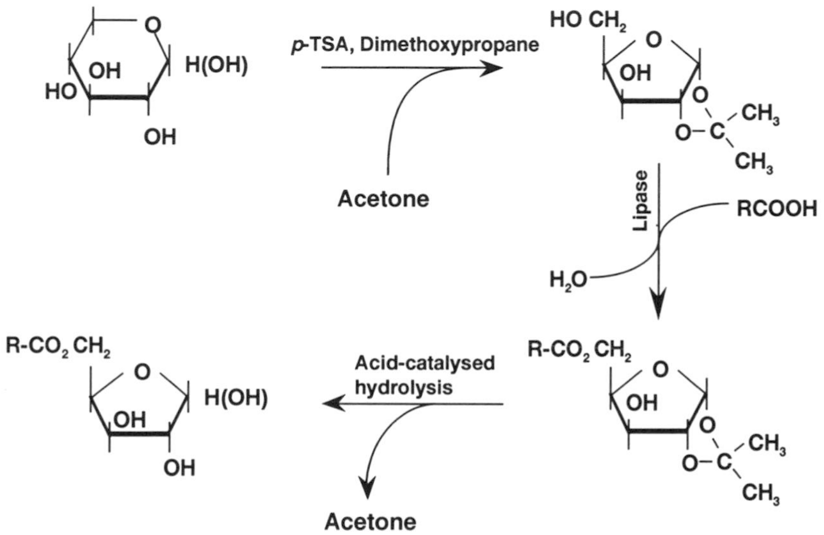

Figure 9.2 'Hydrophobization' of monosaccharide substrates prior to lipase-catalysed esterification in the synthesis of sugar-based surfactants.

tively, whilst when ethyl-, *n*- and iso-propyl- or butyl-glucosides were used, the
reaction was completed in a few hours. A range of 1-*O*-alkyl-6-*O*-acyl-gluco-
pyranosides was prepared in up to 90% yield, and the process has recently
undergone pilot-scale trials. The products are claimed by Novo-Nordisk to be
non-toxic and rapidly biodegradable and are expected to find applications as
industrial and/or household detergents (51) if certain difficulties with the
product formulations are successfully overcome.

   Alternatively, sugar acetals (Figure 9.2b) have been used as starting materials
to increase the miscibility of the reactants (47–50). The final products, mono-

and di-saccharide fatty acid esters, were obtained in good yields *via* the lipase-catalysed, solvent free esterification of the acetals, followed by mild acid hydrolysis of the resulting sugar acetal esters. Although large-scale acetalization and acetal deprotection do not present serious technological difficulties, as demonstrated by the industrial production of ascorbic acid, the overall manu-facturing sequence would be more complicated. It remains to be seen, therefore, whether these additional steps can be justified by the improved quality and superior emulsifying properties of the products. At the same time, this method-ology is probably more versatile as compared to that employing 1-*O*-alkyl-glycosides, since it provides an efficient route to the synthesis of (oligo)-saccharide fatty acid esters.

It should be noted that the use of isopropylidene groups in these syntheses was designed to improve the miscibility of the reactants, and must be distinguished from the conventional protecting-group strategy employed in regioselective organic synthesis. For example, 6′-*O*-acyl-lactose was obtained as the sole product after removal of the isopropylidene groups, even when a crude mixture of partial lactose acetals was enzymatically esterified (50). Due to the strict enzyme regioselectivity towards the 6′-hydroxyl group, no acylation was observed with 4′,6′-isopropylidenelactose, while the reaction with the 3′,4′-iso-propylidene derivative led to the formation of the same monoester product (Figure 9.3). Hence, no separation of the partial acetals from a complex product mixture was required, and the kinetics of acylation were actually better with the crude mixture of acetals than with pure crystalline substrate (Figure 9.4). The desired regioselectivity of esterification was achieved solely through the specificity of the lipase.

Evidently simple chemical modification of the substrate, its 'conditioning' in terms of its solubility or miscibility with the reaction mixture, can lead to a significant improvement in the kinetics and overall productivity. However, it is not always possible to achieve complete miscibility of the reagents. In this case, the synthesis can be performed in a heterogenous reaction mixture, e.g. a suspension of one substrate in the other. It appears that even under these conditions enzyme kinetics and selectivity can be maintained at a satisfactory level, as shown in a comparative study of the lipase-catalysed solvent-free esterification of 1,2-*O*-isopropylidene-xylose (IPX) and -glucose (IPG) with myristic acid (49).

Despite the chemical similarity of these two carbohydrate substrates, a considerable difference in the kinetics and regioselectivity was observed when the reaction was performed in the absence of added solvent. It was found that esterification of IPX (melting point 70°C; completely miscible with molten fatty acids at the reaction temperature of 75°C) under solvent-free conditions proceeded much faster and with greater regioselectivity than that of IPG (melting point 159°C; only sparingly soluble in the oil phase under the same conditions). This was due to: (i) some mass-transfer limitations which inevitably arise in a bioreactor containing a high proportion of solid substrate (IPG); and

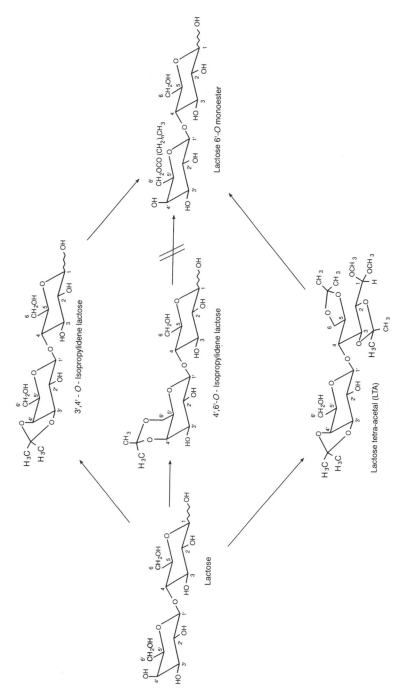

Figure 9.3 Regioselectivity of Lipozyme™-catalysed esterification of lactose acetals. See text for further details.

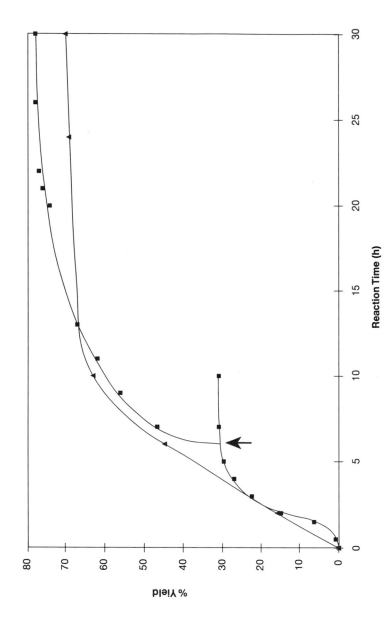

Figure 9.4 Kinetics of Lipozyme™-catalysed esterification of crude lactose acetal (▲) and pure lactose tetra-acetal (■) with myristic acid. When crystalline lactose tetra-acetal (■) was used a small amount of toluene was added at the start of the reaction to aid initial miscibility of the reactants. The arrow indicates the time at which the solvent (and water of esterification) was allowed to evaporate, thus shifting the equilibrium in the direction of synthesis.

(ii) accumulation of the significantly more oil-soluble product (IPG monoester) and its efficient competition with the substrate (IPG). The latter led to considerable formation of diesters and the resulting loss of regioselective control. However, it was possible to improve the enzyme kinetics and to 'restore' the regioselectivity by the addition of a small quantity of suitable organic solvent to enhance the miscibility of the starting materials. A dramatic increase in the reaction rate and a lack of diester formation was observed in the presence of cyclohexanone, *t*-butyl acetate and other solvents, even when the reaction mixture contained a considerable amount of insoluble IPG (49). It should be stressed that the amounts of solvent used in these experiments were insufficient by far to solubilize the substrates completely.

Similar conclusions can be drawn from the results obtained by Yamane and coworkers, who carried out a detailed investigation of the solvent-free glycerolysis of fats and oils (52–55). The enzymatic reaction was performed in a nearly stoichiometric mixture (emulsion) of glycerol and various triglycerides at ambient temperatures using a range of 1,3-specific lipases. The equilibrium was shifted towards accumulation of the final product by decreasing the reaction temperature to below the melting point of the monoglyceride product. Yields of up to 90% were obtained with a range of animal and plant lipids including beef tallow, lard, and rapeseed, olive and palm oils. Although the reaction rates decreased significantly after cooling due to the high solids content of the resulting reaction mixture, it is interesting to note that no loss of regioselectivity (as determined by the accumulation of 2-monoacyl-glycerol) was observed. Unfortunately, no detailed investigation of the factors affecting the reaction kinetics under heterogeneous conditions has yet been reported for this system.

An important issue, where reactions performed in neat mixtures of substrates are concerned, is that of driving the equilibrium in the desired direction. This is usually achieved by removal of the product (e.g. water, alcohol or fatty acid formed during (trans)esterification in the above examples). Approaches include the direct application of vacuum, the use of selective adsorbents such as silica gel or molecular sieves, and selective product precipitation/crystallization (30, 56). However, isobaric removal of water and azeotropic distillation are probably the most suitable processes for industrial production, as exemplified by the synthesis of sugar fatty acid esters (51) and short chain esters (27), respectively.

When implementing any such synthesis, one has to bear in mind the pivotal role that water plays in enzymatic catalysis, and more specifically, in the maintenance of the unique catalytically competent conformation of the enzyme. It has been amply demonstrated that enzyme activity and stability depend critically on the level of biocatalyst hydration. In practical terms, the introduction of thermodynamic water activity (a_w) as a generic parameter (57, 58) has significantly simplified the optimization and control of biotransformations performed in non-aqueous media. Obviously, esterification reactions are favoured by a low a_w, and removal of the water produced provides a key driving force. However, excessive dehydration of the reaction mixture can lead to a dramatic deterioration of enzyme performance. The dual significance of water, in terms of both the reaction equilibrium and the biocatalyst performance, is clearly seen in lipase-catalysed esterifications, which, when initiated at a very low a_w, display an initial autocatalytic phase due to 'self hydration' of the enzyme by the water released (57–61). However, unless water is subsequently removed to maintain the optimal a_w, the product yield levels off, due to the establishment of an equilibrium, a reduction in the overall synthetic rate due to side-product formation, and in some cases phase-separation effects (61).

The control of thermodynamic water activity becomes more important with reactions performed at elevated temperatures, since one also has to consider the effect of a_w on enzyme thermostability. Recently, the thermostability of several lipases and chymotrypsin has been studied in detail by using differential scanning calorimetry (62). In accordance with earlier results (63), the denatura-tion temperature (T_m) of the enzymes was found to be 30–50°C higher in an anhydrous environment than in aqueous solution. As shown in Figure 9.5, an almost linear decrease in T_m as a function of water activity was observed in these experiments. It is evident, therefore, that the highest productivity in a bioreactor operated at elevated temperatures would be obtained at a hydration level significantly below the optimum for catalytic activity. It can readily be seen from the above that the control of a_w has fundamental implications for biocatalyst kinetics and operational stability, and ultimately for the rational optimization of biocatalyst performance and productivity in industrial reactors. Thermodynamic water activity sensors based on modified RH probes, and suitable for vapour-phase or direct-contact use, are now commercially available, thus making it possible to integrate a_w monitoring and control in bioprocesses (57, 58, 64, 65). It should be pointed out, however, that considerable complications can be expected, especially when implementing a_w monitoring/control in heterogeneous systems (60, 61). Thus, the existence of distinct phases, and changes in their composition/distribution, as well as partition effects, can result in drastic temporal and spatial variations in a_w and transient non-equilibrium phenomena during the course of the reaction.

9.3 Enzymatic catalysis in eutectic mixtures

It is apparent from numerous studies performed with lipases that a homogeneous solution of substrates is not a prerequisite for reactions to occur. It has also been established in the last two years that the concept of enzymatic transformation in heterogeneous substrate mixtures is equally applicable to other reactions, and lipases (by virtue of their ability to work in heterogeneous media and their affinity for interfaces (66, 67)) are not the only enzymes suitable for application in multiphase reaction mixtures. This was clearly demonstrated by an extensive investigation into the enzyme-catalysed synthesis of short peptides using a range of proteases from different classes (67–71).

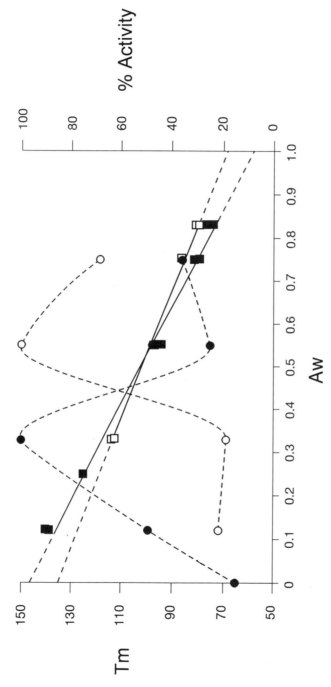

Figure 9.5 Effect of hydration on (i) the denaturation temperature (squares) and (ii) catalytic rate (circles) of *Candida rugosa* lipase (closed symbols) and α-chymotrypsin (open symbols).

In the initial experiments, the solvent-free condensation of an equimolar mixture of L-phenylalanine ethyl ester (L-Phe–OEt, an oil) and L-leucine amide (L-Leu–NH$_2$, a solid) was studied by using immobilized subtilisin and α-chymotrypsin as catalysts. Both enzymes readily catalysed the reaction and, in this particular instance, the highest yield of 83% (0.8 g per g of reaction mixture) was obtained with subtilisin (68). The addition of a small amount of organic solvent, although somewhat increasing the reaction rate, did not significantly influence the yield (Figure 9.6). It is interesting to note that the same reaction gave a yield of only 32 and 33% when carried out using the substrates dispersed (at a concentration of 0.25 M) in ethyl acetate or dichloromethane (66). Thus, it was also possible to dispense with the bulk solvent in the protease-catalysed synthesis of dipeptides.

However, the presence of one of the substrates in a liquid form was not essential for the reaction to proceed. In many other cases, product formation was observed even when the two substrates were solids in the pure state. In these cases, a liquid phase was spontaneously formed on mixing the substrates together, thus implying that from a physico-chemical standpoint the reaction mechanism differed from the previously described catalysis which had suspensions of substrates. This spontaneous formation of liquid phases was further investigated and led to the description of a new phenomenon, enzymatic catalysis in eutectic mixtures.

A eutectic is defined as a mixture of two or more compounds (either both solids, or a solid and a liquid) which displays a pronounced minimum value for its melting point (the eutectic temperature, T_e) at a certain ratio of the constituents. Indeed, it was unambiguously demonstrated that amino acid derivatives could readily form eutectic mixtures with a T_e as low as 36°C (70). Furthermore, the formation of the eutectic and its corresponding T_e could be dramatically affected by the addition of small amounts of organic solvents, termed 'adjuvants' (68, 69). The inclusion of low concentrations of adjuvants typically resulted in a dramatic expansion of the liquid phase, where the reaction was monitored by FT-IR microscopy (70), due to the formation of a ternary eutectic. It was found that a number of solvents acted as eutectic modifiers, although their relative performance varied with the reaction and enzyme studied. In general, the best results were obtained with solvents (or their combinations) displaying Hildebrand (d) values between 8.5 and 10.0, and log P values between -1.5 and -0.5. 2-Methoxyethyl acetate, 2-methoxyethyl ether, water, ethanol, glycerol and ethanol–water mixtures were particularly effective in providing stable liquid or semi-liquid eutectics when added to amino acid derivatives in amounts of 10–30% w/w (70). It is important to note that at the low concentrations used, these solvents functioned solely as eutectic modifiers and not as bulk solvents, since a considerable excess (typically over ten-fold) was required to obtain a true solution. A number of proteases (subtilisin, chymotrypsin, thermolysin, papain and related enzymes) retained their catalytic activity in eutectic media, and high

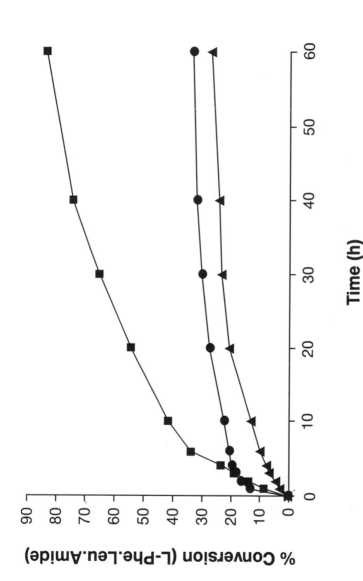

Figure 9.6 Solvent-free synthesis of L-Phe–Leu–NH$_2$ catalysed by immobilized proteases: ■,subtilisin; ●, proteinase K; ▲, α-chymotrypsin.

yields of products (up to 0.75 g per g of reaction mixture) were obtained in the synthesis of several model and bioactive peptides (71).

In many instances water and simple alcohols (ethanol, glycerol) can be used as efficient eutectic modifiers, substantially decreasing the eutectic temperature and consequently improving the kinetics (Figure 9.7). This observation is significant where the synthesis of products intended for use as food additives is concerned. Clearly, the addition of small amounts of water, glycerol and/or ethanol does not present any problems in this respect. The utility of this approach can be illustrated by the preparative scale synthesis of several flavour peptides in eutectics formed in the presence of 20–25% of these adjuvants (72).

It is interesting to note that, because water alone can act as an efficient promoter of eutectic formation, this may offer at least a partial explanation for the recently described 'solid-to-solid' transformations of peptides (73–75). Thus, enzymes including thermolysin and α-chymotrypsin catalysed di- and tri-peptide synthesis in heterogeneous mixtures consisting of largely insoluble substrates dispersed in a small quantity of water, and even in 'microaqueous' solvent-free media composed of the solid substrates together with a salt hydrate or traces of water. A fluidized bed reactor was successfully developed for implementing the latter approach on a preparative scale (74).

The crucial role of N-protecting groups in the process of eutectic formation suggests an interesting possibility of 'tailoring' substrates for enzymatic trans-formations in eutectics or in the absence of bulk solvents. Indeed, rather than using conventional protecting groups one may attempt to design new ones that would impart specific physico-chemical properties. The validity of this approach for enzymatic peptide synthesis has been elegantly demonstrated by Fischer et $al.$ (76). The use of acyl donors with charged N-protecting groups allowed reactions to be performed at equimolar ratios of substrates at concentrations as high as 1.2 M in the presence of aqueous buffers. This approach was successfully applied to the preparation of N-maleyl-kyotorpin ethyl ester on a multikilogram scale (76). In addition, a continuous bioreactor was constructed, and provided this dipeptide with a space time yield of 13.4 kg l^{-1} d^{-1}. Regardless of whether the reaction mixtures employed in these studies were eutectics or not, the results clearly show that a dramatic increase in productivity can be achieved by the appropriate selection of the substrates and solvent.

In practical terms, high productivity is one of the most interesting features of biocatalysis in eutectic mixtures. It often allows a much higher concentration of product to be obtained in a bioreactor of the same size, compared to the same reaction performed in an organic solvent. For example, in dipeptide syntheses, 0.2–0.8 g of product is typically formed per gram of reaction mixture against 0.015–0.035 g g^{-1} obtained in solution even at substrate concentrations as high as 0.25 M. Similarly, the appropriate use of adjuvants in the reaction mixture allows a significant reduction in the reaction time. This can be illustrated by the subtilisin-catalysed synthesis of Phe–Leu–NH$_2$, where the inclusion of 30% w/w

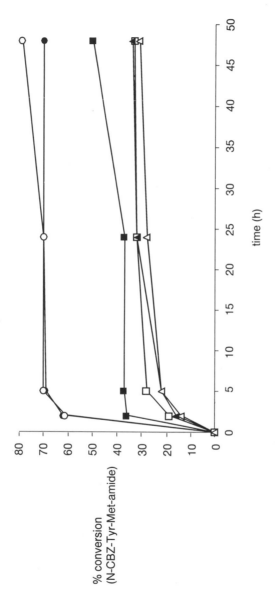

Figure 9.7 Kinetics of protease-catalysed synthesis of N-CBZ-Tyr-Met–NH$_2$ in eutectic media. △, subtilisin with no adjuvant; □, subtilisin with no adjuvant; ●, 2-methoxyethyl-acetate (MEA); ▲, subtilisin with 15% triethyleneglycol dimethyl ether (TGDME); ■, α-chymotrypsin with no adjuvant; ○, α-chymotrypsin with 15% MEA; ○, α-chymotrypsin with 15% TGDME.

of 2:1 ethanol:water reduced the reaction time from 40 h to 4 h, while at the same time improving the yield from 78% to 89%.

It should be pointed out that the final yield of product obtained in eutectic media is generally determined by the solidification of the reaction mixture. This is due to the formation of the peptide product, which at some point crystallizes out of the liquid eutectic phase, eventually leading to the complete solidification of the reaction mixture. This process and, as a result, the product yield can be controlled by the appropriate inclusion of adjuvants, depending on the amount and the nature of the solvent used (70). However, the solidification of the reaction mixture does not present a practical difficulty for product separation or the recovery and re-use of an immobilized enzyme. A mild solvent extraction of the reaction mixture is usually sufficient for recovery of the product, leaving behind an immobilized enzyme in an active form. For example, the subtilisin immobilized on Celite used for the synthesis of Phe–Leu–NH$_2$, was recycled four times (after extraction of the reaction mixture with ethanol, followed by filtration), giving successive dipeptide yields of 86, 82, 81 and 79%.

Enzymatic catalysis in heterogeneous eutectic mixtures is a new and interesting phenomenon which is currently being investigated in more detail. As a bioreaction system it offers a potentially high productivity achieved by dispensing with bulk solvents and allowing the transformation of substrates for which there is no satisfactory common solvent. Unfortunately, little is known at present about the general behaviour of enzymes in these systems. It would be interesting to investigate enzyme kinetics in heterogeneous eutectic media and to determine what factors affect it.

Similarly, no information is available on the enzyme specificity, which may well be somewhat compromised by the extremely high substrate concentrations present in eutectic media. In principle, the same methods of 'media engineering' and 'biocatalyst engineering', successfully used for the optimization of biocatalysis in organic solvents (8–12), should be equally applicable to enzymatic catalysis in eutectic mixtures. Indeed, the ability to manipulate the molecular interactions (hydrogen bonding, polar, hydrophobic, etc.), critical for eutectic formation by varying the structure of the substrates/adjuvants, has two important consequences. Firstly, it allows one to tailor the physico-chemical profiles of eutectic systems, and thereby control enzyme activity. Secondly, it should be possible to extend eutectic catalysis to other classes of biotransformations, providing that an adequate spectrum of functionalized substrates is available.

9.4 Design and implementation of continuous bioreactors

The third approach for improving the productivity of enzymatic transformations is the design of continuous bioreactors, which allows exploitation of the increased operational stability of enzymes in low-water systems compared to conventional aqueous media, as well as bringing the obvious benefits of efficient process control and automation. This point can be demonstrated by the enzymatic conversion of phospholipids to lysophospholipids. The large-scale batch hydrolysis of phospholipids by phospholipase A_2 suffers from operational difficulties and reduced productivity, which arise from the nature of the biocatalyst and the emulsified reaction medium employed (77). Furthermore, the very nature of the biotransformation makes it ill suited for implementation as a continuous process. On the other hand, despite the attraction of performing the reaction continuously in a homogeneous milieu, the low activity and stability of phospholipases in organic solvents have precluded the development of any such technology. In contrast, it has been shown that *sn-1* lysolecithins can be easily synthesized *via* an alternative route based on the lipase-catalysed transesterification of phospholipids in alcohols (78). The high stability of *M. miehei* lipase under the conditions employed allowed the construction of a continuous packed-bed reactor, which was operated for 1180 h with no appreciable loss of enzyme activity, and with a steady state space–time yield of up to 0.15 kg l^{-1} d^{-1} (78). Similar considerations of improved biocatalyst performance and increased productivity can be cited for the thermolysin-catalysed synthesis of aspartame precursor in ethyl acetate (79, 80).

The benefits of continuous processes are most pronounced for multistep/sequential syntheses, where the close integration of process modelling, control and automation allows complex biotransformations to be performed in, effectively, one step. However, the design of the synthetic strategy becomes of paramount importance since, in addition to the obvious concern of interfacing successive transformation modules with regard to their different solvent and water requirements and kinetic profiles, severe problems are encountered with protocols requiring intermediate steps for the manipulation of reactive functional groups. This issue is clearly illustrated by enzymatic peptide synthesis, where the traditional reliance on N-/C-protected substrates has necessitated the inclusion of corresponding deprotection steps prior to each coupling cycle (81). The difficulties faced when integrating chemical/enzymatic deprotection regimes into continuous processes has made it technologically and economically impractical to translate batch protocols into continuous bioreactors. Thus, the successful implementation of multi-step syntheses critically depends on avoiding intermediate modification steps and on simplification of the overall strategy.

The advantages of this approach, when successful, have been recently demonstrated by the design of an enzymatic oligopeptide synthesizer for the continuous production of a precursor of [Leu^5]-enkephalin pentapeptide (82). A simplified three-step approach was formulated, based on the synthesis of di- and tri-peptide intermediates and their subsequent direct coupling without the need for intermediate deprotection steps. This strategy was readily translated into a three-stage column bioreactor employing immobilized proteases as catalysts (82). When operated under steady-state conditions the bioreactor was capable of a space–time yield of up to 3.7 g l^{-1} d^{-1} (Figure 9.8).

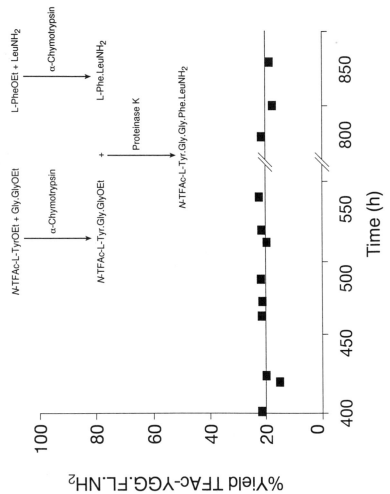

Figure 9.8 Three step synthesis of N-TFAc-enkephalinamide in a continuous bioreactor under steady-state conditions.

9.5 Conclusions

We trust that this review demonstrates some of the progress made in recent years regarding non-aqueous enzymatic catalysis, and more specifically the introduction of new systems and methods allowing improvements in the productivity of biotransformations. It is noticeable that industry, especially food manufacturers and producers of bulk products, tends to avoid using large quantities of solvents. Enzymatic solvent-free processing and the use of eutectic mixtures meet this requirement to a large extent, although the need for organic solvents cannot be completely eliminated (synthesis of substrates and downstream processing). An increased effort has been directed towards the design and optimization of continuous bioreactors containing immobilized enzymes. Since the cost of biocatalysts still remains a major obstacle for the commercialization of many laboratory-scale, enzyme-mediated syntheses, this area of research is likely to receive more attention in the future. It should be stressed that improved productivity is not the only driving force behind this research. The exciting possibility of combining several enzymatic reactions in a single robust module, which can perform a multi-step synthesis in effectively one step, is an industrially attractive prospect. Similarly, increased attention is being paid to the integration of biotransformations with downstream processing to facilitate recovery of the final products. Although the subject has not been discussed in this chapter, it is an important issue and the development of membrane- and supercritical fluids-based reactors may prove to be the key to achieving enhanced productivity and reducing the cost of biotransformations to make them more industrially attractive.

References

1. Davies, H.G., Green, R.H., Kelly, D.R. and Roberts, S.M. (1990) *CRC Crit. Rev. Biotechnol.*, **10**, 129–153.
2. Faber, K. and Franssen, M.C.R. (1993) *Trends Biotechnol.*, **11**, 461–469.
3. Klibanov, A.M. (1990) *Acc. Chem. Res.*, **23**, 114–120.
4. Margolin, A.L. (1993) *Enzyme Microb. Technol.*, **15**, 266–280.
5. Sih, C.J. and Wu, S.H. (1989) *Topics Stereochem.*, **19**, 63–89.
6. Vulfson, E.N. (1993) *Trends Food Sci. Technol.*, **4**, 209–215.
7. Wong, C.-H. and Whitesides, G.M. (1994) *Enzymes in Synthetic Organic Chemistry*. Tetrahedron Organic Chemistry Series, Vol. 12. Pergamon, Oxford.
8. Dordick, J.S. (1989) *Enzyme Microb. Technol.*, **11**, 1194–1211.
9. Halling, P. (1994) *Enzyme Microb. Technol.*, **16**, 178–206.
10. Klibanov, A.M. (1989) *Trends Biochem. Sci.*, **14**, 141–144.
11. Mattiasson, B. and Adlercreutz, P. (1991) *Trends Biotechnol.*, **9**, 394–398.
12. Zaks, A. and Russell, A.J. (1988) *J. Biotechnol.*, **8**, 259–270.
13. Khmelnitsky, Y.L., Levashov, A.V., Klyachko, N.L. and Martinek, K. (1988) *Enzyme Microb. Technol.*, **10**, 710–724.
14. Zaks, A. and Klibanov, A.M. (1988) *J. Biol. Chem.*, **263**, 3194–3198.
15. Deetz, J.S. and Rozzell, J.D. (1988) *Trends Biotechnol.*, **6**, 15–19.
16. Cambou, B. and Klibanov, A.M. (1984) *J. Amer. Chem. Soc.*, **106**, 2687–2692.

17. Wang, Y.F., Lalonde, J.J., Momongan, M., Bergbreiter, D.E. and Wong, C.-H. (1988) *J. Amer. Chem. Soc.*, **110**, 7200–7205.
18. Snijder, A.M., Vulfson, E.N. and Doddema, H.J. (1991) *Recl. Trav. Chim. Pays-Bas*, **110**, 226–230.
19. Macrae, A.R. and How, P. (1983) European patent application 0093602.
20. Nakamura, K., Yokomichi, H., Okisaka, K., Nishide, T., Kawahara, Y. and Nomura, S. (1987) European patent application 0257388.
21. Matsuo, T., Sawamura, N., Hashimoto, Y. and Hashida, W. (1979) UK patent application GB 2035359.
22. Casey, J. and Macrae, A. (1992) *INFORM*, **3**, 203–207.
23. Macrae, A.R. and Hammond, R.C. (1985) *Biotech. Genet. Eng. Rev.*, **3**, 193–217.
24. Arishima, T. and Mori, H. (1994) *Foods Food Ingred. J. Japan*, **159**, 110–119.
25. King, D.M. and Padley, F.B. (1990) European Patent No. 0 209 327 (Unilever plc).
26. Ikeda, M. (1992) *Toxicol. Lett.*, **64/65**, 191–201.
27. Hills, G.A., Macrae, A.R. and Poulina, R.R. (1990) European patent application 0383405.
28. Armstrong, D.W. and Brown, L.A. (1994) In *Bioprocess Production of Flavour, Fragrance and Colour Ingredients*, (ed. A. Gabelman), Wiley, New York, pp. 41–94.
29. Fonteyn, F., Blecker, C., Lognay, G., Marlier, M. and Severin, M. (1994) *Biotechnol. Lett.*, **16**, 693–696.
30. Stevenson, D.E., Stanley, R.A. and Furneaux, R.H. (1994) *Enzyme Microb. Technol.*, **16**, 478–484.
31. Welsh, F.W., Murray, W.D. and Williams, R.E. (1989) *Crit. Rev. Biotechnol.*, **2**, 7–10.
32. Yamane, T., Suzuki, T. and Hoshino, T. (1993) *J. Amer. Oil Chem. Soc.*, **70**, 1285–1287.
33. Yamane, T., Suzuki, T., Sahashi, Y., Vikersveen, L. and Hoshino, T. (1992) *J. Amer. Oil Chem. Soc.*, **69**, 1104–1107.
34. Gross, A. (1991) *Food Technol.*, **45**, 96–100.
35. Young, V. (1990) *Lipid Technol.*, **2**, 7–10.
36. Therisod, M. and Klibanov, A.M. (1986) *J. Amer. Chem. Soc.*, **108**, 5638–5640.
37. Therisod, M. and Klibanov, A.M. (1987) *J. Amer. Chem. Soc.*, **109**, 3977–3981.
38. Riva, S. and Secundo, F. (1990) *Chimicaoggi*, **6**, 9–16.
39. Drueckhammer, D.G., Hennen, W.J., Pederson, R.L., Barbas, C.F. III, Gautheron, C.M., Krach, T. and Wong, C.-H. (1991) *Synthesis*, 499–525.
40. Vulfson, E.N. (1992) In *Surfactants in Lipid Chemistry* (ed. J. Tyman), Royal Society of Chemistry, Cambridge, pp. 16–37.
41. Chopineau, J., McCafferty, F.D., Therisod, M. and Klibanov, A.M. (1988) *Biotechnol. Bioeng.*, **31**, 209–214.
42. Janssen, A.E.M., Klabbers, C., Franssen, M.C.R. and Van't Riet, K. (1991) *Enzyme Microb. Technol.*, **13**, 565–572.
43. Khaled, N., Montet, D., Pina, M. and Graille, J. (1991) *Biotechnol. Lett.*, **13**, 167–172.
44. Schlotterbeck, A., Lang, S., Wray, V. and Wagner, F. (1993) *Biotechnol. Lett.*, **15**, 61–64.
45. Oguntimein, G.B., Erdmann, H. and Schmid, R.D. (1993) *Biotechnol. Lett.*, **15**, 175–180.
46. Adelhorst, K., Bjorkling, F., Godtfredsen, S.E. and Kirk, O. (1990) *Synthesis*, 112–115.
47. Fregapane, G., Sarney, D.B. and Vulfson, E.N. (1991) *Enzyme Microb. Technol.*, **13**, 796–800.
48. Fregapane, G., Sarney, D.B., Greenberg, S.G., Knight, D.J. and Vulfson, E.N. (1994) *J. Amer. Oil Chem. Soc.*, **71**, 87–91.
49. Fregapane, G., Sarney, D.B. and Vulfson, E.N. (1994) *Biocatalysis*, **11**, 9–18.
50. Sarney, D.B., Kapeller, H., Fregapane, G. and Vulfson, E.N. (1994) *J. Amer. Oil Chem. Soc.*, **71**, 711–714.
51. Bjorkling, F., Godtfredsen, S.E. and Kirk, O. (1991) *Trends Biotech.*, **9**, 360–363.
52. McNeill, G.P., Shimizu, S. and Yamane, T. (1991) *J. Amer. Oil Chem. Soc.*, **68**, 6–10.
53. McNeill, G.P. and Yamane, T. (1991) *J. Amer. Oil Chem. Soc.*, **68**, 1–5.
54. McNeill, G.P., Borowitz, D. and Berger, R.G. (1992) *J. Amer. Oil Chem. Soc.*, **69**, 1098–1103.
55. Yamane, T., Kang, S.T., Kawahara, K. and Koizumi, Y. (1994) *J. Amer. Oil Chem. Soc.*, **71**, 339–342.
56. Cabral, J.M.S., Best, D., Boross, L. and Tramper, J. (eds) (1994) *Applied Biocatalysis*, Harwood Academic, Reading.
57. Halling, P. (1987) *Biotechnol. Adv.*, **5**, 7–84.
58. Halling, P. (1994) *Enzyme Microb. Technol.*, **16**, 178–206.

59. Goldberg, M., Thomas, D. and Legoy, M. (1990) *Enzyme Microb. Technol.*, **12**, 976–981.
60. Boyer, J.L., Gilot, B. and Guiraud, R. (1990) *Appl. Microb. Biotechnol.*, **33**, 372–340.
61. Goldberg, M., Thomas, D. and Legoy, M.-D. (1990) *Enzyme Microb. Technol.*, **12**, 976–983.
62. Turner, N.A., Duchateau, D.B. and Vulfson, E.N. (1995) *Biotechnol. Lett.*, **17**, 371–376.
63. Volkin, D.B., Staubli, A., Langer, R. and Klibanov, A.M. (1991) *Biotechnol. Bioeng.*, **37**, 843–853.
64. Khan, S.A., Halling, P. and Bell, G. (1990) *Enzyme Microb. Technol.*, **12**, 453–458.
65. Prior, B.A. (1979) *J. Food Sci.*, **42**, 668–674.
66. Brzozowski, A.M. *et al.* (1991) *Nature*, **351**, 491–494.
67. Tilbeurgh, H.V., Egloff, M.P., Martinez, C., Rugani, N., Verger, R. and Cambillau, C. (1993) *Nature*, **362**, 814–820.
68. Gill, I. and Vulfson, E.N. (1993) *J. Amer. Chem. Soc.*, **115**, 3348–3349.
69. López-Fandiño, R., Gill, I. and Vulfson, E.N. (1994) *Biotechnol. Bioeng.*, **43**, 1024–1030.
70. Gill, I. and Vulfson, E.N. (1994) *Trends Biotechnol.*, **12**, 118–122.
71. López-Fandiño, R., Gill, I. and Vulfson, E.N. (1994) *Biotechnol. Bioeng.*, **43**, 1016–1023.
72. Jorba, X., Gill, I. and Vulfson, E.N. (1995) *J. Agric. Food Chem.* (submitted).
73. Halling, P., Eichorn, U., Kuhl, P. and Jakubke, H.-D. (1995) *Enzyme Microb. Technol.*, in press.
74. Kuhl, P., Eichorn, U. and Jakubke, H.-D. (1995) *Biotechnol. Bioeng.*, **45**, 276–278.
75. Cerovsky, V. (1992) *Biotechnol. Lett.*, **6**, 155–160.
76. Fischer, A., Bommarius, A.S., Drauz, K. and Wandrey, C. (1994) *Biocatalysis*, **8**, 289–307.
77. Novo-Nordisk Enzyme Process Division (1992) *Lecithase: Product Specification Sheet*, Bagsvaerd, Denmark.
78. Sarney, D.B., Fregapane, G. and Vulfson, E.N. (1994) *J. Amer. Oil Chem. Soc.*, **71**, 93–96.
79. Cheetham, P.S.J. (1994) In *Applied Biocatalysis* (eds J.M.S. Cabral *et al.*) Harwood Academic, Reading, pp. 79–108.
80. Oyama, K. (1987) In *Biocatalysis in Organic Media* (eds C. Laane, J. Tramper and M.D. Lilly), Elsevier Science, Amsterdam, pp. 209–224.
81. Kullmann, W. (1987) *Enzymatic Peptide Synthesis*, CRC Press, Boca Raton, Florida.
82. Richards, A.O., Gill, I. and Vulfson, E.N. (1993) *Enzyme Microb. Technol.*, **15**, 928–935.

10 Large-scale enzymatic conversions in non-aqueous media

R.A. SHELDON

10.1 Introduction

During the last decade the organic chemist's and the chemical industry's perception of enzymes has changed dramatically. The traditional view was based on the conventional wisdom that the use of enzymes is restricted to aqueous media and temperatures close to ambient. Moreover, enzymes were perceived as being very expensive, fragile catalysts that present substantial problems in handling and downstream processing, i.e. hardly the sort of reagent one would consider for use in industrial organic synthesis. Conventional wisdom also held that enzymes are very substrate- and reaction-specific (the one enzyme–one reaction–one substrate paradigm).

The pioneering studies of Klibanov and coworkers, from 1984 onwards (1–3), on the use of enzymes in organic solvents containing little or no added water, constitute a turning point in this traditional organic chemist's perception of enzymes. Not only were they able to function perfectly well in organic solvents, many enzymes proved to be more thermally stable in organic media than in water. Since the majority of organic compounds are readily soluble in organic solvents and sparingly soluble in water, organic chemists quickly recognized the synthetic potential of enzymes. The intervening decade, following these seminal studies, has witnessed an explosive growth in the applications of enzymes in organic synthesis (4–11), many of which involve reactions in non-aqueous media (12–14). During the same period the industrial use of enzymes has proliferated (15–20). The purpose of this chapter is to review the current status of the large-scale applications of enzymes in non-aqueous media. Our definition of the latter encompasses hydrolytic processes when the water is present as a reactant rather than as a solvent. We have also included enzymatic conversions which in our opinion have definite industrial potential but have not yet been commercialized.

10.2 General aspects

10.2.1 Why enzymatic processes?

A major driving force in the application of enzymatic processes in the chemical process industries is environmental concern and regulation. There is a rapidly growing demand for cleaner, alternative technologies that produce less waste, e.g. high-atom utilization, low-salt processes (21, 22), which are less energy-intensive and avoid the use of toxic reagents and solvents.

The characteristic features of enzymatic processes—high selectivities under mild reaction conditions, often without the need for protection and deprotection of functional groups—make them highly attractive from an environmental view-point.

Similarly, there is a growing demand for products that are more targeted in their action and more environmentally benign, i.e. have limited environmental impact. This manifests itself, for example, in the marked trend towards the use of enantiomerically pure pharmaceuticals and agrochemicals (23, 24), and polymers that are biodegradable. Many biologically active substances, e.g. pharmaceuticals, herbicides, fungicides, insecticides and flavors and fragrances, are chiral molecules. In general, only one of the enantiomers (the eutomer) of a racemic mixture is responsible for the desired biological effect. At best the other isomer (the distomer) constitutes unnecessary 'isomeric ballast', but more often than not it inhibits the desired effect of the eutomer and/or exhibits adverse side effects. The high stereoselectivities inherent in many enzymatic transformations make them eminently suitable for the synthesis of pure enantiomers.

In the flavor and fragrance industry, comprising food ingredients and personal-care products, there is an increasing consumer demand for natural, 'environmentally friendly' and 'healthy' products, preferably made from renewable resources (25). Natural flavors and fragrances are more readily accepted by the customer and, hence, have significant labeling and marketing advantages. Stringent regulations require that for a product to qualify for the approbrium 'natural' it must be synthesized from natural raw materials using only natural methods. Enzymatic processes, which utilize 'natural' catalysts, generally fall into this category. Nature-identical materials, in contrast, have the same structure as the natural product but are obtained by synthesis or from natural raw materials through chemical processes. A third category, 'synthetic' flavors and fragrances, applies to substances that are not naturally occurring.

10.2.2 Why non-aqueous media?

Before discussing the merits of using organic solvents for enzymatic processes, it is worth while considering recent trends in the use of organic solvents in chemical processes in general. Environmental legislation is severely restricting the use of many traditional organic solvents, such as chlorinated hydrocarbons. Indeed, so many of the solvents favored by organic chemists are now on the environmental blacklist that the whole question of solvents in organic synthesis has to be re-examined. Not only do organic chemists generally use too much solvent they, more often than not, choose the wrong ones. In the first place, the best solvent is no solvent. If the substrate, or the product, is a liquid why not use it as the solvent (diluent)? Thus, many of the enzymatic processes discussed later

Table 10.1 Potential advantages of enzymatic processes in non-aqueous media

1. Increased solubility of apolar substrates.
2. Enhanced thermostability.
3. Easier product recovery from low boiling solvents.
4. Immobilization is often unnecessary as the enzyme is insoluble.
5. Easier recovery of the enzyme by simple filtration or centrifugation.
6. If immobilization is required for better performance, simple adsorption onto non-porous surfaces may be sufficient. Enzymes do not desorb from these surfaces in non-aqueous media.
7. Alteration in substrate specificity.
8. Suppression of water-dependent side reactions, e.g. nonenzymatic blank reaction in enzymatic hydrocyanations of aldehydes which leads to racemic product (see later).
9. Shifting thermodynamic equilibria to favor synthesis over hydrolysis, e.g. esterification and peptide formation.
10. The possibility of carrying out reactions with other nucleophiles which would not be feasible in water, e.g. transesterification and ester ammoniolysis.
11. Easier integration of enzymatic with conventional chemical steps in a multistep synthesis (chemoenzymatic processes).

can be carried out on the neat substrate, and probably should be. If a solvent is needed it should preferably be water or, possibly, supercritical CO_2 (26). Indeed, an important advantage of biocatalysis is that reactions which would normally be carried out in, say, chlorinated hydrocarbons can be performed in water. Hence, we conclude that in order for enzymatic processes in non-aqueous media to be attractive they must offer definite advantages that are unattainable in water. In this context it is worth noting that the use of an organic solvent may disqualify a product prepared by an enzymatic process for the 'natural' label (see above).

Despite the aforementioned caveats, the use of enzymes in non-aqueous media can offer significant processing advantages (12). Several of these advantages (Table 10.1) are only realized when the bulk water phase is eliminated, i.e. in monophasic organic solvents where the enzyme is not in direct contact with a bulk aqueous phase.

Obviously an important advantage can accrue from the shifting of thermo-dynamic equilibria in favor of synthesis over hydrolysis. However, we note that the same result can sometimes be achieved by precipitation of a water-insoluble product from an aqueous solution. All other things being equal, in such cases the latter would probably be the method of choice. For example, the artificial sweetener, aspartame, can be prepared via regio- and stereoselective coupling of protected L-aspartic acid with racemic phenylalanine methyl ester (Figure 10.1) using immobilized thermolysin in ethyl acetate as solvent (27). However, the commercial process, operated by DSM-Tosoh, actually employs the soluble enzyme in aqueous solution, whereby the product precipitates in virtually quantitative yield (28).

10.2.3 Factors influencing industrial utility

In addition to the environment-driven, market-pull factors mentioned above, several technology-push factors have been instrumental in stimulating the appli-

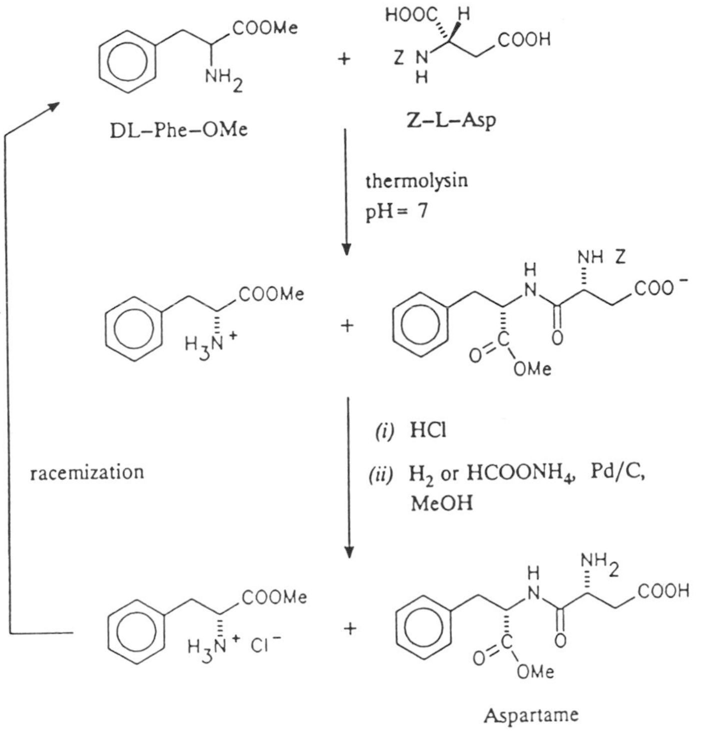

Figure 10.1 Preparation of aspartame using immobilized thermolysin.

cation of enzymes in the chemical industry. Advances in genetic engineering have provided the basis for more efficient, i.e. cheaper, production of enzymes. When a useful enzyme is identified the genetic material encoding for its structure is transferred into a suitable host microbe, e.g. *Bacillus* bacteria and *Aspergillus* fungi, which grows quickly, is safe to handle and produces high yields of the enzyme. Between 40 and 50% of the enzymes currently manufactured by Novo Nordisk, for example, are produced in such host microbes, according to a recent report (29).

By using protein-engineering techniques, such as site-directed mutagenesis (30–34), enzymes can be 'redesigned' to improve their performance or endow them with new characteristics such as altered substrate specificity, pH optimum and tolerance to elevated temperatures and organic solvents. Moreover, advances in computer-aided molecular modeling of enzyme structures have provided the basis for predicting the changes needed in the redesigning processes. Once a pure crystalline sample of an enzyme is available and its amino acid sequence is known, its three-dimensional structure can be determined by high-resolution X-ray crystallography. This information is then used in computer-aided modeling to predict the structural changes resulting from replacement of one or more amino acids. Finally, the gene is manipulated so that it encodes for the redesigned enzyme. An alternative approach involves post-translational modification of enzymes (35).

Improved techniques for the stabilization of enzymes, e.g. via immobilization, subsequent to their production have also played an important role in stimulating their large-scale application. In this context it is worth mentioning the recently developed technique for producing crosslinked enzyme crystals by reacting small enzyme crystals with glutaraldehyde (36). These so-called CLECs are much more robust than conventional immobilized enzymes.

10.2.4 The industrial enzyme market

Enzymes are classified according to the International Enzyme Commission into six major classes (Table 10.2). Although about 2500 enzymes have been classified only 300–350 are commercially available and only a small proportion of these are actually used in significant quantities.

The world market for industrial enzymes is estimated at around $900 million in 1990 (15), of which 74% is accounted for by three major applications: household detergents, dairy products and starch processing (Table 10.3). It is not surprising, therefore, that enzyme manufacturers devote much more effort to developing better enzymes for these applications than for organic synthesis. Moreover, we note that these and other important applications of enzymes, such as in leather processing, textiles and paper manufacture, all involve the use of enzymes in aqueous media and, hence, fall outside the scope of this book. Nevertheless, applications of enzymes in organic chemical processing can benefit from the development of enzymes for other industries. Indeed, many of the enzymes currently used in organic synthesis were developed for other applications, e.g. *Mucor miehei* lipase for cheesemaking and *Candida antarctica* lipase for detergents.

Applications of biotransformations in organic chemicals manufacture are largely confined to the use of enzymes that do not require stoichiometric quantities of expensive cofactors, e.g. hydrolases, lyases and, to a lesser extent,

Table 10.2 Classification of enzymes

Enzyme class	Known	Available	Reaction type(s)
1 Oxidoreductases	650	90	Redox reactions
2 Transferases	720	90	Group transfer reactions (e.g. methyl, glycosyl, phosphate)
3 Hydrolases	636	125	Hydrolysis of esters, amides, epoxides, etc.
4 Lyases	255	35	Addition to or formation of C=C, C=N and C=O bonds
5 Isomerases	126	6	Racemizations, epimerizations, etc.
6 Ligases	83		Formation of C–C, C–N, C–O and C–S bonds with consumption of ATP

Table 10.3 Industrial enzyme market

Conversion/enzyme type	Sales (1994) ($Million)	Main applications
Protein hydrolysis		
Alkaline proteases (e.g. subtilisin)	370	Detergents
Rennets (calf, microbial)	105	Cheesemaking, dairy
Plant proteases (papain, bromelain)	7	Food processing
Pancreatic proteases	10	Leather processing
	492	
Carbohydrate conversion		
Amylases	167	Starch processing, baking, brewing, textile, detergents
Pectinases	30	Beverages
Cellulases	153	Detergents, textile, baking
Xylanases/β-glucanases	45	Animal feed processing, brewing
Glucose isomerase	20	Food and beverages
Pullulanases	5	Starch
Phytases	12	Animal feed processing
Lactases	9	Dairy
	441	
Fat hydrolysis		
Lipases	70	Fat processing, dairy, detergents, organic synthesis
Total	1003	

isomerases. Cofactor-consuming enzymes, e.g. many oxidoreductases, are generally used in fermentation processes where continuous regeneration of the cofactor from a cheap carbon source (e.g. glucose) is provided for. Transferases and ligases generally require complex cofactors and have found little use in organic synthesis, although the latter are of immense importance in genetic engineering.

The synthesis of enantiomerically pure amino acids (natural and unnatural) constitutes by far the largest application of enzymes in organic chemicals manufacture (38). Most of these processes involve the use of various types of amidases and, since amino acids are generally water-soluble, are performed in aqueous media. On the other hand, the majority of scientific publications on the use of enzymes in organic synthesis concerns reactions with lipases and esterases, often in non-aqueous media. Many lyases and certain oxidoreductases are also eminently suited to operation in non-aqueous media (see later).

10.2.5 Whole cells versus isolated enzymes

Biocatalytic transformations can be carried out using whole microbial cells or isolated cell-free enzymes. Whole cells are often used in the case of labile, intracellular enzymes and/or cofactor-dependent enzymes. Although both types can, in principle, be used in non-aqueous media, e.g. baker's yeast has been used in benzene as solvent (39), most of the examples we shall discuss involve the use of cell-free enzymes. They have the advantage of avoiding contamination from other enzymes present in microbial cells. Fermentations, i.e. reactions carried out with growing microbial cells, fall outside the scope of this review.

Having set the scene for the large-scale applications of enzymes in non-aqueous media, we shall now describe specific areas where they have been applied or, in some cases, where they appear to offer industrial potential.

10.3 Modification of oils and fats

The most important application, from a tonnage viewpoint, is undoubtedly in the modification of oils and fats (20).

10.3.1 Lipases in the oleochemical industry

The scope for the application of enzymes in the oleochemical industry is enormous. The annual world-wide production of oils and fats is approximately 60 million tonnes (20). Many of the processes that are used in the oleochemical industry are energy-intensive, requiring high temperatures (often above 200°C) and pressures. The resulting products generally need to be redistilled to remove impurities formed via thermal degradation. Moreover, highly unsaturated oils are thermally labile and cannot be processed without prior hydrogenation.

Enzymatic processes can provide significant advantages such as energy savings and product quality improvement. Moreover, the regioselectivity of lipase-catalyzed reactions of triglycerides (see below) is an important advantage compared to chemical methods.

The natural function of lipases is to catalyze the (reversible) hydrolysis of fats, i.e. triglycerides of long-chain fatty acids. The latter are water-insoluble and a characteristic feature of lipases is their ability to catalyze the hydrolysis of ester bonds at the interface between the water-insoluble substrate and the aqueous phase containing the enzyme. At very low water levels (e.g. < 0.1%) other reactions, such as transesterification (alcoholysis, glycerolysis and acidolysis) and interesterification become possible (Figure 10.2).

Microbial lipases can be divided into three categories according to their specificity. The first group exhibits no specificity with regard to the three acyl groups in the triglyceride. The second group, the 1,3-specific lipases, catalyzes selective hydrolysis at the 1 and 3 positions affording a mixture of free fatty acids and the 2-monoglyceride. 1,3-Specificity is common among microbial lipases, e.g. *Mucor miehei* lipase, and results from the inability of the sterically hindered 2-glycerol esters to fit into the active site of the lipase. The third group of lipases is specific for a particular type of acyl group. For example, *Geotrichum candidum* lipase exhibits a marked preference for unsaturated fatty acid residues containing a *cis* double bond at the 9 position.

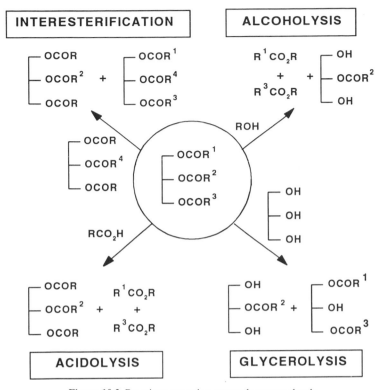

Figure 10.2 Reactions occurring at very low water levels.

The properties of fats and oils, and hence their commercial value, depend on the fatty acid composition of the triglyceride structure. Traditionally, upgrading of low-quality fats was achieved by blending natural fats and oils with different triglyceride compositions or by alkali-catalyzed transesterification. However, the latter method affords mixtures of triglycerides with random fatty acid compositions, which generally precludes the formation of products with the desired physico-chemical properties.

The potential of 1,3-specific lipases for effecting selective transformations of triglycerides was recognized more then ten years ago by Unilever (40–42) and Fuji Oil (43) scientists. The relatively high price of lipases was, however, a major obstacle to their commercialization. In recent years advances in biotechnology have led to significant reductions in price and improvements in performance of microbial lipases. *Mucor miehei* lipase (Novo's Lipozyme™), for example, has a half-life at 60°C of 1800 h when immobilized on an anion-exchange resin (19). Hence, there is currently a marked trend towards the replacement of conventional chemical processes by more selective enzymatic alternatives.

For example, Unilever (44) and Fuji Oil have commercialized the production of cocoa-butter substitute by lipase-catalyzed conversion of inexpensive palm oil mid-fraction. The latter consists mainly of 1,3-dipalmitoyl-2-mono-oleine (POP). Acidolysis (reaction 10.1) or interesterification (reaction 10.2) with stearic acid or tristearin, respectively, in the presence of a 1,3-specific lipase, affords a product enriched in the more valuable components of cocoa butter, 1(3)-palmitoyl-3(1)-stearoyl-2-mono-oleine (POSt) and 1,3-distearoyl-2-mono-oleine (StOSt).

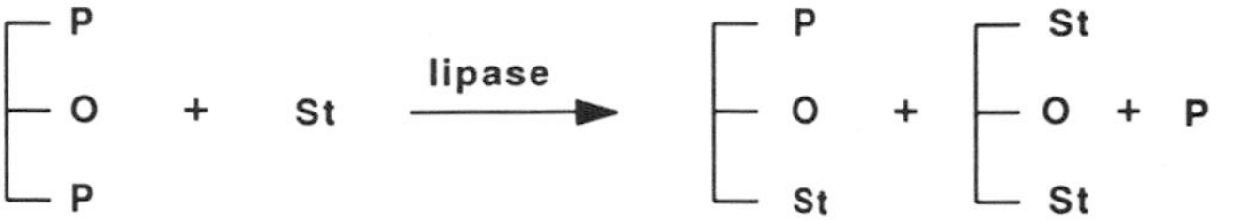

$$
\begin{bmatrix} P \\ O \\ P \end{bmatrix} + \text{St} \xrightarrow{\text{lipase}} \begin{bmatrix} P \\ O \\ St \end{bmatrix} + \begin{bmatrix} St \\ O \\ St \end{bmatrix} + P \qquad (10.1)
$$

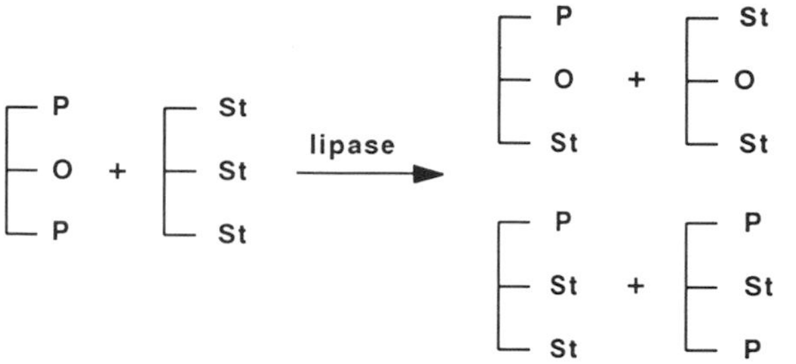

$$
\begin{bmatrix} P \\ O \\ P \end{bmatrix} + \begin{bmatrix} St \\ St \\ St \end{bmatrix} \xrightarrow{\text{lipase}} \begin{bmatrix} P \\ O \\ St \end{bmatrix} + \begin{bmatrix} St \\ O \\ St \end{bmatrix}
$$
$$
\qquad\qquad \begin{bmatrix} P \\ St \\ St \end{bmatrix} + \begin{bmatrix} P \\ St \\ P \end{bmatrix} \qquad (10.2)
$$

P = palmitate ; St = stearate ; O = oleate

In order to minimize non-enzymatic 1,2-acyl migration the process is carried out with an immobilized lipase in fixed-bed reactors with short residence times.

10.3.2 Structured lipids for nutrition

The same approach can also be applied to the synthesis of structured tri-glycerides having valuable nutritional or dietary properties, e.g. milk-fat substitutes. In contrast to plant oils, triglycerides of human milk fat contain palmitic acid almost exclusively in the 2-position, with 1,3-dioleoyl-2-palmitoylglycerol as the major component. The latter is readily obtained by acidolysis of palm top-fraction, which is rich in tripalmitin, with oleic acid in the presence of a 1,3-specific lipase (19, 20, 45).

Similarly, triglycerides with polyunsaturated fatty acids (PUFA) in the 2-position and medium-chain fatty acids (MCFA; mainly C_8 and C_{10}) in the 1 and 3 positions can be produced by 1,3-specific lipase-catalyzed acidolysis. These triglycerides provide rapid delivery of energy via oxidation of the more hydrophilic MCFAs, while at the same time providing an adequate supply of an essential fatty acid from the remaining 2-monoglyceride (46). Such triglycerides are finding applications in clinical nutrition, e..g. as a concentrated form of calories for patients with pancreatic deficiency.

Increasing interest in the medical and nutritional significance of PUFAs is leading to new applications of lipases (47, 48) in the modification of oils rich in PUFAs, such as arachidonic, eicosapentaenoic and docosahexenoic fatty acids. Commercial interest is focused on the development of mild methods for the production of these labile molecules in a form suitable for incorporation into dietary formulations, skin creams, etc.

10.3.3 Fatty acid amides via enzymatic ammoniolysis

The recently discovered (49, 50) enzymatic ammoniolysis of esters to the corresponding amides, which is catalyzed by certain lipases such as *Candida antarctica* and *Humicola* lipases, has been successfully applied (51) to the synthesis of fatty acid amides by ammoniolysis of triglycerides (reaction 10.3) or fatty acid esters (reaction 10.4) in a non-aqueous medium. The latter can be synthesized by esterificaztion of the free fatty acid using the same enzymes.

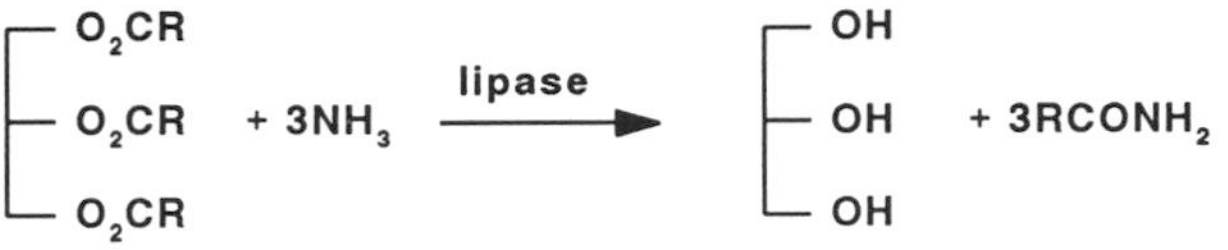

$$
\begin{array}{l}
\left[\begin{array}{l} O_2CR \\ O_2CR \\ O_2CR \end{array}\right. + 3NH_3 \xrightarrow{\text{lipase}} \left[\begin{array}{l} OH \\ OH \\ OH \end{array}\right. + 3RCONH_2
\end{array} \qquad (10.3)
$$

$$
R^1 CO_2 R^2 + NH_3 \xrightarrow{\text{lipase}} R^1 CONH_2 + R^2 OH \qquad (10.4)
$$

Fatty acid amides find a variety of applications, e.g. in the production of water repellents for textiles and as additives in lubricating oils, coatings and polyolefins. Traditionally they are manufactured by high temperature (200°C) reaction of ammonia with the fatty acid or its methyl ester. Enzymatic methods can, in principle, afford substantial energy savings and product quality improvement. Enzymatic synthesis could be particularly useful for the synthesis of more labile amides of unsaturated fatty acids, e.g. erucamide $(CH_3(CH_2)_7CH=CH(CH_2)_{11}CONH_2)$.

10.3.4 Bioesters as ingredients of personal care products

Unilever has recently commercialized the production of simple esters of fatty acids, e.g. isopropyl palmitate, isopropyl myristate and 2-ethylhexyl palmitate for use as emollients in personal-care products such as skin creams, bath oils, etc. Immobilized *Mucor miehei* lipase is used as the catalyst for the esterification under solvent-free conditions, with removal of the water formed in the process by vacuum distillation (52, 53) (reaction 10.5). The enzymatic process apparently affords products of much higher quality, thus obviating the need for downstream refining. The products are marketed under the name 'Bioester'.

$$R^1 CO_2H + R^2 OH \xrightarrow{\text{lipase}} R^1CO_2R^2 + H_2O$$

$$R^1 = n - C_{14}H_{29} \; ; \; n - C_{16}H_{33} \tag{10.5}$$

$$R^2 = (CH_3)_2CH - \; ; \; CH_3(CH_2)_3CH(C_2H_5)CH_2-$$

Similarly, wax esters (esters of fatty acids with fatty alcohols) are also used in personal-care products and are being manufactured by Croda Universal Ltd using *Candida rugosa* lipase in a batch reactor. The slightly higher costs of the enzymatic process compared to the conventional method are offset by the improved product quality.

10.4 Regioselective acylation of carbohydrates and steroids

Growing consumer demand for 'green' products has, in recent years, focused attention on the utilization of carbohydrates as raw materials for specialty chemicals. Carbohydrates are abundantly available, renewable resources and carbohydrate-based products are, generally speaking, non-toxic, biocompatible and biodegradable. Moreover, carbohydrate-based products would qualify for the accolade 'natural', assuming that the methods used for their production are natural, i.e. enzymatic. This is particularly important in food and personal-care applications. Another advantage of enzymatic processes for carbohydrate conversions accrues from the fact that carbohydrates are polyhydroxy compounds. Hence, the high regioselectivity of many enzymatic processes can provide an important advantage over less selective more conventional chemical processes.

Examples of applications where these various attributes of carbohydrate raw materials and enzymatic processes converge are detergents and emulsifiers. Interestingly, these applications also constitute a convergence of the two major classes of renewable raw materials: carbohydrates and oleochemicals. Moreover, since many of these applications involve regioselective acylation of carbohydrates this also dictates processing in non-aqueous media.

10.4.1 Carbohydrate-based surfactants by lipase-catalyzed acylation of glucose and its derivatives

Alkyl polyglucosides (APGs), derived from starch, are already produced on a commercial scale by Henkel (54) for use in detergents. There is also considerable interest in the selective acylation of alkylglucosides and underivatized mono- and disaccharides for use, *inter alia*, in detergent formulations (55, 56). Initial studies of enzymatic acylations of carbohydrates were carried out with lipases or proteases in polar, aprotic solvents such as pyridine (57) and dimethylformamide (58) because of the low solubility of carbohydrates in nonpolar solvents. Such solvents are, however, unattractive for large-scale reactions and unacceptable for the manufacture of food additives and personal-care products (see later).

Björkling and coworkers at Novo (59, 60) showed that immobilized *Candida antarctica* lipase catalyzes the regioselective 6-*O*-acylation of ethyl glucose with fatty acids (reaction 10.6). Yields exceeding 90% were obtained by performing the reaction at 70°C, in a solvent-free system, and continuously removing the water of reaction by vacuum distillation.

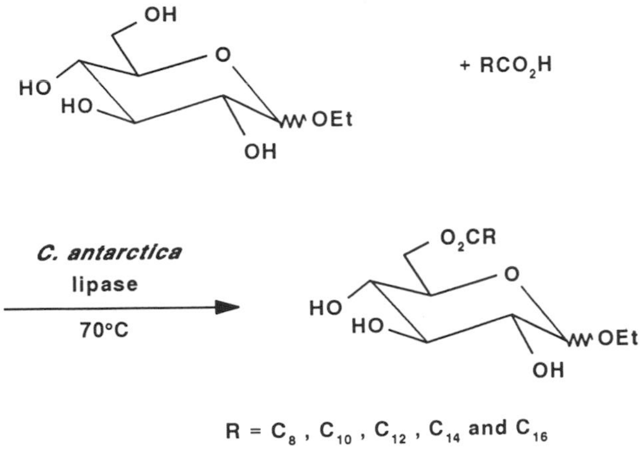

$$R = C_8, C_{10}, C_{12}, C_{14} \text{ and } C_{16}$$

The products exhibit surfactant properties and have potential applications in cosmetics and household detergents and as degreasing agents for metals, electronics and leather (19). The process has been developed to pilot-plant scale by Novo (19).

Similarly, alkyl glucosides, galactosides and fructosides are regioselectively acylated by transesterification with an ethyl ester in the presence of an immobilized *Candida antarctica* lipase (61). An excess of the ester substrate was used as the solvent and zeolite CaA was added to remove the ethanol formed in the reaction and drive the equilibrium to the right. The zeolite also removed the last traces of (free) water in the system. Reaction of ethyl acrylate with octyl-α-D-glucopyranoside afforded the mono-6-*O*-acrylate in 99% selectivity at 99% conversion (reaction 10.7).

Thus, even in the presence of a large excess of the acyl donor the monoester is formed virtually exclusively. Other lipases e.g. from *Candida rugosa*, *Mucor miehei* and porcine pancreas gave much inferior conversions and selectivities. The product is of interest as a raw material for synthesizing water-soluble polyacrylates (see section 10.7).

tert-Butanol was also shown to be a good solvent for lipase-catalyzed acylations of carbohydrates (61, 62). Due to its steric bulk, *tert*-butanol does not act as an acyl donor in lipase-catalyzed reactions. For example, glucose underwent slow regioselective 6-monoacylation by reaction with excess ethyl

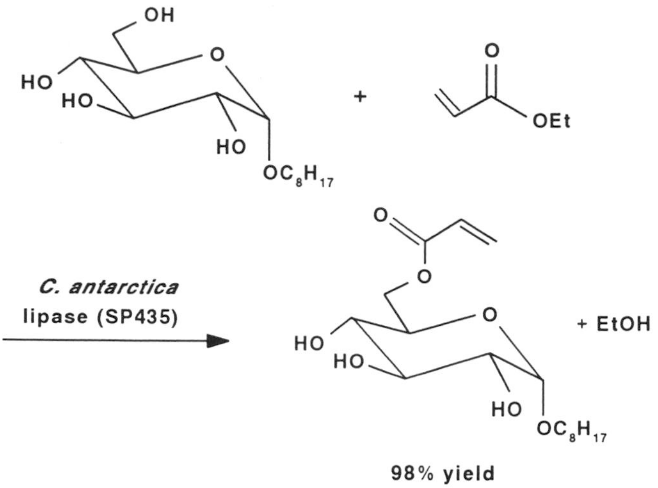

(10.7)

C. antarctica
lipase (SP435)

98% yield

butanoate in the presence of *Candida antarctica* (SP435) lipase and zeolite CaA at 40°C (reaction 10.8) (56, 62). The formation of di- and tri-esters was also suppressed in *tert*-butanol, compared to excess ethyl butanoate as solvent.

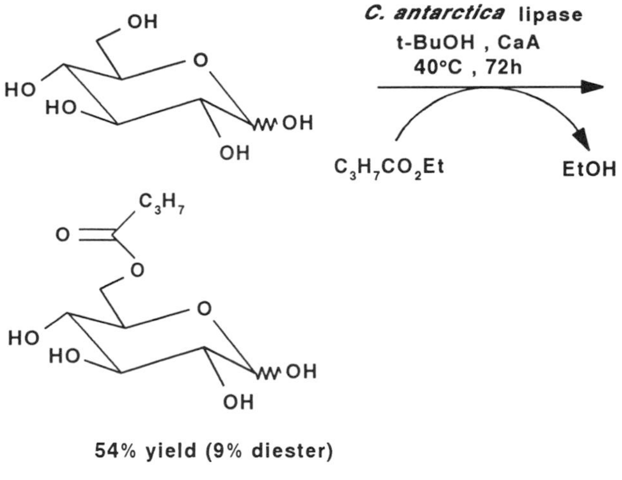

(10.8)

54% yield (9% diester)

10.4.2 *Sucrose fatty acid esters by enzymatic acylation*

Sucrose monoesters of fatty acids have recently attracted interest as emulsifiers in food and personal-care applications. They are currently manufactured by Sisterna B.V., a joint venture of Suiker Unie and Dai-ichi Kogyo Seiyaku, via a chemical procedure. An enzymatic alternative could have significant selectivity and labeling advantages. Unfortunately, acylation of sucrose is even more difficult than that of glucose, due to its extreme low solubility in organic media.

Enzymatic acylation of sucrose was first achieved using a protease (subtilisin) together with an activated acyl donor, trichloroethyl butanoate, in dimethylformamide (58). Obviously this is unattractive for large-scale production. Recently, it has been shown (62) that *Candida antarctica* lipase catalyzes the slow acylation of sucrose by ethyl butanoate in refluxing *tert*-butanol. A roughly equimolar mixture of the 6- and 6′-monoesters was obtained in 66% yield after 48 h (reaction 10.9). With ethyl dodecanoate, the monoesters were formed in 32% yield in 72 h.

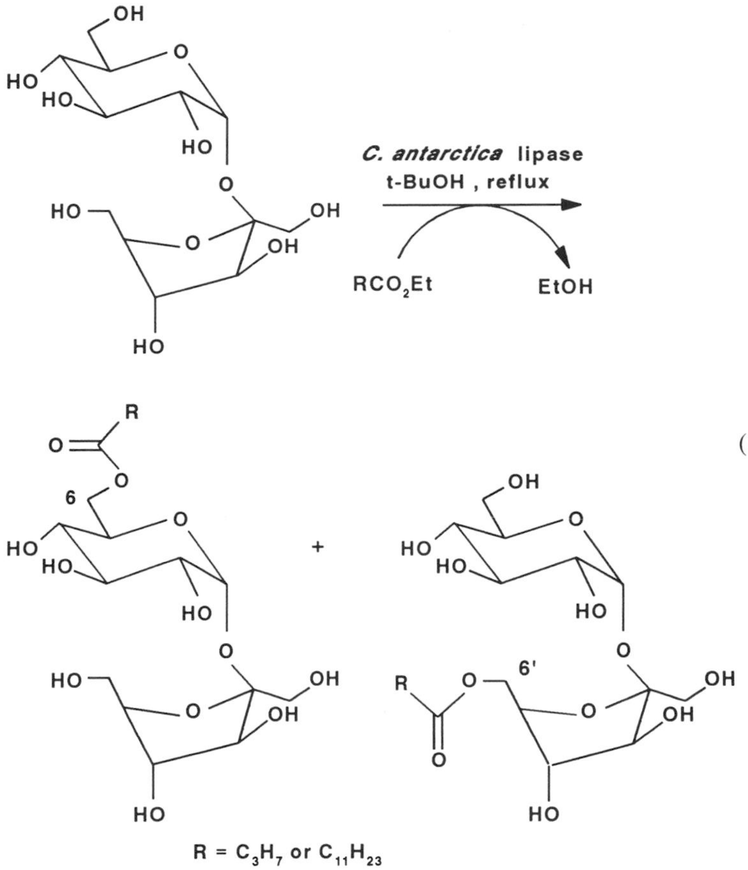

R = C$_3$H$_7$ or C$_{11}$H$_{23}$

(10.9)

Other disaccharides underwent more facile acylation using this procedure, e.g. maltose afforded the 6′-*O*-butanoate in 93% yield after 24 h (62). Hence, we tentatively conclude that enzymatic acylations in non-aqueous media are potentially attractive methods for the large-scale preparation of monoesters of sucrose and other disaccharides.

10.4.3 Selective acylation of steroids

Another class of commercially important compounds where regioselective reactions play an essential role are the steroids. Thus, many steroids contain two or more hydroxyl groups that are difficult to differentiate using chemical procedures. Regioselective acylation of steroids in non-aqueous media is a useful method for selective protection of a hydroxyl group in a steroid molecule (63). Alternatively, regioselective deacylation of a fully acylated steroid can be used, e.g. *Candida rugosa* lipase-catalyzed transesterification of 3β,17β-diacetoxy-5α-androstane to 17β-acetoxy-5α-androstan-3β-ol (reaction 10.10) (64).

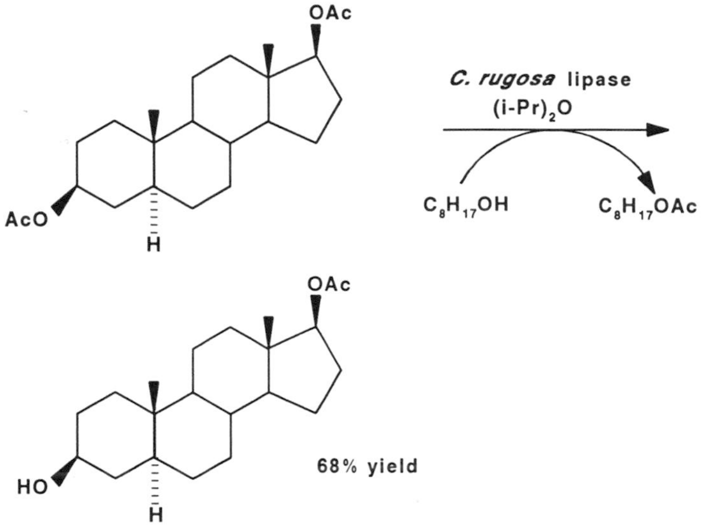

$$(10.10)$$

It is not clear if such techniques have been reduced to commercial practice but they appear to have industrial potential.

10.5 Flavors and fragrances

Increasing consumer demand for natural ingredients (see section 10.2) has resulted in a marked trend towards the use of biotransformations for the manufacture of flavors and fragrances (25, 65). However, many of these transformations involve fermentation processes based on the same biogenetic pathways that are responsible for the production of such materials in nature. There is a simple reason for this: in order to qualify for the label 'natural', the total synthesis should not involve any 'chemical' steps. Thus, although enzymatic conversions have been widely applied to the synthesis of flavor and fragrance molecules the substrate in many of these examples is prepared via conventional 'chemical' synthesis. Consequently, the product would not have the added value associated with the 'natural' label. Nevertheless, the high chemo-, regio- and enantioselectivities of enzymatic transformations could offer significant processing advantages over conventional synthetic methods. In this context it should be emphasized that the use of an organic solvent in an enzymatic step could also disqualify the product for the 'natural' label. The relevant regulations governing such matters are not completely clear on what is allowed in this respect. As noted earlier, supercritical carbon dioxide could possibly constitute an attractive alternative to an organic solvent.

10.5.1 Esters of terpene alcohols

Many natural fragrances contain terpene alcohols and their esters (usually acetates) as major constituents. Geranyl acetate, for example, is one or the most important natural fragrances. Lipase-catalyzed transesterification of geraniol in organic media is a potentially attractive route to geranyl acetate and related terpene esters (66, 67). Thus, reaction of *n*-propyl acetate (20 equivalents) with geraniol in *n*-hexane (reaction 10.11), in the presence of immobilized *Mucor miehei* lipase, afforded geranyl acetate in 85% yield after 3 days (67). Methyl or ethyl acetate could also be used, but gave slightly lower rates than *n*-propyl acetate.

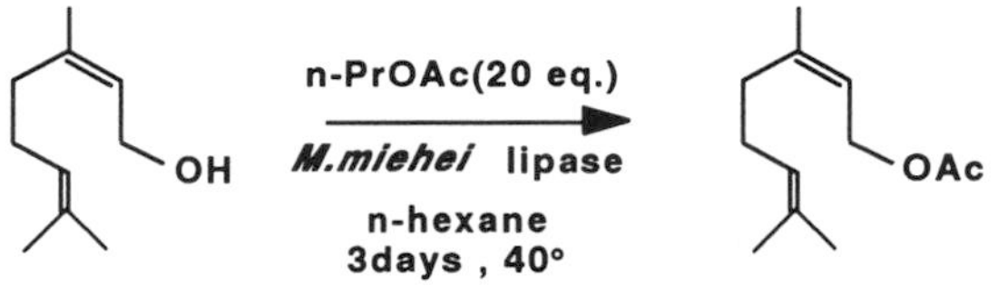

$$\text{geraniol} \quad \xrightarrow[\substack{\textit{M.miehei} \text{ lipase} \\ \text{n-hexane} \\ \text{3days , 40°}}]{\text{n-PrOAc(20 eq.)}} \quad \text{geranyl acetate} \tag{10.11}$$

Similarly, citronellol acetate was prepared by *Candida antarctica* lipase-catalyzed esterification of acetic acid (reaction 10.12) in a solvent-free medium (68). The addition of molecular sieves, to remove the water formed, afforded a 10% yield enhancement.

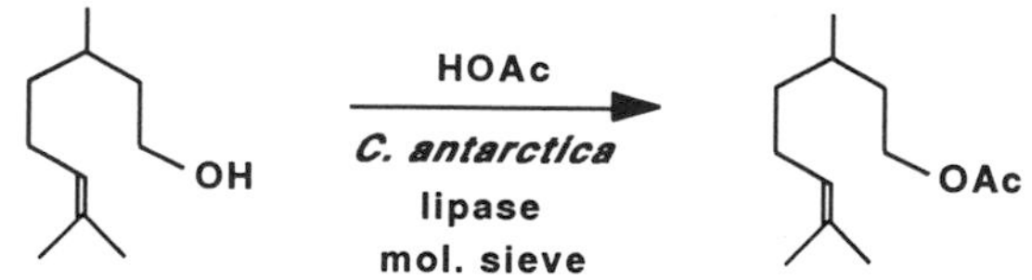

$$\text{citronellol} \quad \xrightarrow[\substack{\textit{C. antarctica} \\ \text{lipase} \\ \text{mol. sieve}}]{\text{HOAc}} \quad \text{80% yield} \tag{10.12}$$

In the above examples the mild conditions of enzymatic processes could have selectivity advantages over conventional esterifications catalyzed by strong mineral acids. Terpenes tend to be sensitive to acid-catalyzed rearrangements under strongly acidic conditions.

In this context it is worth noting that many flavor constituents are simple aliphatic and aromatic esters, e.g. isoamyl acetate (banana), isoamyl butyrate (chocolate), benzyl acetate (jasmine), cinnamyl acetate (guava) and methyl anthranilate (grape, honey). In such cases conventional mineral acid-catalyzed esterification works perfectly well. Nevertheless, if the alcohol and the acid are available from natural sources (e.g. acetic acid via fermentation), enzymatic esterification would qualify the product for the approbium 'natural'.

10.5.2 Chiral alcohols, carboxylic acids, esters and lactones

Chiral secondary alcohols, carboxylic acids, esters and lactones occur widely in nature as food flavors. In most cases, as one would expect, only one of the enantiomers is responsible for the characteristic taste and aroma. Examples include (R)-1-octen-3-ol (mushroom odor) (69) and ethyl(5)-2-methylbutanoate (pineapple) (70):

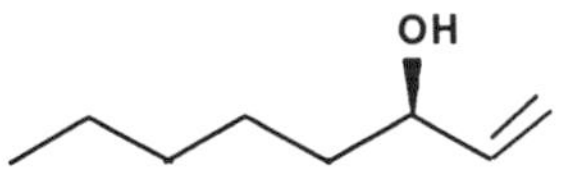

mushroom 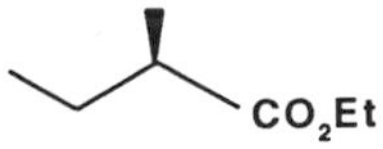   pineapple

Lipase-mediated esterifications and transesterifications in organic media have been widely used for the kinetic resolution of racemic alcohols and acids (4–9). For example, an extract of porcine pancreas (PPE) was used for the kinetic resolution of various chiral saturated, unsaturated and benzylic alcohols (71), via esterification with an achiral acid, such as dodecanoic acid, in n-hexane, e.g. reactions 10.13 and 10.14.

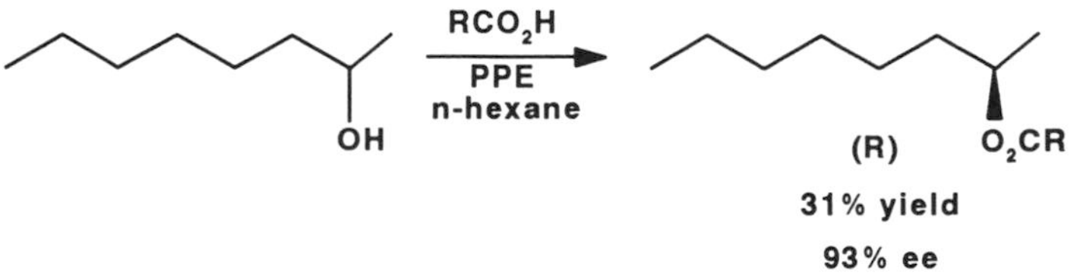

(10.13)

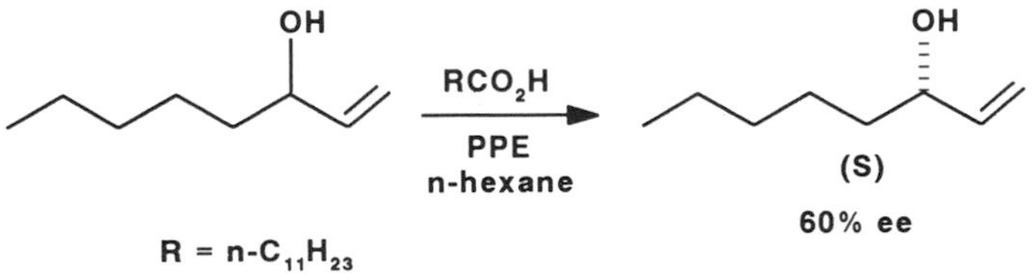

31% yield
93% ee

(10.14)

R = n-C$_{11}$H$_{23}$

60% ee

Similarly, 2-methyl-substituted carboxylic acid esters were resolved via *Candida rugosa* lipase-catalyzed acidolysis (reaction 10.15) in n-heptane

LIVERPOOL
JOHN MOORES UNIVERSITY
AVRIL ROBARTS LRC

(70, 72). The observed enantiomeric ratios (E) were poor (c. 6 with 2-methyl-butanoate) to moderate (13–52 with 2-methylhexanoate) and need to be improved to render the method industrially viable.

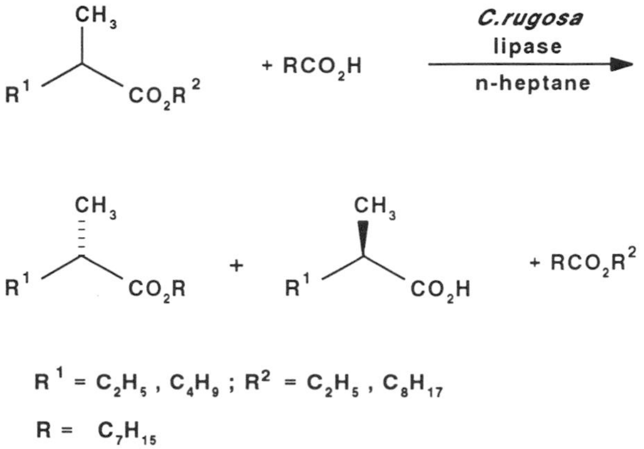

$$R^1 = C_2H_5, C_4H_9 ; R^2 = C_2H_5, C_8H_{17}$$

$$R = C_7H_{15}$$

(10.15)

More recently, *Chromobacterium viscosum* lipase immobilized in micro-emulsion-based organogels was used (73) for the enantioselective esterification of 2-methylbutyric acid with ethanol in *n*-heptane (reaction 10.16). The (S)-ester was obtained in moderate enantioselectivity (32% ee).

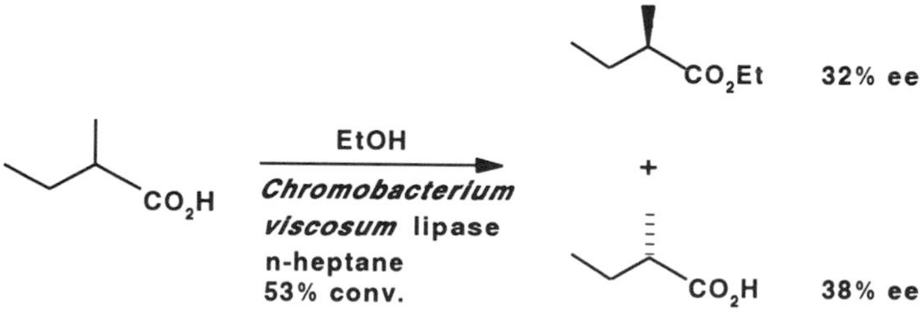

(10.16)

Although the enantioselectivity needs to be significantly improved this technique appears to be worthy of further investigation.

A variety of chiral hydroxy acids are of commercial interest as prominent flavor components of tropical fruits (74). Lipase-catalyzed transesterification (alcoholysis or acidolysis) appears to be a promising strategy for the synthesis of the enantiomerically enriched materials (75).

Chiral aliphatic γ- and δ-lactones are also widely distributed in nature as volatile components of several fruits, e.g. peaches, apricots, nectarines and strawberries (76). The naturally occurring lactones are optically active and the enantiomeric composition can vary with the source. Commercial interest in these products has stimulated the development of biotechnological methods for their

LIVERPOOL
JOHN MOORES UNIVERSITY
AVRIL ROBARTS LRC
TEL. 0151 231 4022

production (25, 76). Many of these processes are fermentations, based on the biogenetic pathways for the *in vivo* synthesis of these lactones via β-oxidation of unsaturated fatty acids, and fall outside the scope of this chapter. Nevertheless, enzymatic lactonization of hydroxyacids (esters) can constitute an interesting chemoenzymatic method for the synthesis of nature-identical materials.

For example, porcine pancreatic lipase-catalyzed lactonization of γ-hydroxy-acid methyl esters, in dry ether, afforded optically active γ-lactones (reaction 10.17) (77).

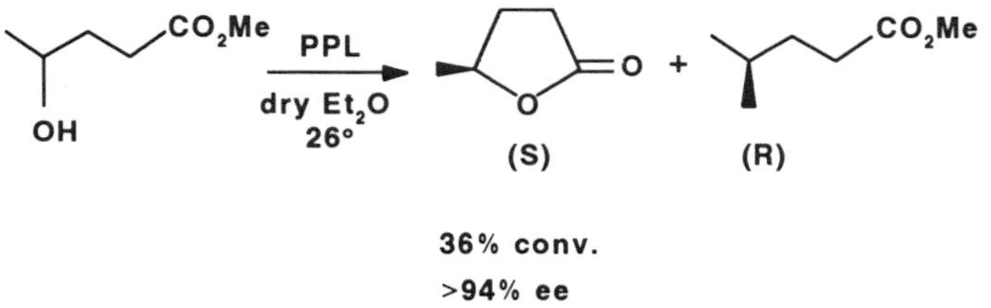

$$\text{(10.17)}$$

36% conv.
>94% ee

Naturally occurring macrocyclic lactones are commercially important fragrance ingredients. Many of these compounds are chiral and enantioselective lipase-catalyzed lactonization in organic media has been used for the synthesis of specific enantiomers (78, 79), e.g. reaction (10.18). Yields were generally low due to competing intermolecular processes.

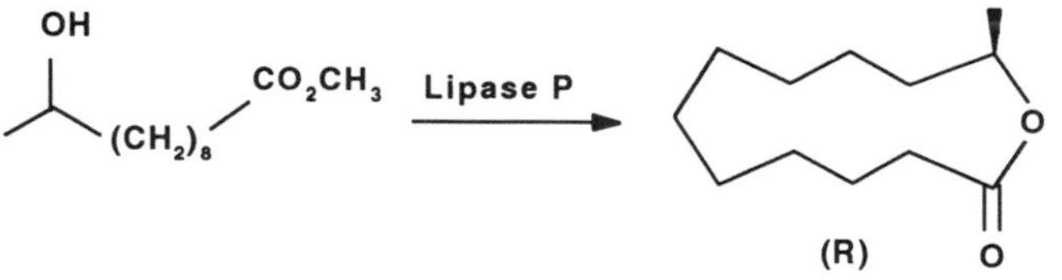

$$\text{(10.18)}$$

10.6 Optically active pharmaceuticals and pesticides

As was already noted in section 10.2 there is currently a marked trend towards marketing chiral pharmaceuticals and pesticides as the pure, biologically active enantiomers (23, 80–82). In contrast to flavor and fragrance intermediates, in the synthesis of pharmaceuticals and pesticides 'natural' methods do not command a premium compared to 'chemical' procedures. In these industries enzymatic processes have to compete solely on the basis of process economics. Nevertheless, enzymatic methods are finding increasing application in the synthesis of enantiomerically pure drugs and pesticides (83–86). Many of these processes involve the use of hydrolytic enzymes, such as lipases and other esterases, in nonaqueous media.

10.6.1 Lipases in the synthesis of chiral C_3 synthons for beta blockers

The beta blockers have been a very successful group of antihypertensive agents for more than two decades. They are chiral molecules having the general structure **1**, but the commercially most important ones—propranolol, atenolol and metoprolol—are all marketed as racemic mixtures.

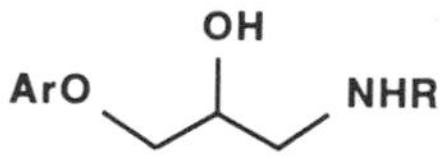

Structure **1**

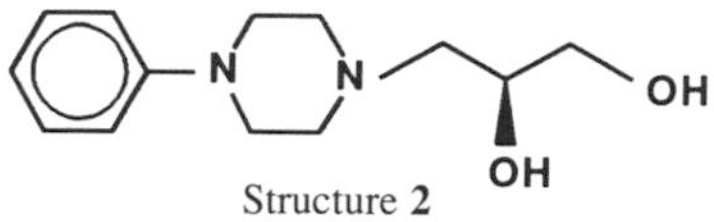

Structure **2**

An enormous effort has been devoted to the development of methods for the synthesis of the active (S)-enantiomers of beta blockers. Most of the routes investigated involve enantioselective syntheses of chiral C_3-synthons, often using an enzymatic kinetic resolution (23, 87). For example, DSM-Andeno commercialized the production of optically active glycidyl derivatives via lipase-catalyzed hydrolysis of racemic glycidyl butyrate (reaction 10.19) (88). The reaction is performed in the absence of solvent with relatively small amounts of water.

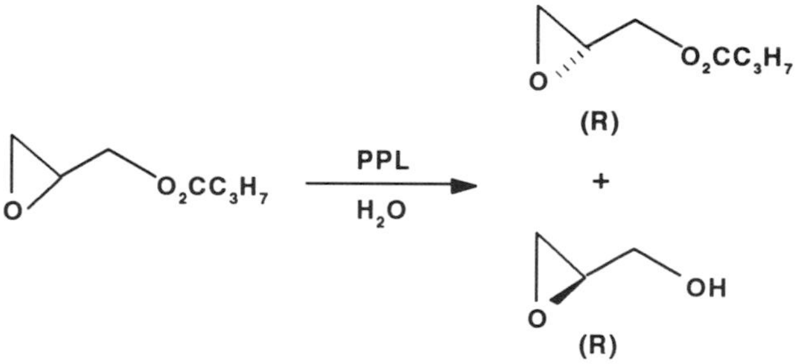

(10.19)

Similarly, chiral C_3 synthons have been made via lipase-catalyzed trans-esterifications in organic media (89–93), e.g. reactions 10.20 and 10.21 (89).

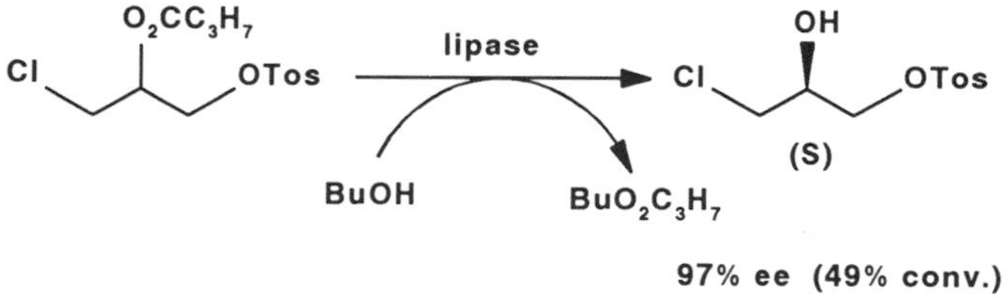

(10.20)

97% ee (49% conv.)

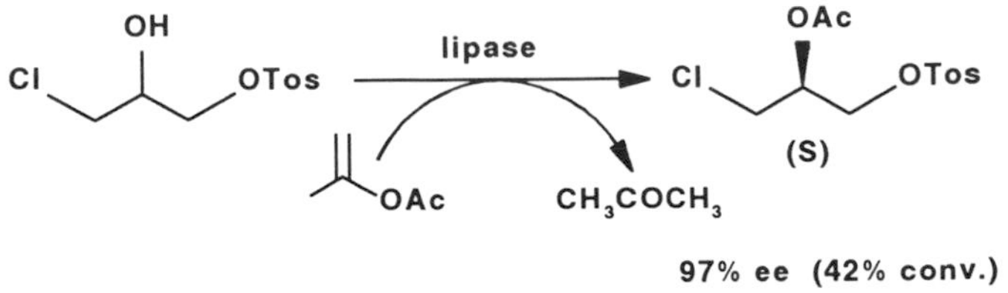

$$(10.21)$$

97% ee (42% conv.)

Kanegafuchi in Japan has also commercialized the production of such chiral C_3 synthons (23). Unfortunately, the expectation that racemic beta blockers would be replaced by the (S)-enantiomers has not materialized. Fortunately, many of these chiral C_3 synthons have commercial utility in the synthesis of other classes of optically active drugs such as the antitussive drug, ledol (2). Moreover, the valuable experience gained in research on enzymatic processes for beta-blocker intermediates was, in many companies, applied to analogous syntheses of intermediates for other chiral drugs, e.g. diltiazem.

10.6.2 Enzymatic synthesis of chiral glycidic esters

Diltiazem, a so-called calcium antagonist, is one of the ten best selling drugs worldwide. A key intermediate in its synthesis is a chiral glycidic ester. The original synthetic route developed by Tanabe, the discoverer of diltiazem, involved a classical resolution, via diastereomeric salt crystallization, at a later stage in the synthesis (see Figure 10.3). This route has now largely been super-seded by a more economical process, developed by DSM-Andeno, involving enzymatic kinetic resolution of the arylglycidate ester (Figure 10.3). The latter bears a distinct resemblance to the glycidyl butyrate hydrolysis mentioned in the preceding section. Thus, *Mucor miehei* lipase-catalyzed hydrolysis of the arylglycidate ester in an organic medium, such as toluene or methyl *tert*-butyl ether, afforded the (2R,3S) isomer in 100% de at c. 50% conversion (94).

The DSM-Andeno route is more economical as it involves an early resolution step (23) and the overall route is two or three steps shorter than the original process.

A transesterification variant of the key enzymatic step has also been reported

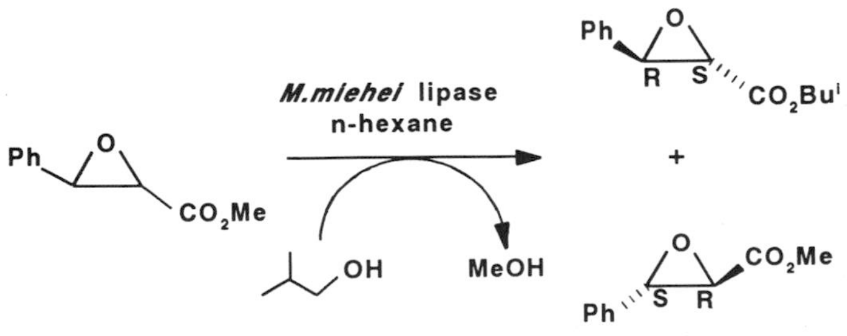

$$(10.22)$$

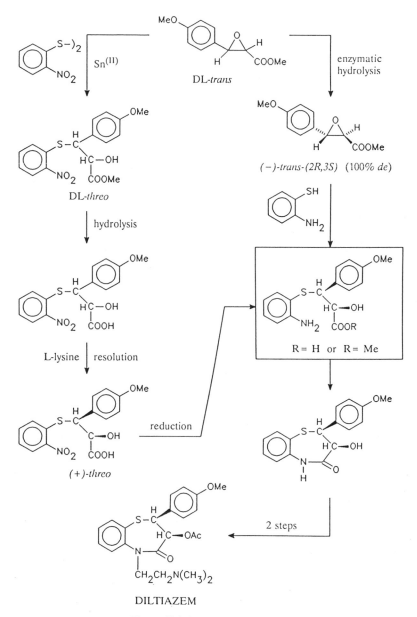

Figure 10.3 Two routes to diltiazem.

(95). Interestingly, a lipase-catalyzed transesterification in n-hexane was used for the kinetic resolution of the analogous methyl-*trans*-β-phenylglycidate (reaction 10.22) (96). The product is an intermediate in the synthesis of the side-chain of the anticancer drug, taxol.

10.6.3 α-Arylpropionic and α-aryloxypropionic acids

Two structurally similar groups of commercially important chiral molecules are the α-arylpropionic acids and the α-aryloxypropionic acids (Figure 10.4). The former are nonsteroid anti-inflammatory drugs (NSAIDs), and the (S)-enantiomer is responsible for the desired therapeutic effect. The latter constitute an important group of herbicides, the active enantiomer of which has the (R)-configuration, which is stereochemically equivalent to the (S)-enantiomer of an α-arylpropionic acid (see below).

The two commercially most important examples of α-arylpropionic acids are naproxen and ibuprofen, which are marketed as the single (S)-enantiomer and the racemate, respectively. A wide variety of strategies, including enzymatic transformations, has been applied to the synthesis of (S)-naproxen and related NSAIDs (97). For example, a continuous process has been described (98) for the enantiospecific hydrolysis of the ethoxyethyl ester of naproxen using a fixed bed of *Candida rugosa* lipase immobilized on a polyacrylate resin. (S)-Naproxen was obtained in > 95% ee. Similarly, (S)-ibruprofen was synthesized by *C. rugosa* lipase-catalyzed hydrolysis of racemic esters in a multiphase membrane reactor (99). Extensive screening by Gist Brocades scientists (100) led to the isolation of a *Bacillus subtilis* strain that produced an esterase (carboxylesterase NP) capable

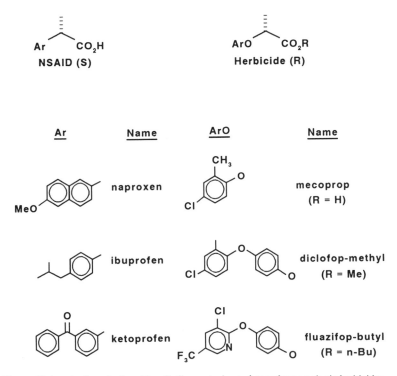

Figure 10.4 α-Arylpropionic acid antiinflammatories and α-aryloxypropionic herbicides.

of catalyzing the hydrolysis of ibuprofen and naproxen methyl esters in very high (> 99% ee) enantioselectivities (reaction 10.23).

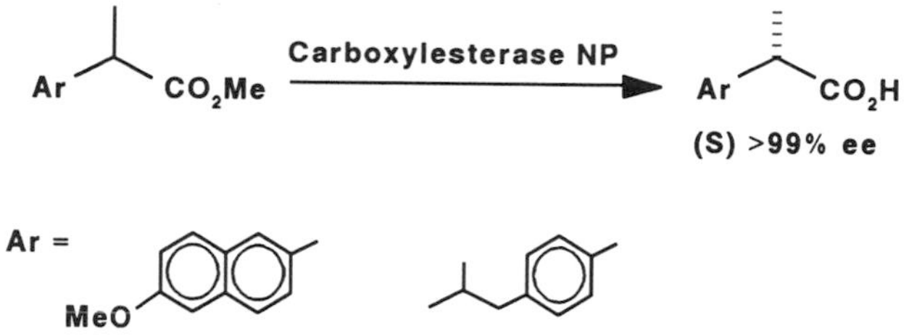

$$\text{Ar} \diagdown \text{CO}_2\text{Me} \xrightarrow{\textbf{Carboxylesterase NP}} \text{Ar} \diagdown \text{CO}_2\text{H}$$

(S) >99% ee

(10.23)

Ar =

However, to our knowledge, enzymatic resolution methods for the synthesis of (S)-naproxen have not been commercialized. Apparently they offer no distinct economic advantage compared to the well-entrenched classical resolution process operated by Syntex or the chiral auxiliary route of Zambon (101). Similarly, if (S)-ibuprofen ever reaches the market place, enzymatic resolution is unlikely to be competitive with catalytic asymmetric syntheses which start from precursors of the racemic acid (23).

Enzymatic methods have also been widely used in the synthesis of inter-mediates for optically active α-aryloxypropionic acid herbicides (23). Many of these processes are conducted in non-aqueous media. For example, α-chloro- and α-bromopropionic acid were resolved via *Candida rugosa*-catalyzed esterification with *n*-butanol in *n*-hexane (reaction 10.24) (102). Chemie Linz (Austria) has reportedly scaled this process up to pilot plant level. The (S)-bromopropionic acid is converted via an inversion into the required (R)-α-phenoxypropionic acid. Alternatively, the latter can be prepared via lipase-catalyzed hydrolysis, or transesterification of its methyl ester (reaction 10.25) (103).

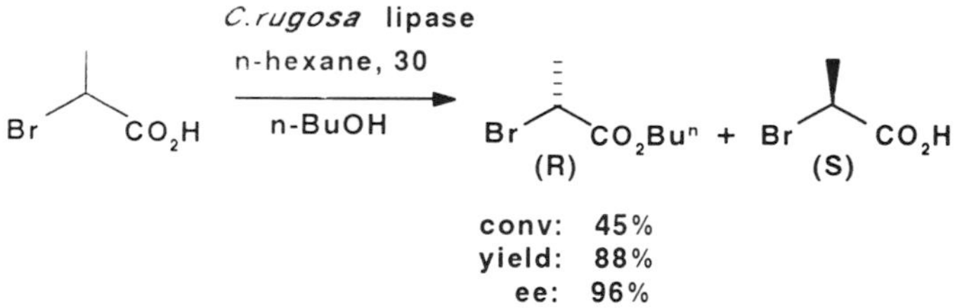

$$\text{Br} \diagdown \text{CO}_2\text{H} \xrightarrow[\text{n-BuOH}]{\substack{\textit{C.rugosa} \text{ lipase} \\ \text{n-hexane, 30}}} \text{Br} \diagdown \text{CO}_2\text{Bu}^n + \text{Br} \diagdown \text{CO}_2\text{H}$$

(R)   (S)

conv: 45%
yield: 88%
ee: 96%

(10.24)

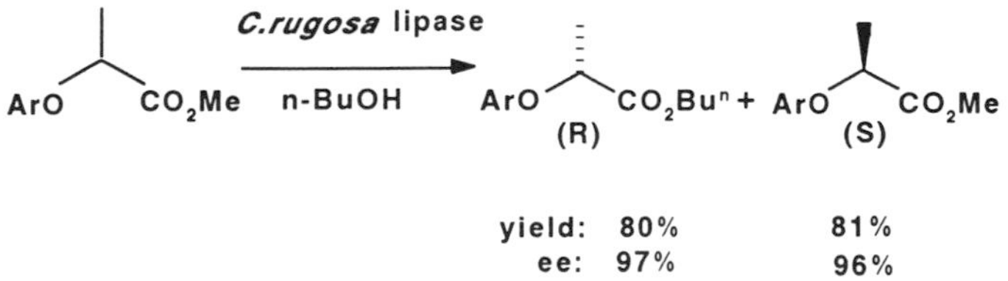

$$\text{ArO} \diagdown \text{CO}_2\text{Me} \xrightarrow[\text{n-BuOH}]{\textit{C.rugosa} \text{ lipase}} \text{ArO} \diagdown \text{CO}_2\text{Bu}^n + \text{ArO} \diagdown \text{CO}_2\text{Me}$$

(R)   (S)

yield: 80%   81%
ee: 97%   96%

(10.25)

The unwanted (S)-enantiomer of the methyl ester was readily racemized by heating with a catalytic amount of NaOMe (103). It was concluded that asymmetric hydrolysis was the method of choice because of a substantially higher rate compared to transesterification. α-Aryloxypropionic acids, such as mecoprop, are now marketed in Europe as the single (R)-enantiomer, and the key intermediate for their synthesis is (S)-α-chloropropionic acid. Interestingly, lipase-catalyzed processes cannot compete with the two routes that are used for commercial production: reaction of (R)-lactic acid ester (derived from fermentation) with SOCl$_2$ (BASF), and kinetic resolution of racemic α-chloropropionic acid using an (R)-specific dehalogenase (ICI) (23).

10.6.4 ACE-Inhibitor intermediates

(R)-2-Hydroxy-4-phenylbutanoate (3) and (R)-β-acetylmercaptoisobutyric acid (4) are key intermediates for the synthesis of ACE inhibitors, a highly successful group of cardiovascular drugs (104). Lipase and esterase based processes have been described for the synthesis of (3) (reaction 10.26) (105) and (4) (reaction 10.27) (106), respectively. The latter reaction is particularly enantioselective and may be able to compete with the established methods for producing (4) (23, 104).

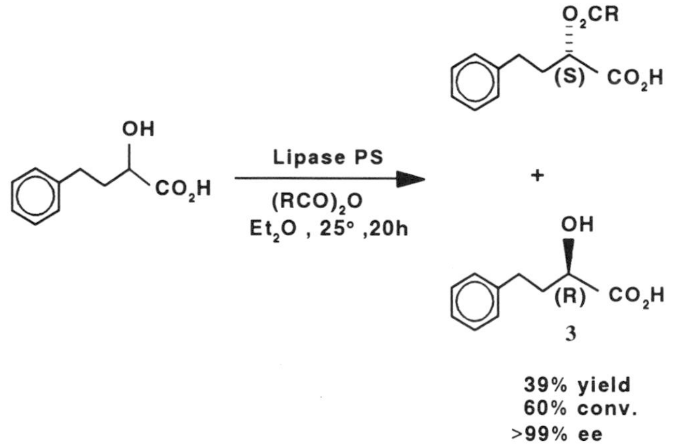

(10.26)

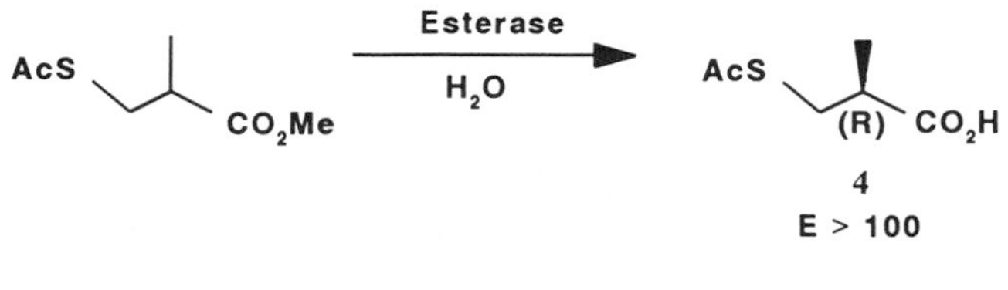

39% yield
60% conv.
>99% ee

(10.27)

E > 100

10.6.5 Lipases in the synthesis of pyrethroid intermediates

Lipase-catalyzed kinetic resolutions have been extensively applied to the synthesis of intermediates for optically active pyrethroid insecticides (107). The Sumitomo company has been particularly active in this area. For example, (S)-m-phenoxybenzaldehyde cyanohydrin, the alcohol moiety of most pyrethroid esters, was prepared by lipase-catalyzed hydrolysis of the racemic acetate (reaction 10.28) and subsequently converted to S,S-fenvalerate (107). The remaining (R)-acetate was racemized by heating with a catalytic amount of triethylamine.

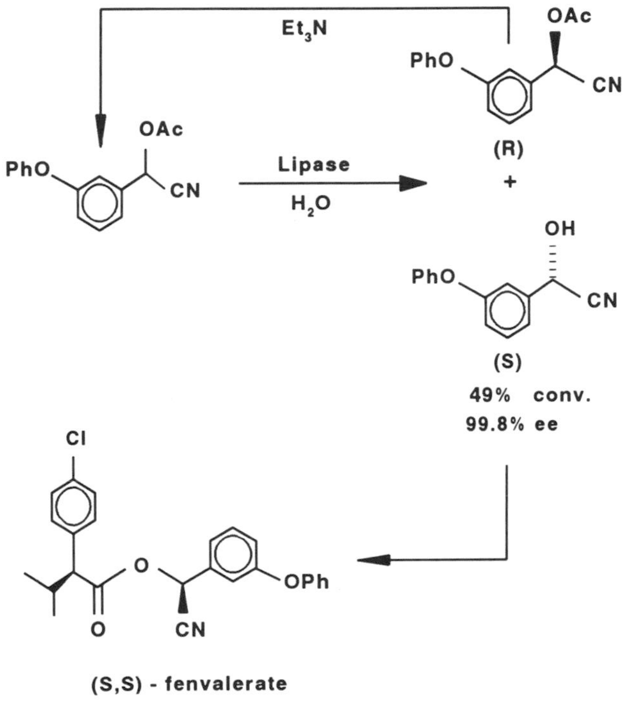

(10.28)

(S,S) - fenvalerate

It is not clear whether or not reaction 10.28 is used commercially. An alternative method for the synthesis of the (S)-cyanohydrin involves enzymatic asymmetric hydrocyanation of the aldehyde (see later).

Similarly, a *Pseudomonas* lipase-catalyzed transesterification with vinyl acetate in *n*-hexane was used for the synthesis of (S)-4-methyl-1-heptyn-4-en-3-ol (reaction 10.29), the alcohol moiety of the synthetic pyrethroid S-2852 (108).

10.6.6 Enzymatic ester ammoniolysis

Lipases are known to catalyze a wide variety of reactions involving 'unnatural' acyl acceptors, such as alcohols, amines, oximes and hydrogen peroxide, in

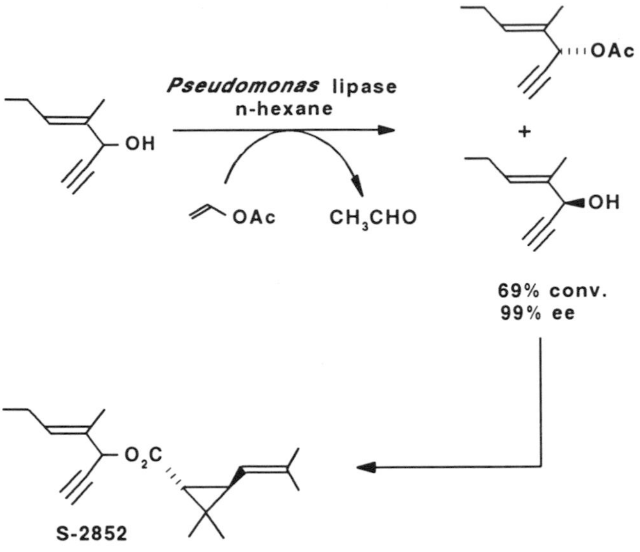

(10.29)

organic media (109). Recently yet another acyl acceptor was added to this repertoire: ammonia. Thus, ethyl octanoate was shown to undergo smooth lipase-catalyzed ammoniolysis, to the corresponding amide (reaction 10.30), by reaction with ammonia at 40°C in *tert*-butanol (110–112). Lipases from *Candida antarctica* (SP435) and *Humicola* sp. (SP398) were particularly effective as they were able to tolerate high concentrations (2.5 M) of ammonia. Enzymatic ammoniolysis in organic media could be a useful method for the production of thermally labile amides. As noted in section 10.3, it could be useful for the production of fatty acid amides from fatty acid esters or triglycerides. Many amides are sparingly soluble in organic solvents and precipitate during the ammoniolysis reaction, thus facilitating their isolation.

$$\text{C}_7\text{H}_{15}\text{CO}_2\text{Et} + \text{NH}_3 \xrightarrow[\substack{\text{t-BuOH} \\ 40°}]{\text{lipase}} \text{C}_7\text{H}_{15}\text{CONH}_2 + \text{EtOH} \quad (10.30)$$

The corresponding carboxylic acids do not undergo ammoniolysis, since they form the ammonium salt which cannot act as a substrate for the lipase. However, since lipases also mediate the esterification of the carboxylic acid, the latter can be converted to the amide in a two-step process using the same lipase to catalyze the formation of the ester and its subsequent ammoniolysis (112).

Enzymatic ammoniolysis is also an excellent method for the kinetic resolution of chiral carboxylic acids or chiral alcohols. For example, the *C. antarctica* lipase-catalyzed ammoniolysis of the 2-chloroethyl ester of ibuprofen (reaction 10.31) was *c.* ten times as enantioselective as the corresponding hydrolysis ($E = 28$ and 4, respectively) (110). High enantioselectivity ($E > 100$) was also observed in the ammoniolysis of α-methylbenzyl-*n*-butyrate (reaction 10.32) (112). Lipase-mediated ammoniolysis was also applied to the conversion of α-amino acid esters to the corresponding amino amides (113). For example, the ethyl ester of racemic phenylglycine was converted to the (*R*)-amide (reaction 10.33) with high enantioselectivity ($E = 38$). Moreover, the remaining (*S*)-ester racemized under the reaction conditions which means that, in principle, both enantiomers of the ester can be converted to the (*R*)-amide. The latter is potentially a key intermediate in second-generation enzymatic processes for the manufacture of semisynthetic penicillins and cephalosporins.

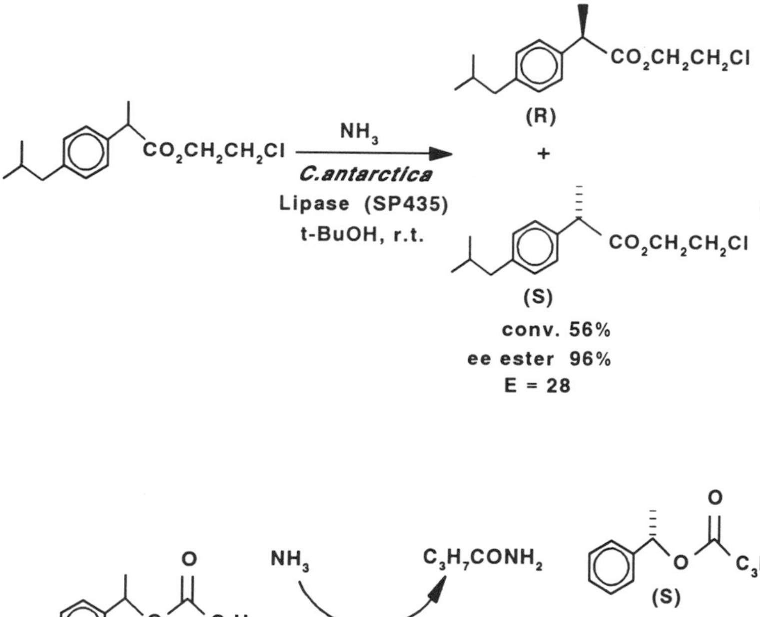

$$(10.31)$$

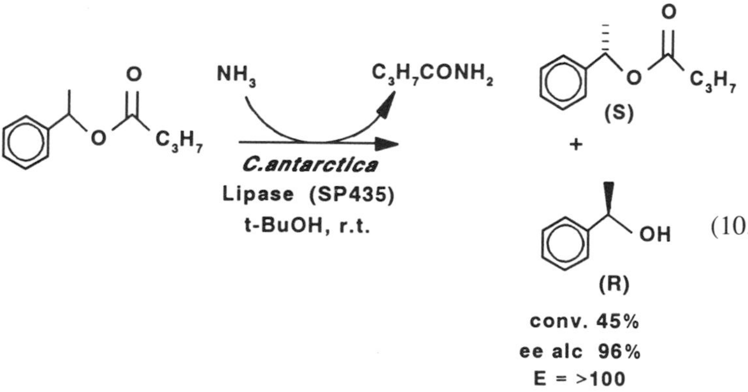

conv. 56%
ee ester 96%
E = 28

$$(10.32)$$

conv. 45%
ee alc 96%
E = >100

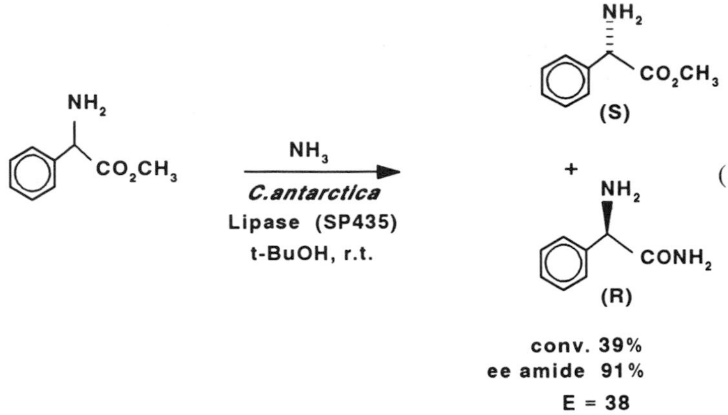

$$\text{(10.33)}$$

conv. 39%
ee amide 91%
E = 38

10.6.7 Enzymatic hydrocyanation in organic media

All of the examples discussed up till now involved the use of hydrolytic enzymes in non-aqueous media. As mentioned in section 10.2, lyases constitute another group of enzymes that are of particular interest for large-scale production. Lyase-catalyzed additions of N, O, and C nucleophiles to double bonds have considerable potential in (enantio)selective synthesis. Many of these reactions can, in principle, be performed in non-aqueous media.

Oxynitrilases, for example, occur widely in nature, where they catalyze the conversion of cyanogenic glycosides to aldehydes and HCN (114). Pioneering studies by Becker and Pfeil (115) in the 1960s demonstrated that the (R)-oxynitrilase (E.C. 4.1.2.10), present in bitter almonds, catalyzes the enantioselective formation of (R)-cyanohydrins by reaction of HCN with aldehydes (reaction 10.34). Subsequently, oxynitrilase (E.C. 4.1.2.11) from sorghum extract was shown to catalyze the selective formation of (S)-cyanohydrins (reaction 10.35) (116). The two enzymes differ in structure and substrate specificity: (S)-oxynitrilase is effective only with aromatic aldehydes, while (R)-oxynitrilase accommodates a wide variety of aromatic, heterocyclic and aliphatic aldehydes (114).

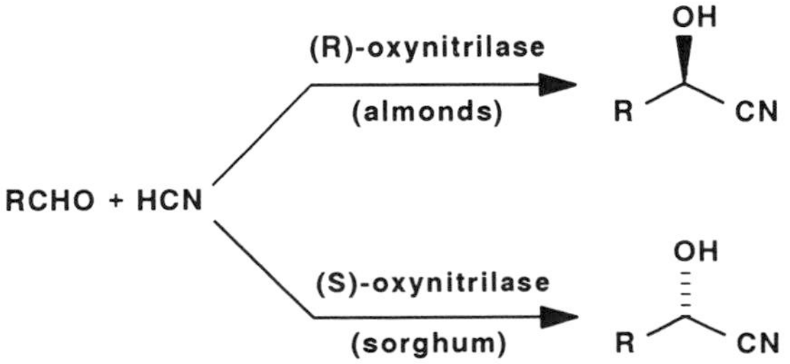

$$\text{(10.34)}$$

$$\text{(10.35)}$$

Achieving high enantioselectivities in aqueous media is hampered by competing non-enzymatic addition of HCN to the substrate and by facile racemization of the product in aqueous buffer. Dramatic improvements in enantioselectivity are observed when the reaction is carried out in a water-immiscible solvent, such as ethyl acetate or di-isopropyl ether (116, 117). In these solvents the enzymatic reaction is hardly affected whereas the chemical reaction is substantially retarded. For example, on changing from aqueous ethanol to ethyl acetate as solvent, the ee of the (R)-cyanohydrin obtained from m-phenoxybenzaldehyde increased from 10.5 to 98% (117).

(R)-Oxynitrilase is readily available and highly efficient: milligram quantities are enough to catalyze the formation of kilograms of cyanohydrin. It has been immobilized on cellulose-based supports (117) or by encapsulation in lyotropic liquid crystals embedded in a porous support (118). Alternatively, Solvay Duphar has developed a process which operates in a biphasic aqueous–organic system (114).

Asymmetric hydrocyanation is of considerable commercial interest as the optically pure cyanohydrins can be subsequently converted to a variety of chiral building blocks, e.g. α-hydroxy acids, α-hydroxyamines, etc. (114).

The (S)-cyanohydrin of m-phenoxybenzaldehyde, a key intermediate for optically active pyrethroid insecticides (see earlier), was prepared in 92% ee by hydrocyanation of the aldehyde (reaction 10.36), in the presence of (S)-oxynitrilase bound to a porous membrane, in dibutyl ether (119).

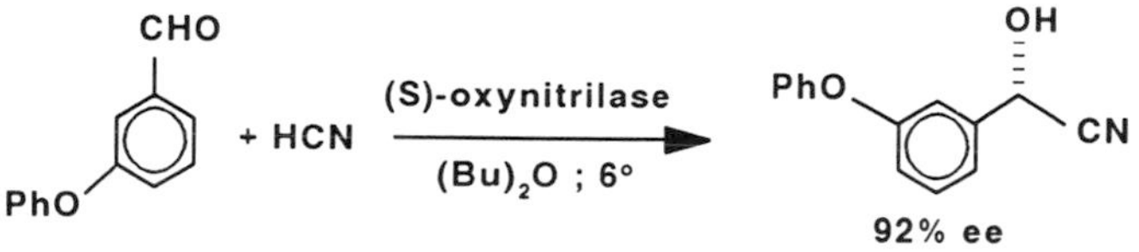

$$(10.36)$$

A process for the synthesis of (R)-pantolactone, a precursor of pantothenic acid (vitamin B$_5$), via (R)-oxynitrilase-catalyzed addition of HCN to hydroxy-pivalaldehyde (reaction 10.37) has been described (120).

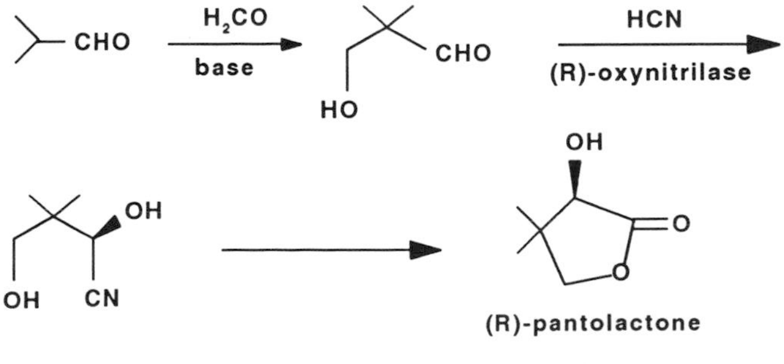

$$(10.37)$$

An alternative means of circumventing the non-enzymatic process has been described: enzymatic transhydrocyanation with acetone cyanohydrin. For example, benzaldehyde was converted to (R)-mandelonitrile (reaction 10.38) (120) or (S)-mandelonitrile (reaction 10.39) (122) using (R)- or (S)-oxynitrilase, respectively.

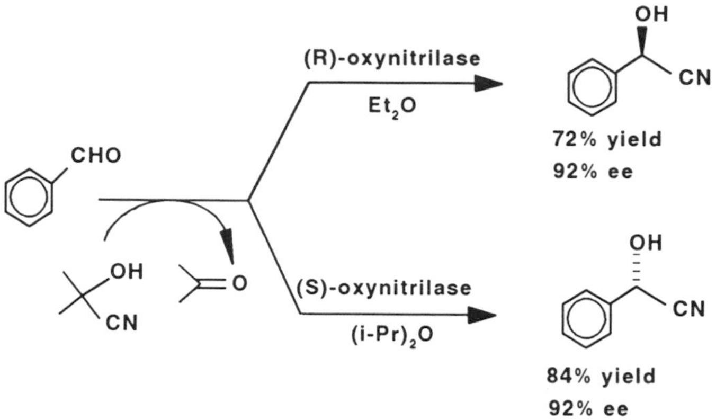

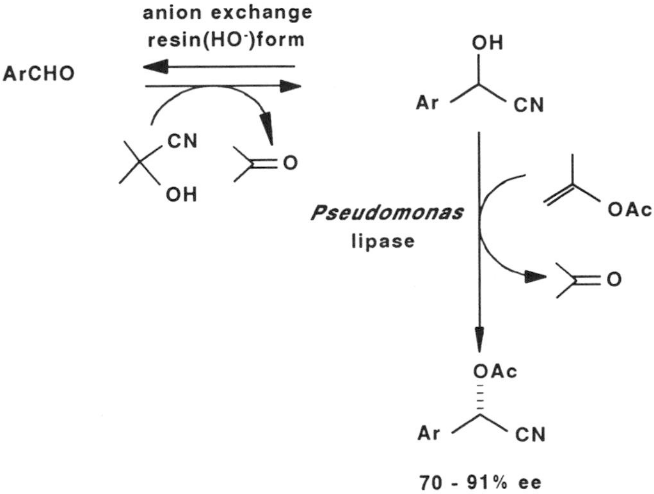

Yet another approach to the synthesis of optically active cyanohydrins is the lipase-catalyzed hydrolysis of cyanohydrin acetates (123). In a further elaboration of this approach, advantage was taken of the reversibility of cyanohydrin formation to devise an elegant method for obtaining a theoretical yield of 100% in one step (124). A variety of aromatic aldehydes underwent transhydrocyanation with acetone cyanohydrin in the presence of an anion exchange resin (HO⁻ form) in di-isopropyl ether (reaction 10.40). The resulting racemic cyanohydrin was subjected to in situ, lipase-catalyzed transesterification with isopropenyl acetate, yielding the (S)-cyanohydrin acetate in high ee. Because of the reversible nature of the base-catalyzed transhydrocyanation, the remaining (R)-cyanohydrin undergoes *in situ* racemization. (S)-*m*-Phenoxy-benzaldehyde cyanohydrin acetate, for example, was obtained in 80% yield and 89% ee. The success of the method depends on the fact that acetone cyanohydrin, because of its steric bulk, cannot be accommodated by the lipase and is not acylated.

Asymmetric hydrocyanation of 3-phenylpropanal, readily available from hydrogenation of cinnamaldehyde or hydroformylation of styrene, is a potentially attractive route to the key ACE-inhibitor intermediate (**5**) (reaction 10.41).

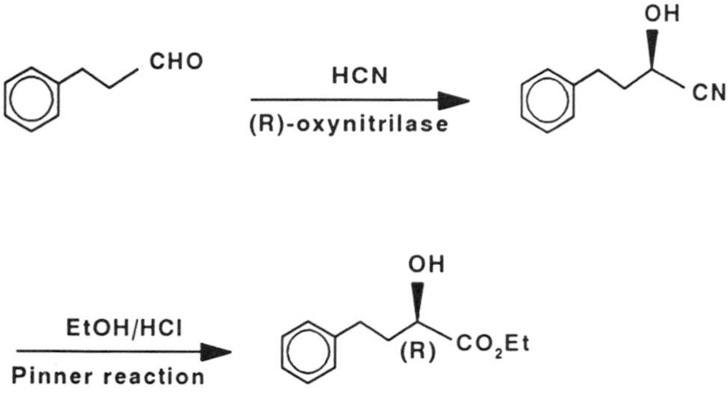

(10.41)

In short, enzymatic hydrocyanation in organic media constitutes a potentially attractive method for the industrial production of fine chemicals via enantio-selective C–C bond formation. The industrial utility of oxynitrilases would be substantially enhanced if the genes encoding for these enzymes were cloned into a suitable host micro-organism.

10.7 Enzymatic polymer synthesis

In living cells enzymes catalyze the formation of polysaccharides, proteins and nucleic acids with unparalleled selectivities. The chemo-, regio- and enantio-selectivities characteristic of enzymes can also be utilized for the production of specialty polymers with desirable features such as chirality and biodegradability. Hence, enzymatic methods are finding increasing use in the synthesis of novel polymeric materials with applications such as water adsorbents, hydrogels, chiral adsorbents, liquid crystals and permselective membranes (125). In non-aqueous media, hydrolytic enzymes can catalyze the polymerization of bifunctional monomers. Moreover, improved productivity can accrue from the increased solubility of hydrophobic monomers in organic solvents.

10.7.1 Polyester synthesis

Lipases, esterases and proteases have been used (125) for the synthesis of poly-esters via condensation of diols with diesters or self-condensation of hydroxy-carboxylic acid esters. Although such condensation polymerizations are trivial to perform chemically, the use of enzymes allows for the enantioselective polymerization of racemic monomers, as illustrated in reaction 10.42. Carbohydrates can be used as the diol monomer, e.g. to produce a sucrose–adipate copolymer via *Bacillus* alkaline protease-catalyzed condensation of di-(2,2,2-trifluoroethyl) adipate with sucrose in anhydrous pyridine (reaction 10.43) (126). The product was readily biodegradable and has potential appli-cations as a water adsorbent.

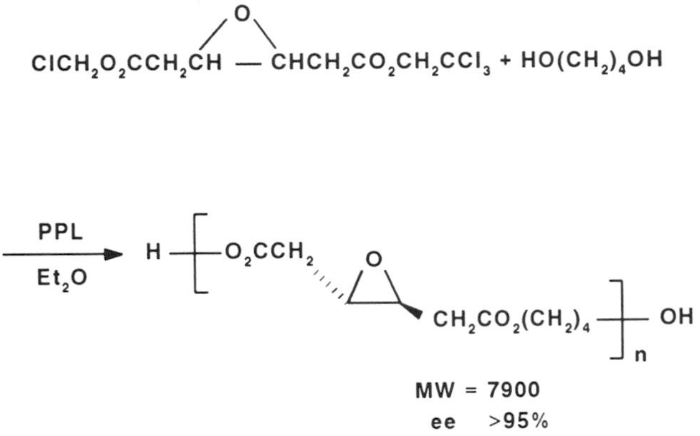

$$ \text{ClCH}_2\text{O}_2\text{CCH}_2\text{CH} - \text{CHCH}_2\text{CO}_2\text{CH}_2\text{CCl}_3 + \text{HO(CH}_2)_4\text{OH} $$

(10.42)

$$ \xrightarrow[\text{Et}_2\text{O}]{\text{PPL}} \quad \text{H} \left[-\text{O}_2\text{CCH}_2 \cdots \triangle \cdots \text{CH}_2\text{CO}_2\text{(CH}_2)_4 \right]_n \text{OH} $$

MW = 7900
ee >95%

A serious drawback to enzymatic polycondensations is the low molecular weight that is universally observed. The enzyme must continually generate an acyl-enzyme intermediate by reaction with the terminal ester moiety of the growing polymer. The efficiency of this process drops rapidly with increasing molecular weight due to the inherent diffusional limitations.

10.7.2 Peroxidase-catalyzed synthesis of polyphenols

Because of the health and environmental hazards associated with formaldehyde, alternatives are being sought for the widely used phenol–formaldehyde resins. The polymerization of phenols can be accomplished in the absence of formaldehyde by using enzymes. For example, horseradish peroxidase (HRP), in the presence of H_2O_2, catalyzes the free-radical polymerization of phenols. Because HRP catalyzes only the formation of chain-initiating phenoxy radicals, polymer chain growth proceeds non-enzymatically. The formation of high molecular weight polymers is hampered, however, by the poor solubility of

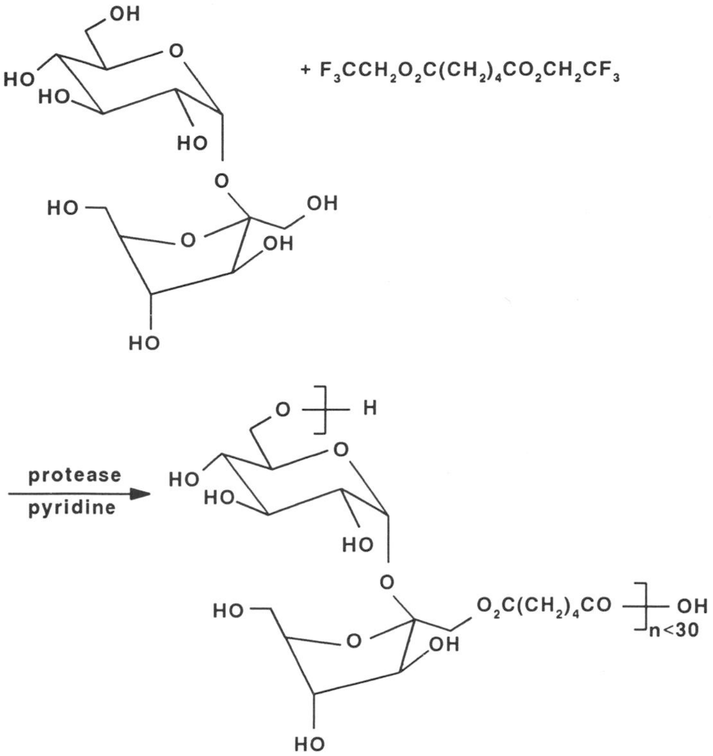

+ F3CCH2O2C(CH2)4CO2CH2CF3
protease
pyridine
O2C(CH2)4CO
OH
n<30

(10.43)

the growing polymer in water. In organic media, in contrast, high molecular weight polymers can be produced. Polymers of *p*-phenylphenol, for example, were produced in aqueous dioxan and the molecular weight of the polymer inreased with decreasing water content (127). Moreover, the productivity of the polymer formation was much higher due to the high solubility of the phenol in dioxane. The resulting polymer is being commercialized for application as a photographic developer resin (125).

Another biocatalytic alternative to phenol–formaldehyde resins is enzymatic modification of natural phenolic polymers such as lignin. For example, HRP was used to catalyze the copolymerization of phenols with lignins (128). In this way polyphenol resins can be produced using a natural, renewable resource.

10.7.3 Enzymatic synthesis of specialty monomers

Another approach to the synthesis of novel polymers is to use enzymatic transformations for the synthesis of novel monomers which are subsequently polymerized using standard techniques. For example, optically active acrylates and methacrylates were synthesized by lipase or protease-catalyzed reactions of (meth)acrylates with racemic alcohols or amines in methyl-*tert*-butyl ether, e.g. reaction 10.44 (129). Subsequent free-radical polymerization using azo-bisisobutyronitrile (AIBN) afforded optically active polymers of high molecular weight. Similarly, PPL was used for the enantioselective acylation of racemic 2-ethylhexanol with *O*-acryloyloxime in tetrahydrofuran (reaction 10.45) (130). Homopolymerization or copolymerization with styrene afforded optically active polymers.

As discussed in section 10.4, lipase-catalyzed transesterification in organic media can also be used for the regioselective synthesis of 6-acrylate esters of glucosides and galactosides, e.g. with *Candida antarctica* lipase and ethyl acrylate in *tert*-butanol (61) or *Pseudomonas cepacia* lipase and vinyl acrylate in pyridine (131). Subsequent free radical polymerization using a conventional chemical initiator led to the formation of high molecular weight poly(carbo-hydrateacrylate) (reaction 10.46) (131). In the presence of 0.3% ethylene glycol dimethylacrylate, a 6-acryloyl-β-galactoside afforded a copolymer which retained 50 times its own weight of water. Such materials have potential as biocompatible hydrogels in biomedical and membrane applications.

Analogous polymers have also been synthesized from a sucrose acrylate monomer (125). The biocompatible and biodegradable properties of these

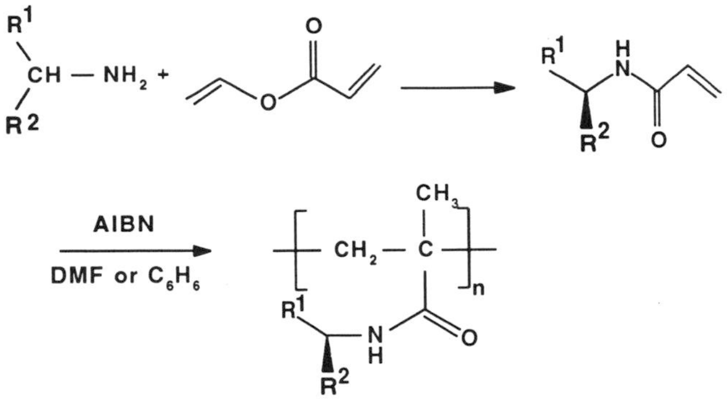

(10.44)

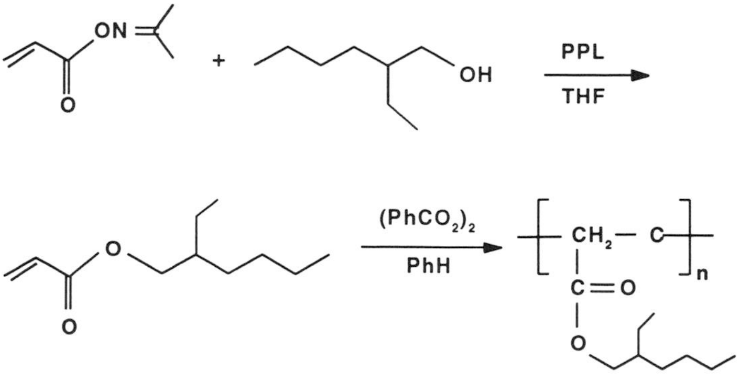

(10.45)

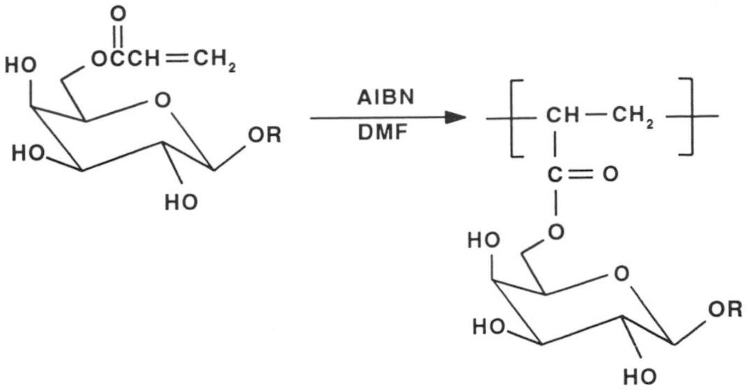

(10.46)

carbohydrate-based synthetic polymers appear to hold considerable promise for commercial application. Indeed, as noted by Dordick (125), chemoenzymatic synthesis of novel polymeric materials is only just beginning to be exploited. For example, stereo- and regioselective (meth)acryloylation of amino acids, steroids and antibiotics could yield polymers containing biologically active pendant groups for use in drug delivery systems.

10.8 Concluding remarks and future prospects

Although no attempt has been made to be comprehensive in this review, hopefully it has shown that enzymatic conversions in non-aqueous media have many (potential) industrial applications. The areas of application vary widely, from food additives, flavors and fragrances to pharmaceuticals, pesticides and specialty polymers. Stimulated by increasing consumer demands for more natural, healthy and environmentally benign products, the use of enzymatic processes is expected to achieve even further penetration in these industry segments in the future.

Up till now most applications involve the use of hydrolytic enzymes, particularly lipases. In the future we expect the use of lyases and oxidoreductases in non-aqueous media to find applications in large-scale syntheses. The report (37) that cross-linked enzyme crystals (CLECs) of an aldolase render it suitable for use in non-aqueous media, for example, appears to hold considerable promise for synthetic applications.

Moreover, further advances in enzyme engineering are expected to make enzymes even more amenable for large-scale organic synthesis. Indeed, we may conclude that considerable progress has been made since the pioneering studies of the groups of Klibanov and Macrae on lipase-catalyzed transformations in organic solvents. In looking back at this review in ten years' time we shall probably conclude that it provided a mere glimpse of the industrial potential of enzymatic conversions in non-aqueous media.

References

1. Zaks, A. and Klibanov, A.M. (1984) Enzymatic catalysis in organic media at 100°C. *Science* **224**, 1249–1251.
2. Klibanov, A.M. (1986) Enzymes that work in organic solvents. *Chemtech*, 354–359.
3. Klibanov, A.M. (1990) Asymmetric transformations catalyzed by enzymes in organic solvents. *Acc. Chem. Res.*, **23**, 114–120.
4. Wong, C.H. and Whitesides, G.M. (1994) *Enzymes in Synthetic Organic Chemistry*, Pergamon, Elsevier, Amsterdam.
5. Faber, K. (1992) *Biotransformations in Organic Chemistry*, Springer-Verlag, Berlin.
6. Davies, H.G., Green, R.H., Kelly, D.R. and Roberts, S.M. (1989) *Biotransformations in Preparative Organic Chemistry*, Academic Press, New York.
7. Abramowicz, D.A. (ed.) (1990) *Biocatalysis*, Van Nostrand Reinhold, New York.
8. Halgas, J. (1992) *Biocatalysts in Organic Synthesis*, Elsevier, Amsterdam.
9. Poppe, L. and Novak, L. (1992) *Selective Biocatalysis*, Verlag Chemie, Weinheim, Germany.
10. Holland, H.L. (1992) *Organic Synthesis with Oxidative Enzymes*, Verlag Chemie, Weinheim, Germany.
11. Faber, K. and Franssen, M.C.R. (1993) Prospects for the increased application of biocatalysts in organic transformations, *Tibtech*, **11**, 461–470.
12. Dordick, J.S. (1991) Principles and applications of nonaqueous enzymology. In *Applied Biocatalysis*, vol. 1 (eds H.W. Blanch and D.S. Clark), Marcel Dekker, New York, pp. 1–51.
13. Tramper, J., Vermue, M.H., Beeftink, H.H. and von Stockar, U. (eds) (1992) *Biocatalysis in Non-Conventional Media*, Elsevier, Amsterdam.
14. Laane, C., Tramper, J. and Lilly, M.D. (eds) (1987) *Biocatalysis in Organic Media*, Elsevier, Amsterdam.
15. Gerhartz, W. (ed.) (1991) *Enzymes in Industry. Production and Application*, Verlag Chemie, Weinheim, Germany.
16. Dordick, J.S. (ed.) (1991) *Biocatalysts for Industry*, Plenum Press, New York.
17. Falch, E.A. (1991) Industrial enzymes—developments in production and application, *Biotech. Advan.*, **9**, 643–658.
18. Arbige, M.V. and Pitcher, W.H. (1989). Industrial enzymology: a look towards the future, *Trends Biotech.*, **7**, 330–335.
19. Björkling, F., Godtfredsen, S.E. and Kirk, O. (1991) The future impact of industrial lipases, *Trends Biotech.*, **9**, 360–363.
20. Vulfson, E.N. (1994) Industrial applications of lipases. In *Lipases, their Structure, Biochemistry and Application* (eds P. Woolley and S.B. Petersen), Cambridge University Press, Cambridge.
21. Sheldon, R.A. (1994) Consider the environmental quotient, *Chemtech*, 38–47.
22. Sheldon, R.A. (1992) Organic synthesis—past, present and future, *Chem. Ind., London*, 903–906.
23. Sheldon, R.A. (1993) *Chirotechnology: Industrial Synthesis of Optically Active Compounds*, Marcel Dekker, New York.
24. Collins, A.N., Sheldrake, G.N. and Crosby, J. (eds) (1992) *Chirality in Industry*, Wiley, New York.
25. Cheetham, P.S.J. (1993) The use of biotransformations for the production of flavours and fragrances, *Tibtech*, **11**, 478–488.
26. Randolph, T.W., Blanch, H.W. and Clark, D.S. (1991) Biocatalysis in supercritical fluids. In *Biocatalysts for Industry* (ed. J.S. Dordick) Plenum Press, New York.
27. Oyama, K. and Kihara, K. (1984) A new horizon for enzyme technology, *Chemtech*, 100–104.
28. Oyama, K. (1992) The industrial production of aspartame. In *Chirality in Industry* (eds A.N. Collins, G.N. Sheldrake and J. Crosby) Wiley, New York, pp. 237–247.
29. Samdani, G., Moore, S. and Ondrey, G. (1993) Enzymes move from nature to the plant, *Chem. Eng.*, December, 35–39.
30. Estell, D.A. (1993) Engineering enzymes for improved performance in industrial applications. *J. Biotechnol.*, **28**, 25–30.
31. Estell, D.A., Gaycar, T.P. and Wells, J.A. (1985) Engineering an enzyme by site-directed mutagenesis to be resistant to chemical oxidation. *J. Biol. Chem.*, **260**, 6518–6521.

32. Bott, R., Shield, J.W. and Poulose, A.J. (1994) Protein engineering of lipases. In *Lipases, Their Structure, Biochemistry and Application* (eds P. Woolley and S.B. Petersen) Cambridge University Press, Cambridge.
33. Hwang, J.Y. and Arnold, F.H. (1991) Enzyme design for nonaqueous solvents. In *Applied Biocatalysis*, vol. 1 (eds H.W. Blanch and D.S. Clark), Marcel Dekker, New York, pp. 53–86.
34. Arnold, F.H. (1990) Engineering enzymes for non-aqueous solvents. *Tibtech*, **8**, 244–249.
35. Hilvert, D. (1991) Extending the chemistry of enzymes and abzymes, *Tibtech*, **9**, 11–17.
36. St. Clair, N.L. and Navia, M.A. (1992) Cross-linked enzyme crystals as robust biocatalysts. *J. Amer. Chem. Soc.*, **114**, 7314–7316.
37. Sobolov, S.B., Bartoszko-Malik, A., Oeschger, T.R. and Montelbano, M.M. (1994) Cross-linked enzyme crystals of fructose diphosphate aldolase: development as a biocatalyst for synthesis, *Tetrahedron Lett.*, **35**, 7751–7754.
38. Chibata, I., Tosa, T. and Shibatani, T. (1992) The industrial production of optically active compounds by immobilized biocatalysts. In *Chirality in Industry* (eds A.N. Collins, G.N. Sheldrake and J. Crosby), Wiley, New York, pp. 351–370.
39. Nakamura, K., Kondo, S., Kawai, Y. and Ohno, A. (1991) Reduction by baker's yeast in benzene, *Tetrahedron Lett.*, **32**, 7075–7078.
40. Macrae, A.R. (1983) Lipase-catalyzed interesterification of oils and fats, *J. Amer. Oil. Chem. Soc.*, **60**, 291–294.
41. Macrae, A.R. and Hammond, R.C. (1985) Present and future applications of lipases, *Biotech. Genet. Eng. Rev.*, **3**, 193–217.
42. Coleman, M.H. and Macrae, A.R. (1980) Fat process and composition. UK Patent 1577993, to Unilever.
43. Matsuo, T., Sawamura, N., Hashimoto, Y. and Hashida, W. (1981) Method for enzymatic transesterification of lipid and enzyme used therein. European Patent 0035883, to Fuji Oil.
44. Quinlan, P. and Moore, S. (1993) Modification of triglycerides by lipases: process technology and its application to the production of nutritionally improved fats, *Inform*, **4**, 580–585.
45. King, D.M. and Padley, F.B. (1990) Milk fat substitutes, European Patent 0209327, to Unilever.
46. Jandacek, R., Whiteside, J.A., Holcombe, B.N., Volpenheim, R.A. and Taulbee, J.D. (1987) The rapid hydrolysis and efficient absorption of triglycerides with octanoic acid in the 1 and 3 positions and long-chain fatty acid in the 2 position, *Amer. J. Clin. Nutr.*, **45**, 940–945.
47. Mukherjee, K.D. and Kiewitt, I. (1991) Enrichment of γ-linoleic acid from fungal oil by lipase-catalysed reactions. *Appl. Microbiol. Biotechnol.*, **35**, 579–584.
48. Zaks, A. and Bross, A.T. (1990) Production of glycerides rich in omega-3 fatty acids by lipase-catalysed transesterification and recrystallization, International Patent WO 90/13656, to Enzytech Inc.
49. De Zoete, M.C., Kock-van Dalen, A.C., van Rantwijk, F. and Sheldon, R.A. (1993) Ester ammoniolysis: a new enzymatic reaction, *J. Chem. Soc., Chem. Commun.*, 1831–1832.
50. De Zoete, M.C., Kock-van Dalen, A.C., van Rantwijk, F. and Sheldon, R.A. (1994) A new enzymatic reaction: enzyme catalyzed ammoniolysis of carboxylic esters, *Biocatalysis*, **10**, 307–316.
51. De Zoete, M.C., van Rantwijk, F. and Sheldon, R.A. (1995) submitted foir publication.
52. Hills, G.A., Macrae, A.R. and Poulina, R.R. (1990) Esters preparation from acids and alcohols with lipase catalyst and azeotropic distillation of alcohol and obtained water, European Patent 0383405, to Unichema B.V.
53. Macrae, A., Roehl, E.L. and Brand, H.M. (1990) Bioesters in cosmetics, *Drug Cosmet. Ind.*, **147**, 36–39.
54. Koch, M., Beck, R. and Röper, H. (1993) Starch-derived products for detergents, *Starch/Stärke*, **45**, 2–7, and references cited therein.
55. Riva, S. and Secundo, F. (1990) Selective enzymatic acylations and deacylations of carbohydrates and related compounds, *Chimicaoggi*, June, 9–16.
56. De Goede, A.T.J.W., Woudenberg-van Oosterom, M. and Van Rantwijk, F. (1994) Selective lipase-catalyzed esterification of carbohydrates, *Carbohydrates in Europe*, May, 18–20.
57. Therisod, M. and Klibanov, A.M. (1986) Facile enzymatic preparation of monoacylated sugars in pyridine, *J. Amer. Chem. Soc.*, **108**, 5638–5640.
58. Riva, S., Chopineau, J., Kieboom, A.P.G. and Klibonov, A.M. (1988) Protease-catalyzed regioselective esterification of sugars and related compounds in anhydrous dimethylformamide, *J. Amer. Chem. Soc.*, **110**, 584–589.

59. Björkling, F.M., Godtfredsen, S.E. and Kirk, O. (1989) A highly selective enzyme catalyzed esterification of simple glucosides, *J. Chem. Soc., Chem. Commun.*, 924–925.
60. Adelhorst, K., Björkling, F., Godtfredsen, S.E. and Kirk, O. (1990) Enzyme-catalyzed preparation of 6-O-acylglucopyranosides, *Synthesis*, 112–115.
61. De Goede, A.T.J.W., van Oosterom, M., van Deurzen, M.P.J., Sheldon, R.A., van Bekkum, H. and van Rantwijk, F. (1994) Selective lipase-catalyzed esterification of alkylglycosides, *Biocatalysis*, **9**, 145–155.
62. Sheldon, R.A., van Rantwijk, F. and Woudenberg-van Oosterom, M. (1995) Regioselective acylation of disaccharides in *tert*-butyl alcohol catalyzed by *Candida antarctica*, *Biotechnol. Bioeng.*, accepted for publication.
63. Riva, S. (1991) Enzymatic modification of steroids. In *Applied Biocatalysis*, vol. 1 (eds H.W. Blanch and D.C. Clark) Marcel Dekker, New York, pp. 179–220.
64. Njar, V.C.O. and Caspi, E. (1987) Enzymatic transesterification of steroid esters in organic solvents, *Tetrahedron Lett.*, **28**, 6549–6552.
65. Teranishi, R., Takeoka, G.R. and Güntert, M. (eds) (1992) *Flavor Precursors, ACS Symposium Series*, vol. 490, American Chemical Society, Washington, DC.
66. Langrand, G., Rondot, N., Triantaphylides, C. and Baratti, J. (1990) Short-chain flavor esters synthesis by microbial lipases. *Biotechnol. Lett.*, **12**, 581–586.
67. Chulaksananukul, W., Condoret, J.S. and Combes, D. (1992) Kinetics of geranyl acetate synthesis by lipase-catalysed transesterification in *n*-hexane, *Enzym. Microb. Technol.*, **14**, 293–298.
68. Fonteyn, F.M., Blecker, C., Lognoy, G., Marlier, M. and Severin, M. (1994) Optimization of lipase-catalyzed synthesis of citronellol acetate in solvent-free medium, *Biotechnol. Lett.*, **16**, 693–696.
69. Mosandi, A., Heusinger, G. and Gessner, M. (1986) Analytical and sensory differentiation of 1-octen-3-ol enantiomers, *J. Agric. Food Chem.*, **34**, 119–122.
70. Engel, K-H. (1992) Lipases: useful biocatalysts for enantioselective reactions of chiral flavor compounds. In *Flavor Precursors* (R. Teranishi, G.R. Takeoka and M. Güntert, eds) American Chemical Society, Washington, DC, pp. 20–31.
71. Lutz, D., Huffer, M., Gerlach, D. and Schreier, P. (1992) Carboxylester-lipase mediated reactions: a versatile route to chiral molecules. In *Flavor Precursors* (R. Teranishi, G.R. Takeoka and M. Güntert, eds) American Chemical Society, Washington, DC, pp. 32–45.
72. Engel, K-H. (1992) Lipase-catalyzed enantioselective acidolysis of chiral 2-methylalkanoates, *J. Amer. Oil Chem. Soc.*, **69**, 146–150.
73. Uemasu, I. and Hinze, W.L. (1994) Enantioselective esterification of 2-methylbutyric acid catalyzed via lipase immobilized in microemulsion-based organogels, *Chirality*, **6**, 649–653.
74. Engel, K-H., Heidlas, J., Albrecht, W. and Tressl, R. (1989). Biosynthesis of achiral flavor and aroma compounds in plants and microorganisms. In *Flavor Chemistry, Trends and Developments* (eds R. Teranishi, R.G. Buttery and F. Shahidi) ACS Symposium Series, vol. 388, American Chemical Society, Washington, DC, pp. 8–22.
75. Engel, K-H., Bohnen, M. and Dobe, M. (1991) Lipase-catalyzed reactions of chiral hydroxy-acid esters: competition of esterification and transesterification, *Enzym. Microb. Technol.*, **13**, 655–660.
76. Albrecht, W., Heidlas, J., Schwarz, M. and Tressl, R. (1992) Biosynthesis and biotechnological production of aliphatic γ- and δ-lactones. In *Flavor Precursors* (R. Teranishi, G.R. Takeoka and M. Güntert, eds) American Chemical Society, Washington, DC, pp. 46–58.
77. Gutman, A.L., Zuobi, K. and Boltansky, A. (1987) Enzymatic lactonisation of γ-hydroxyesters in organic solvents. Synthesis of optically pure γ-methylbutyrolactones and γ-phenylbutyrolactone, *Tetrahedron Lett.*, **28**, 3861–3864.
78. Makita, A., Nihira, I. and Yamada, Y. (1987) Lipase catalysed synthesis of macrocylic lactones in organic solvents, *Tetrahedron Lett.*, **28**, 805–808.
79. Yamada, H., Ohsawa, S., Sugai, T., Ohta, H. and Yashikawa, S. (1989) Lipase-catalyzed highly enantioselective macrolactonization of hydroxyacid esters in an organic solvent, *Chem. Lett.*, 1775–1776.
80. Stinton, S. (1994) Chiral drugs, *Chem. Eng. News*, September 19, 38–72.
81. Ramos Tombo, G.M. and Bellus, D. (1991) Chiralität und Pflanzenschutz, *Angew. Chem.*, **103**, 1219–1241.

82. Ariens, E.J., van Rensen, J.J.S. and Welling, W. (eds) (1988) *Stereoselectivity of Pesticides*, Elsevier, Amsterdam.

83. Sih, C.J., Gu, Q.-M., Fülling, G., Wu, S.-H. and Reddy, D.R. (1988) The use of microbial enzymes for the synthesis of optically active pharmaceuticals, *Dev. Ind. Microb.*, **29**, 221–229.

84. Margolin, A.L. (1993) Enzymes in the synthesis of chiral drugs, *Enzym. Microb. Technol.*, **15**, 266–280.

85. Ader, U., Andersch, P., Berger, M., Goergens, U., Seemayer, R. and Schneider, M. (1992) Hydrolases in organic synthesis: preparation of enantiomerically pure compounds, *Pure Appl. Chem.*, **64**, 1165–1170.

86. Bianchi, D., Cesti, P., Golini, P., Spezia, S., Garavaglia, C. and Mirenna, L. (1992) Enzymatic preparation of optically active fungicide intermediates in aqueous and organic media, *Pure Appl. Chem.*, **64**, 1073–1078.

87. Kloosterman, M., Elferink, V.H.M., van Iersel, J., Roskam, J.H., Meyer, E.M., Hulshof, L.A. and Sheldon, R.A. (1988) Lipases in the preparation of beta-blockers, *Tibtech.*, **6**, 251–256.

88. Ladner, W.E. and Whitesides, G.M. (1984) Lipase-catalyzed hydrolysis as a route to esters of chiral epoxy alcohols, *J. Amer. Chem. Soc.*, **106**, 7250–7251.

89. Chen, C.S., Liu, Y.C. and Marsella, M. (1990) A convenient chemoenzymatic synthesis of (*R*)- and (*S*)-chloromethyloxirane, *J. Chem. Soc., Perkin Trans. I*, 2559–2561.

90. Ader, U. and Schneider, M. (1992) Enzyme assisted preparation of enantiomerically pure beta-adrenergic blockers I. A facile screening method for suitable biocatalysts, *Tetrahedron Assymmetry*, **3**, 201–204.

91. Ader, U. and Schneider, M. (1992) Enzyme assisted preparation of enantiomerically pure beta-adrenergic blockers II. Building blocks of high optical purity and their synthetic conversions, *Tetrahedron Asymmetry*, **3**, 205–208.

92. Ader, U. and Schneider, M. (1992) Enzyme assisted preparation of enantiomerically pure beta adrenergic blockers III. Optically active chlorohydrin derivatives and their conversion, *Tetrahedron Asymmetry*, **3**, 531–524.

93. Bevinakatti, H.S. and Banerji, A.A. (1991) Practical chemoenzymatic synthesis of both enantiomers of propranolol, *J. Org. Chem.*, **56**, 5372–5375.

94. Hulshof, L.A. and Roskam, J.H. (1989) Phenylglycidate stereoisomers, conversion products thereof with e.g. 2-nitrophenol and preparation of diltiazem, European Patent Application 0343714, to Stamicarbon.

95. Kanerva, L.T. and Sundholm, O. (1993) Lipase catalysis in the resolution of racemic intermediates of diltiazem synthesis in organic solvents, *J. Chem. Soc. Perkin Trans. I*, 1385.

96. Gou, D.M., Liu, Y.C. and Chen, C.S. (1993) A practical chemoenzymatic synthesis of the taxol C-13 side chain N-benzoyl-(2R,3S)-3-phenylisoserine, *J. Org. Chem.*, **58**, 1287–1289.

97. Sonawane, H.R., Bellur, N.S., Ahuja, J.R. and Kulkarni, D.G. (1992) Recent developments in the synthesis of optically active α-arylpropionic acids: an important class of non-steroidal anti-inflammatory agents, *Tetrahedron Asymmetry*, **3**, 163–192.

98. Battistel, E., Bianchi, D., Cesti, P. and Pina, C. (1991) Enzymatic resolution of (*S*)-(+)-naproxen in a continuous reactor, *Biotechnol. Bioeng.*, **38**, 659–664.

99. McConville, F.X., Lopez, J.L. and Wald, S.A. (1990) Enzymatic resolution of ibuprofen in a multiphase membrane reactor. In *Biocatalysis* (ed. D.A. Abramowicz) Van Nostrand Reinhold, New York, pp. 167–177.

100. Smeets, J.W.H. and Kieboom, A.P.G. (1992) Enzymatic enantioselective ester hydrolysis by carboxylesterase NP, *Recl. Trav. Chim. Pays-Bas*, **111**, 490–495.

101. Giordano, C., Villa, M. and Panossian, S. (1992) Naproxen: industrial asymmetric synthesis. In *Chirality in Industry* (eds A.N. Collins, G.N. Sheldrake and J. Crosby) Wiley, New York, pp. 303–312.

102. Klibanov, A.M. and Kirchner, G. (1986) Enzymatic production of 2-halopropionic acids, US Patent 4,601,987, to Massachusetts Institute of Technology.

103. Cambou, B. and Klibanov, A.M. (1984) Comparison of different strategies for the lipase-catalyzed preparative resolution of racemic acids and alcohols: asymmetric hydrolysis, esterification and transesterification, *Biotechnol. Bioeng.*, **26**, 1449–1454.

104. Sheldon, R.A., Zeegers, H.J.M., Houbiers, J.P.M. and Hulshof, L.A. (1991) The synthesis of angtiotensin-converting enzyme (ACE) inhibitors, *Chimica oggi (Chemistry Today)*, May, 35–47.

105. Sugai, T. and Ohta, H. (1991) A simple preparation of (*R*)-2-hydroxy-4-phenylbutanoic acid,

Agric. Biol. Chem., **55**, 293–294.

106. Sakimae, A., Hosoi, A., Kobayashi, E., Ohsuga, N., Numazawa, R., Watanabe and Ohnishi, H. (1992) Screening of microorganisms producing D-β-acetylmercaptoisobutyric acid from methyl-DL-β-acetylthioisobutyrate, *Biosci. Biotech. Biochem.*, **56**, 1252–1256.

107. Hirohara, H., Mitsuda, S., Ando, E. and Komaki, R. (1985) Enzymatic preparation of optically active alcohols related to synthetic pyrethroids insecticides. In *Biocatalysts in Organic Synthesis* (eds J. Tramper, H.C. van der Plas and P. Linka), Elsevier, Amsterdam, pp. 119–134.

108. Mitsuda, S. and Nabeshima, S. (1991) Enzymatic optical resolution of a synthetic pyrethroid alcohol. Enantioselective transesterification by lipase in organic solvent, *Tecl. Trav. Chim. Pays-Bas*, **110**, 151–154.

109. De Zoete, M.C., van Rantwijk, F. and Sheldon, R.A. (1994) Lipase-catalyzed transformations with unnatural acyl acceptors, *Catalysis Today*, **22**, 563–590.

110. De Zoete, M.C., Kock-van Dalen, A.C., van Rantwijk, F. and Sheldon, R.A. (1993) Ester ammoniolysis: a new enzymatic reaction, *J. Chem. Soc., Chem. Commun.*, 1831–1832.

111. De Zoete, M.C., van Rantwijk, F. and Sheldon, R.A. (1994) Enzymatic ammoniolysis, International Patent Application PCT/EP94/03038.

112. De Zoete, M.C., Kock-van Dalen, A.C., van Rantwijk, F. and Sheldon, R.A. (1994) A new enzymatic reaction: enzyme-catalyzed ammoniolysis of careboxylic esters, *Biocatalysis*, **10**, 307–316.

113. De Zoete, M.C., Ouwehand, A.A., van Rantwijk, F. and Sheldon, R.A. (1995) Enzymatic ammoniolysis of amino acid derivatives, *Recl. Trav. Chim. Pays-Bas*, **114**, 171–174.

114. Kruse, C.G. (1992) Chiral Cyanohydrins—their manufacture and utility as chiral building blocks. In *Chirality in Industry* (eds A.N. Collins, G.N. Sheldrake and J. Crosby), Wiley, New York, pp. 279–299.

115. Becker, W. and Pfeil, E. (1966) The flavine enzyme, D-hydroxynitrilase, *Biochemistry*, **346**, 301–321.

116. Effenberger, F., Hörsch, B., Förster, S. and Ziegler, T. (1990) Enzyme-catalyzed synthesis of (*S*)-cyanohydrins and subsequent hydrolysis to (*S*)-a-hydroxycarboxylic acids, *Tetrahedron Lett.*, **31**, 1249–1252.

117. Effenberger, F., Ziegler, T. and Förster, S. (1987) Enzyme catalyzed cyanohydrin synthesis in organic solvents, *Angew. Chem. Int. Ed. Engl.*, **26**, 458–459.

118. Kula, M.R., Stürtz, I.M., Wandrey, C. and Krag, U. (1991) European Patent 446826, to Forschungszentrum Julich.

119. Andruski, S.W. and Goldberg, B. (1991) Process for preparing optically active cyanohydrins with enzymes, US Pat. 5177242, to FMC Corp.

120. Beisswengerm, T., Huthmacher, K. and Klenk, H. (1991) Preparation of D-2,4-dihydroxy-3,3-dimethylbutyronitrile as synthetic intermediate in D-pantolactone synthesis, Ger. Offen. DE 4126580, to Degussa.

121. Ognyabov, V.I., Datcheva, V.K. and Kyler, K.S. (1991) Preparation of chiral cyanohydrins by an oxynitrilase-mediated transcyanation, *J. Amer. Chem. Soc.*, **113**, 6992–6996.

122. Kiljunen, E. and Kanerva, L.T. (1994) Sorghum bicolor shoots in the synthesis of (*S*)-mandelo-nitrile, *Tetrahedron Assymmetry*, **5**, 311–314.

123. van Almsick, A., Buddrus, J., Hönicke-Schmidt, Laumen, K. and Schneider, M.P. (1989) Enzymatic preparation of optically active cyanohydrin acetates, *J. Chem. Soc., Chem. Commun.*, 1391–1393.

124. Inagaki, M., Hiratake, J., Nishioka, T. and Oda, J. (1991) Lipase-catalyzed kinetic resolution with *in-situ* racemization: one-pot synthesis of optically active cyanohydrin acetates from aldehydes, *J. Amer. Chem. Soc.*, **113**, 9360–9361.

125. Dordick, J.S. (1992) Enzymatic and chemoenzymatic approaches to polymer synthesis, *Tibtech*, **10** (August) 187–293, and references cited therein.

126. Patil, D.R., Redhwisch, D.G. and Dordick, J.S. (1991) Enzymatic synthesis of a sucrose-containing linear polyester in nearly anhydrous media, *Biotechnol. Bioeng.*, **37**, 639–646.

127. Dordick, J.S., Marletts, M.A. and Klibanov, A.M. (1987) Polymerization of phenols catalyzed by peroxidase in nonaqueous media, *Biotechnol. Bioeng.*, **13**, 964–968.

128. Popp, J.L., Kirk, T.K. and Dordick, J.S. (1991) Incorporation of *p*-cresol into lignins via peroxidase-catalyzed copolymerization in nonaqueous media, *Enzyme Microb. Technol.*, **13**, 964–968.

129. Margolin, A.L., Fitzpatrick, P.A., Dubin, P.L. and Klibanov, A.M. (1991) Chemoenzymatic synthesis of optically active (meth)acrylic polymers, *J. Amer. Chem. Soc.*, **113**, 4693–4694.
130. Ghogare, A. and Kumar, S. (1990) Novel route to chiral polymers involving biocatalytic transesterification of *O*-acryloyl oximes, *J. Chem. Soc., Chem. Commun.*, 134–135.
131. Martin, B.D., Ampofo, S.A., Linhardt, R.J. and Dordick, J.S. (1992) Biocatalytic synthesis of sugar-containing polyacrylate-based hydrogels, *Macromolecules*, **25**, 7081–7085.

LIVERPOOL
JOHN MOORES UNIVERSITY
AVRIL ROBARTS LRC
TITHEBARN STREET
LIVERPOOL L2 2ER
TEL. 0151 231 4022

Epilogue: Prospects and challenges of biocatalysis in organic media

A.M. KLIBANOV

Following our first publication on enzymatic catalysis in anhydrous solvents (Zaks and Klibanov, 1984), I still remember how skeptical colleagues would approach me at scientific meetings and voice their doubts of this being a geneal phenomenon. This book provides a striking illustration of how far such research has progressed since that time. Numerous examples of enzymatic reactions in anhydrous or nearly anhydrous media are described in several chapters. In addition, novel enzyme properties in nonaqueous media compared to those in water, such as the ability to catalyze reactions virtually impossible in aqueous solution, enhanced stability, and altered selectivity, are convincingly documented. Furthermore, useful synthetic applications of nonaqueous biocatalysis are presented, including industrial applications and those moving toward that status.

Professor Ari Koskinen, my co-editor, ably outlined synthetic opportunities of nonaqueous biocatalysis, in particular with respect to the production of chiral synthons, in his introductory chapter. He also introduced the contributors to this book in terms of the facets of nonaqueous enzymology covered by their chapters. In these concluding remarks, to avoid repetition, I will share some thoughts on the mechanistic aspects of this area of research with the focus on future requirements.

The question is often asked as to how enzymatic activity in organic solvents compares to that in water. While this question is hardly relevant from the practical viewpoint—wherein the only thing of real concern is a given enzymatic process's chances of completion within an acceptable period of time (usually, a few hours) and with a reasonably small enzyme sample—it is mechanistically interesting. The emerging answer is that if enzymes function in hydrous organic solvents, such as water-immiscible ones nearly saturated with water (i.e. water activity approaching unity) or water-miscible ones containing several percent of water, then enzyme activities may approach activities found in aqueous solution (Zaks and Klibanov, 1988). On the other hand, in anhydrous organic solvents, enzymes are usually orders of magnitude less reactive, and an understanding of why is forthcoming.

Among the chief factors contributing to the diminished enzymatic activity in anhydrous media are the highly restricted conformational mobility and the less favorable energetics of enzyme–substrate binding compared to those in water. Other factors include reversible, partial enzyme denaturation upon lyophilization (prior to placement in organic solvents) (Prestrelski *et al.*, 1993) and possible diffusional limitations, though this seems to be rare. Note that although these factors may be compelling, all appear avoidable. For example, not only water itself but also nonaqueous water mimics can greatly enhance conformational flexibility of enzymes and, with it, enzymatic activity (Zaks and Klibanov, 1988). While the binding of hydrophobic substrates to lipases and other enzymes designed for hydrophobic interactions will surely be more thermodynamically advantageous in water than in organic solvents, this may not be the case for hydrophilic substrates or other types of enzymes. Lyophilization-induced enzyme damage may be prevented by lyoprotectants (Prestrelski *et al.*, 1993; Dabulis and Klibanov, 1993) or circumvented by suspending cross-linked enzyme crystals (Persichetti *et al.*, 1995) instead of lyophilized enzymes in organic solvents. Consequently, there seems to be no reason why enzymes could not be even more active in organic solvents than in water. However, more mechanistic insights are required to consistently achieve this outcome.

As repeatedly mentioned in this volume, one of the most exciting discoveries of nonaqueous enzymology is the possibility of controlling the enzyme selectivity by the solvent (Wescott and Klibanov, 1994; Carrea *et al.*, 1995). However, our fundamental understanding of this phenomenon is still in its infancy. A great deal of thoughtful study will be necessary to achieve the obvious ultimate objective here—to rationally manipulate enzymatic selectivity simply by changing the solvent. The relationship between enzyme selectivity and physicochemical properties of the solvents/substrates will have to be elucidated.

Recent studies (Fitzpatrick *et al.*, 1993a, 1994; Yennwar *et al.*, 1994) have demonstrated that X-ray crystal structures of the serine proteases subtilisin and chymotrypsin in anhydrous acetonitrile and hexane, respectively, are very similar to those in water. It remains to be determined how general these conclusions are and to fully rationalize them. The apparent lack of strongly deleterious enzyme–solvent interactions is certainly a major surprise yielded by nonaqueous enzymology (Desai and Klibanov, 1995).

Ironically, almost all enzymes examined in organic solvents to date claim water as their natural habitat. What will be the distinctive characteristics of enzymes naturally embedded in biological membranes when placed in organic solvents? At an even higher level of complexity, much is to be gained from using live, functional microbial cells (in contrast to dead cells which can simply be viewed as bags filled with enzymes) to catalyze multi-step transformations in organic solvents. This task will require the knowledge of specific mechanisms of solvent microbial toxicity. Only initial steps are being made in this direction (Inoue and Horikoshi, 1989; Ferrante *et al.*, 1995).

The phenomenon of 'enzyme memory' (Klibanov, 1995), non-existent in aqueous solution, is an exciting enigma of nonaqueous biocatalysis. If truly explained, it promises to provide yet another powerful means of control over the behavior of enzymes in organic media by altering the history and/or the mode of preparation of the biocatalyst.

Finally, up till now, nonaqueous enzymology has been on the receiving side of its aqueous counterpart by benefitting from the wealth of knowledge accumulated by biochemists concerning enzyme structure and function. There are opportunities for us to pay some of this back by affording new approaches to the study of enzyme mechanisms, as illustrated by recent intriguing publications (Woo and Silverman, 1995; Mabrouk, 1995).

The above is an example of new challenges and prospects in nonaqueous enzymology. Over the last decade, this avenue of investigation has grown from an oddity to a fast-moving, fascinating, and useful field of study, akin to protein engineering, catalytic antibodies, and ribozymes. As in these other breakthroughs in biocatalysis, the potential appears immense and only limited by the imagination and creativity of practitioners in the field.

References

Carrea, G., Ottolina, G. and Riva, S. (1995) Role of solvents in the control of enzyme selectivity in organic media. *Trends Biotechnol.*, **13**, 63–70.

Dabulis, K. and Klibanov, A.M. (1993) Dramatic enhancement of enzymatic activity in organic solvents by lyoprotectants. *Biotechnol. Bioeng.*, **41**, 566–571.

Desai, U.R. and Klibanov, A.M. (1995) Assessing the structural integrity of a lyophilized protein in organic solvents. *J. Am. Chem. Soc.*, **117**, 3940–3945.

Ferrante, A.A., Augliera, J., Lewis, K. and Klibanov, A.M. (1995) Cloning of an organic solvent resistant gene in *Escherichia coli*: the unexpected role of alkylhydroperoxide reductase. *Proc. Natl Acad. Sci. USA*, **92**, 7617–7621.

Fitzpatrick, P.A., Steinmetz, A.C.U., Ringe, D. and Klibanov, A.M. (1993) Enzyme crystal structure in a neat organic solvent. *Proc. Natl Acad. Sci. USA*, **90**, 8653–8657.

Fitzpatrick, P.A., Ringe, D. and Klibanov, A.M. (1994) X-ray crystal structure of cross-linked subtilisin Carlsberg in water *vs* acetonitrile. *Biochem. Biophys. Res. Commun.*, **198**, 675–681.

Inoue, A. and Horikoshi, K. (1989) A *Pseudomonas* thrives in high concentrations of toluene. *Nature*, **338**, 264–265.

Klibanov, A.M. (1995) Enzyme memory—What is remembered and why? *Nature*, **374**, 596.

Mabrouk, P.A. (1995) The use of nonaqueous media to probe biochemically significant enzyme intermediates: the generation and stabilization of horseradish peroxidase compound II in neat benzene solution at room temperature. *J. Am. Chem. Soc.*, **117**, 2141–2146.

Persichetti, R.A., St Clair, N.L., Griffith, J.P., Navia, M.A. and Margolin, A.L. (1995) Cross-linked enzyme crystals (CLECs) of thermolysin in the synthesis of peptides. *J. Am. Chem. Soc.*, **117**, 2732–2737.

Prestrelski, S.J., Tedeschi, N., Arakawa, T. and Carpenter, J.F. (1993) Dehydration-induced conformational transitions in proteins and their inhibition by stabilizers. *Biophys. J.*, **65**, 66–671.

Wescott, C.R. and Klibanov, A.M. (1994) The solvent dependence of enzyme specificity. *Biochim. Biophys. Acta*, **1206**, 1–9.

Woo, J.C.G. and Silverman, R.B. (1995) Monoamine oxidase B catalysis in low aqueous medium. Direct evidence for an imine product. *J. Am. Chem. Soc.*, **117**, 1663–1665.

Yennawar, N.H., Yennawar, H.P. and Farber, G.K. 1994. X-ray crystal structure of γ-chymotrypsin in hexane. *Biochemistry*, **33**, 7326–7336.

Zaks, A. and Klibanov, A.M. (1984) Enzymatic catalysis in organic media at 100°C. *Science*, **224**, 1249–1251.

Zaks, A. and Klibanov, A.M. (1988) The effect of water on enzyme action in organic media. *J. Biol. Chem.*, **263**, 8017–8021.

Index